Turn Left at Orion

Hundreds of night sky objects to see in a
home telescope – and how to find them

FIFTH EDITION

Guy Consolmagno

Vatican Observatory, Vatican City State *and* Tucson, Arizona

Dan M. Davis

Stony Brook University, Stony Brook, New York

Illustrated by the authors
Cover and title page:
Mary Lynn Skirvin
Additional illustrations by
Karen Kotash Sepp, Todd Johnson, and Anne Drogin

CAMBRIDGE
UNIVERSITY PRESS

University Printing House, Cambridge CB2 8BS, United Kingdom

One Liberty Plaza, 20th Floor, New York, NY 10006, USA

477 Williamstown Road, Port Melbourne, VIC 3207, Australia

314–321, 3rd Floor, Plot 3, Splendor Forum, Jasola District Centre, New Delhi – 110025, India

103 Penang Road, #05-06/07, Visioncrest Commercial, Singapore 238467

Cambridge University Press is part of the University of Cambridge.

It furthers the University's mission by disseminating knowledge in the pursuit of education, learning, and research at the highest international levels of excellence.

www.cambridge.org
Information on this title: www.cambridge.org/9781108457569
DOI: 10.1017/9781108558464

© Guy Consolmagno and Dan M. Davis 2018

First edition published 1989
Second edition published 1995
Third edition published 2000
Fourth edition published 2011
Fifth edition published 2018
9th printing 2022

Printed in the United Kingdom by Bell & Bain Ltd, Glasgow

A catalog record for this publication is available from the British Library

Library of Congress Cataloging in Publication data
Names: Consolmagno, Guy, 1952– author. | Davis, Dan M. (Dan Michael), 1956– author.
Title: Turn left at Orion : hundreds of night sky objects to see in a home telescope – and how to find them / Guy Consolmagno (Vatican Observatory, Vatican City State, and Tucson, Arizona), Dan M. Davis (Stony Brook University, Stony Brook, New York).
Description: Fifth edition. | Cambridge ; New York, NY : Cambridge University Press, 2018. | Includes bibliographical references and index.
Identifiers: LCCN 2018027240 | ISBN 9781108457569 (alk. paper)
Subjects: LCSH: Astronomy – Amateurs' manuals.
Classification: LCC QB63 .C69 2018 | DDC 523–dc23
LC record available at https://lccn.loc.gov/2018027240

ISBN 978-1-108-45756-9 Spiral bound

Additional resources for this publication at www.cambridge.org/turnleft

Contents

How do you get to Albireo?

A while back I spent a couple of years teaching physics in Africa, as a volunteer with the US Peace Corps. At one point during my service I had to return to the US for a month, and while I was home I visited with my friend Dan in New York City. We got to talking about the beautiful dark skies in Africa, and the boundless curiosity of my students about things astronomical … and so that afternoon we went into Manhattan and, with Dan's advice, I bought a little telescope to take back with me to Kenya.

Dan was far more excited about my purchase than I was. He'd been an avid amateur astronomer since he was a little kid, something of an achievement when you're growing up in the grimier parts of Yonkers and your eyesight is so bad you can start fires with your glasses. And he was just drooling over some of the things I'd be able to see in Africa.

I didn't really understand it, at first. You see, when I was a kid I'd had a telescope, too, a little two-inch refractor that I had bought with trading stamps. I remembered looking at the Moon; and I knew how to find Jupiter and Saturn. But after that, I had sort of run out of things to look at. Those glorious color pictures of nebulae that you see in the glossy magazines? They're all taken with huge telescopes, after all. I knew my little telescope couldn't show me anything like that, even if I knew where to look. And of course I didn't know where to look, anyway.

But now here was Dan getting all worked up about my new telescope, and the thought that I'd be taking it back to Africa, land of dark skies and southern stars. There were plenty of great things to look at, he insisted. He gave me a star atlas, and a pile of books listing double stars and clusters and galaxies. Could it be that I could really see some of these things with my little telescope?

Well, the books he gave me were a big disappointment. At first, I couldn't make heads or tails of their directions. And even when I did figure them out, they all seemed to assume that I had a telescope with at least a six-inch mirror or lens. There was no way of telling which, of all the objects they listed, I might be able to see with my little three-incher.

Finally, Dan went out with me one night. "Let's look at Albireo," he said.
I'd never heard of Albireo.
"It's just over here," he said. "Point it this way, zip, and there you are."
"Neat!" I said. "A double star! You can actually see both of them!"
"And look at the colors," he said.
"Wow … one of them's yellow, and the other's blue. What a contrast."
"Isn't that great?" he said. "Now let's go on to the double-double."
And so it went for the next hour.

Eventually it occurred to me that all of the books in the world weren't as good as having a friend next to you to point out what to look for, and how to find it. Unfortunately, I couldn't take Dan back to Africa with me.

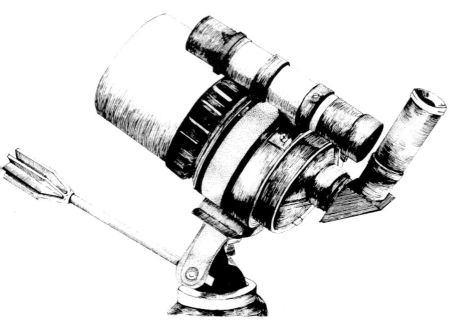

I suspect the problem is not that unusual. Every year, thousands of telescopes are sold, used once or twice to look at the Moon, and then they wind up gathering dust in the attic. It's not that people aren't interested – but on any given night there may be 2,000 stars visible to the naked eye, and 1,900 of them are pretty boring to look at in a small telescope. You have to know where to look, to find the interesting double stars and variables, or the nebulae and clusters that are fun to see in a small telescope but invisible to the naked eye.

The standard observer's guides can seem just incomprehensible. Why should you have to fight with technical coordinate systems? All I wanted to do was point the telescope "up" some night and be able to say, "Hey, would you look at this!"

It's for people who are like I was when I was starting out, the casual observers who'd like to have fun with their telescopes without committing themselves to hours of technical details, that we decided to write *Turn Left at Orion*.

— *Guy Consolmagno (Easton, Pennsylvania; 1988)*
to read more about the 2018 Fifth Edition, see page 242

How to use this book

The facts of life about small telescopes

How night-sky objects look in a telescope depends on the kind of telescope you have. That, in turn, will determine the sort of objects you will want to observe. We talk at great length later on about different kinds of telescopes, but it's worth outlining here at the start a few basic telescope facts of life.

A good telescope can magnify the surface of the Moon, reveal details on the small disks of the planets, and split double stars. But just as importantly, it can also gather light and concentrate it at the eyepiece to make faint nebulae bright enough for the human eye to see. A telescope's light-gathering ability, or *aperture*, is key.

Every telescope has a big lens or mirror to gather in starlight and bring it to a focus near the eyepiece. The bigger the lens or mirror, the more light it can capture. All things being equal, then, bigger is better. But in the real world, all things are never equal. Thus there are a variety of different ways to gather that light. For this book we've divided scopes into three classes: binoculars, small telescopes, and Dobsonian telescopes. They each have their strong points and weaknesses.

Binoculars are relatively inexpensive, very portable, and they give you a wide field of view. Furthermore, by using both eyes most people can pick out more detail in a faint nebula seen with binoculars than you would using a single telescope of the same aperture. Still, their very portability means that they're usually limited to an aperture of a couple of inches, maximum. And most binoculars don't attach easily to a tripod, making it hard to keep them steady and fixed on faint objects.

At the other extreme, a *Dobsonian* is essentially a big mirror at the bottom of a large lightweight tube, with the eyepiece at the top, all fixed in a simple mount. It has fantastic light-gathering power. But Dobs are very awkward to take on trips, or even just to carry into your back yard for a brief look at the sky. And their simplicity comes at a sometimes subtle cost to their optical quality.

In between are the small telescopes, ranging from the classic refractors (a tube with a lens at each end) to the more sophisticated *catadioptric* designs that combine mirrors and lenses in a compact package. "Cats" are portable and powerful; but to gather as much light as a "Dob" you have to spend about four times the money. Still, a small Cat prowling through dark skies can outperform a big Dob fighting city lights. Note also, as we explain below, Cats and refractors use a star diagonal and that gives them a view of the sky that's the mirror image of what's seen in binoculars or a Dob.

We provide two views for each object in this book: one is the view with a small Cat or refractor, the other the view in a larger Dob. For binoculars, use our finder-scope views as a guide. If you're lucky enough to have an 8" Cat, you'll see the greater detail in our Dob views, but with the orientation of the Cat view. Figures drawn with this orientation can be downloaded from our website, www.cambridge.org/turnleft

Finding your way

Once you've got your telescope, where do you point it? Answering that question is what this book is all about.

There are two classes of night-sky objects. The *Moon* and *planets* move about in the sky relative to the stars; fortunately they are bright enough that it's easy to find them. *Seasonal objects* – double stars, clusters, nebulae, and galaxies – stay fixed in the same relative positions to each other, rising and setting together each night, slightly faster than the Sun, thus changing location slowly as the seasons change.

The Moon, Sun, and planets: Finding the **Moon** is never a problem; in fact, it is the only astronomical object that is safe and easy to observe directly even in broad daylight. (Indeed, unless you're up in the wee hours of the morning, daytime is the only time you can see the third quarter Moon. Try it!)

The Moon changes its appearance quite a bit as it goes through its phases, and for each phase there are certain things on the Moon that are particularly fun to look for. We've included pictures and discussions for seven different phases of the Moon, plus what to look for during lunar and solar eclipses. We also give a brief introduction to observing the **Sun** itself… but only try that if you have the proper equipment.

Planets are small bright disks of light in the telescope. In a good telescope with a high powered lens they can be spectacular to look at! But even a pair of binoculars will be able to show you the phases of Venus and the moons of Jupiter.

This book lists our favorite small-telescope objects, arranged by the months when they're best visible in the evening and the places in the sky where they're located. For all these objects, we assume you have a telescope much like one of ours: either a small scope whose main lens or mirror is only 2.4" to 4" (6 to 10 cm) in diameter; or a modest Dobsonian with an aperture of 8" to 10" (20 to 25 cm). Everything in this book can be seen with such small telescopes under ordinary sky conditions: it's how they looked to us.

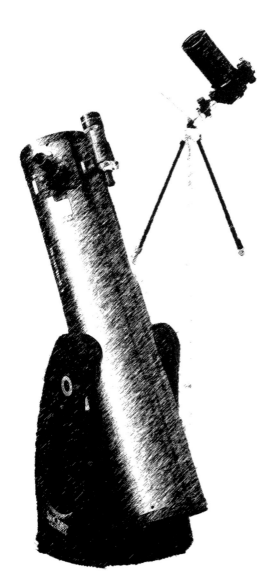

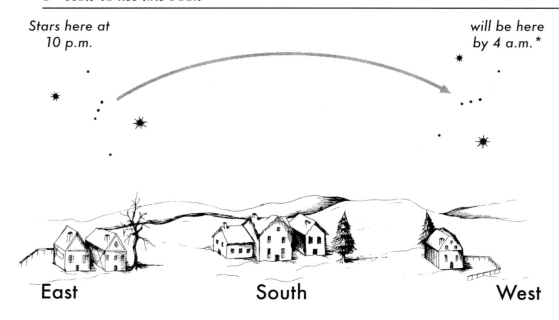

Stars here at 10 p.m.

will be here by 4 a.m.*

East   South   West

*or at 10 p.m., 3 months later

The stars and deep-sky objects stay in fixed positions relative to one another. But which of those stars will be visible during the evening changes with the seasons; objects that are easy to see in March will be long gone by September. Thus we refer to these objects as seasonal. Some objects are visible in more than one season. When we talk about "January skies" we're referring to what you'd see in January at around 10:00 p.m. local standard time. If you are up at four in the morning the sky will look quite different. The general rule of thumb is to advance one season for every six hours, so spring stars will be visible on winter mornings, summer stars on spring mornings, and so forth.

The positions of the planets, relative to the other stars, change from year to year; but if you know in general where to look for them, they're very easy to find. They're generally as bright as the brightest stars. We describe here things to look for when you observe them. Our website **www.cambridge.org/ turnleft** links to tables of when and where to find each planet.

The seasonal objects: The stars, and all the *deep-sky objects* we talk about in this section, stay in fixed positions relative to one another. But which of those stars will be visible during the evening changes with the seasons; objects that are easy to see in March will be long gone by September.

How do you know, on any given night, what objects are up? That's what the **What, where and when** chart at the back of the book is all about. For a given month, and the time when you'll be observing, it lists each of the constellations where seasonal objects are located; the direction you should turn to look for that constellation (e.g. W = west); and if you should be looking low, towards the horizon (e.g. "W – "), or high up (e.g. "W + "). Constellations marked with only a "++" symbol are right overhead. Then you can turn to the seasonal pages to check out the objects visible in each constellation available that night.

But note, you don't need to memorize the constellations in order to use this book. Constellations are merely names that astronomers give to certain somewhat arbitrarily defined regions of the sky. The names are useful for labeling the things we'll be looking at; otherwise, don't worry about them. If you do want to know the constellations, there are a number of good books available (our personal favorite is H. A. Rey's *The Stars*) but for telescope observing, all you need is an idea of where to find the brightest stars, to use them as guideposts.

Note that the brightness of a star, its *magnitude*, is defined in a somewhat counter-intuitive way that goes back to the ancient Greeks: the brighter a star, the lower the numerical value of its magnitude. The common rule is that a star of the *first magnitude* is about 2½ times brighter than a *second-magnitude* star, which is about 2½ times brighter than a *third-magnitude* star, and so forth. The very brightest stars can be zeroth magnitude, or even have a negative magnitude! The star Vega has a magnitude of 0; Sirius, the brightest star of all, has a negative magnitude, –1.4. There are only about 20 stars that are first magnitude or brighter. On a dark night, the typical human eye can see down to about sixth magnitude without a telescope.

We start off each season by describing the location of our **guideposts**, a selection of the brightest and easiest stars to find in the evening skies. Most of our readers are in the northern hemisphere, so we have set our guidepost charts to be appropriate for what you will see at about 10 p.m. in the USA, Canada, Europe, China, Japan … anywhere between latitudes 30° and 60° N. Most of these objects will also be visible from Australia, New Zealand, Africa, or South America, but their positions will all be shifted northwards.

Stars set in the west, just like the Sun, so if you're planning a long observing session you should start by observing the western objects first, before they get too low in the sky. The closer to

Magnitude	Brightness (vs. Polaris)
Sun (-26.7)	330,000,000,000 × Polaris
Full Moon (-12.7)	800,000 × Polaris
First Quarter Moon (-9.5)	40,000 × Polaris
Venus at its brightest (-4.4)	400 × Polaris
Sirius, the brightest star (-1.4)	25 × Polaris
Polaris (2.1)	Polaris
Naked-eye visibility limit (about 6)	1/40 × Polaris
Limit with 4" telescope (about 12)	1/10,000 × Polaris
Limit with 10" telescope (about 14)	1/60,000 × Polaris
Palomar 200" Telescope (about 22)	1/100,000,000 × Polaris
Keck Telescope (about 25)	1/1,500,000,000 × Polaris
Hubble Telescope (about 31)	1/350,000,000,000 × Polaris
Webb Telescope (34, in infrared)	1/5,000,000,000,000 × Polaris

Magnitude scale: -25, -20, -15, -10, -5, 0, 5, 10, 15, 20, 25, 30, 35

the horizon an object sits, the more the atmosphere obscures and distorts it, so you want to catch things when they're as high up in the sky as possible. (However, telescopes with alt-azimuth mounts – see page 19 – have trouble aiming straight up, so try to catch such objects sometime before or after they get to that point.)

Most objects are visible in more than one season. Just because the Orion Nebula is at its best in December doesn't mean you should risk missing it in March. Some of the nicest objects from the previous season that are still visible in the western sky are listed in a table at the beginning of each season.

You'll be able to see objects around your hemisphere's celestial pole at some time during any night, year round; but objects around the opposite pole will always be hidden. Northerners will always be able to find the Little Dipper but never see the Southern Cross; southerners get the Cross but never the Dipper. Thus we have sorted the northernmost and southernmost objects into separate chapters.

Also, remember that when we talk about "January skies," for instance, we're referring to what you'd see then at around 10:00 p.m., standard time. If you are up at 4 in the morning the sky will look quite different. The general rule of thumb is to advance one season for every six hours, so spring stars will be visible in the wee hours of winter, summer stars in early spring morning hours, and so forth.

Who are these guys?

We start with the name and **official designation** of each object. Catalogs and catalog numbers can seem confusing at first, but these are the methods that everyone uses to identify objects in the sky, so you may as well get to know them. It's fun to compare what you see in your telescope with the glossy color pictures that appear in astronomy magazines, where these objects are often identified only by their catalog number.

Brighter stars usually have names – historically, a variety of names. Starting in 2017 the IAU began designating official names and spellings, which we use here.

In addition, stars are also designated by Greek letters or Arabic numerals, followed by the Latin name of their constellation (in the genitive case, for the benefit of Latin scholars). The Greek letters are assigned to the brighter stars in the (very approximate) order of their brightness within their constellation. For example, Sirius is also known as *Alpha Canis Majoris* (or *Alpha CMa* for short) since it's the brightest star in Canis Major, the Big Dog. The next brightest are *Beta*, *Gamma*, and so on. Fainter naked-eye stars are known by their *Flamsteed number*, e.g. *61 Cygni*, assigned by position west to east across the constellation.

Subsequent catalogs have followed for fainter stars. The most common are the *Yale Bright Star Catalog* (numbers beginning with HR, after its predecessor, the *Harvard Revised Catalogue*); the *Henry Draper Catalogue* (numbers beginning with HD); and the *Hipparcos/Tycho Catalogue* (HIP numbers) assembled for the Hipparcos spacecraft, which measured parallax distances to a hundred thousand stars.

Double stars also have catalog numbers. Friedrich Struve and Sherburne Burnham were two nineteenth-century double-star hunters; doubles that filled their catalogs now bear their names. Struve variables are traditionally marked with the Greek letter Σ followed by a number. (Friedrich's son Otto also made a catalog of doubles; his stars are denoted with the letters OΣ.)

Variable stars are given letters. The first known in each constellation were lettered R through Z. As more were discovered, a double-lettering system was introduced: e.g. *VZ Cancri*. After they ran out of letters, they used numbers as in *V1016 Orionis*.

For clusters, galaxies, and nebulae, two catalogs most used in this book are the *Messier Catalog* (with numbers like *M13*) and the *New General Catalog* (with numbers like *NGC 2362*). Charles Messier was a comet-hunter in the 1700s who had no use for galaxies and nebulae. He kept finding them over and over again, and would get confused because many of them looked like comets to him. So he made a list of them, to let him know what *not* to look at while he was searching for comets. In the process he wound up finding and cataloging most of the prettiest objects in the sky. But he managed to number them in a totally haphazard order. The *New*

When you look directly overhead, you don't have to look through as much dirty, turbulent air as you do when you look at something low in the sky. Try to avoid looking at things near the horizon.

If you live in the north, stars to the south never do rise very high; there's nothing you can do about that. (Southern hemisphere residents have the same problem with northern stars, of course.) But stars along the other horizons will appear higher in the sky during different seasons, or at different times of the night.

Open cluster: M35

Galaxies: the Leo Trio

Globular cluster: M13

Diffuse nebula: the Swan Nebula

General Catalog, assembled in the 1880s from a century of observations by William Herschel and his son John, numbers objects from west to east across the sky. Objects that you see in the same area of the sky have similar NGC numbers.

What are these guys?

An *open cluster* is a group of stars, often quite young (by astronomical standards), that are clumped together. Viewing an open cluster can be like looking at a handful of delicate, twinkling jewels. Sometimes they are set against a background of hazy light from the unresolved members of the cluster. On a good dark night, this effect can be breathtaking. We discuss open clusters in more detail when we talk about the clusters in Auriga, on page 71.

Galaxies, globular clusters, and the various types of nebulae will all look like little clouds of light in a small telescope.

A *galaxy* consists of billions of stars in an immense assemblage, similar to our own Milky Way but millions of light years distant from us. It is astonishing to realize that the little smudge of light you see in the telescope is actually another "island universe" so far away that the light we see from any of the galaxies that we talk about (except the Magellanic Clouds) left it before human beings walked the Earth. We discuss galaxies in greater detail on page 109, with the Virgo galaxies.

A *globular cluster* is a group of up to a million stars within our own galaxy, bound together forever in a densely packed, spherical swarm of stars. On a good crisp dark night, you can begin to make out individual stars in some of them. These stars may be among the oldest in our galaxy, perhaps in the Universe. On page 115 (M3 in Canes Venatici) we go into the topic of globular clusters in greater detail.

Diffuse nebulae are clouds of gas and dust from which young stars are formed. Though they are best seen on very dark nights, these delicate wisps of light can be among the most spectacular things to look at in a small telescope. See page 55 (the Orion Nebula) for more information on these nebulae.

Planetary nebulae have nothing to do with planets; they are the hollow shells of gas emitted by some aging stars. They tend to be small but bright; some, like the Dumbbell and the Ring Nebulae, have distinctive shapes. We talk about them in greater detail on page 99 (the Ghost of Jupiter, in Hydra).

If the dying star explodes into a supernova, it leaves behind a much less structured gas cloud. M1, the Crab Nebula, is a *supernova remnant*; see page 67.

A *double star* looks like one star to the naked eye, but in a telescope it turns out to be two (or more) stars. That can be a surprising and impressive sight, especially if the stars have different colors. They're also generally easy to locate, even when the sky is hazy and bright.

Variables are stars that vary their brightness. We describe how to find a few that can change brightness dramatically in a matter of an hour or less.

Ratings and tips

For each object you'll see a little box where we provide a *rating*, and list the *sky conditions, eyepiece power,* and the *best months* of the year to look for them. We also note the objects where a *nebula filter* might help the view. And we point out some of the reasons why this particular object is worth a look.

The rating is our own highly subjective judgement of how impressive each object looks. Since each kind of telescope will see the object in a different way, we repeat the ratings three times, with the appropriate symbols for Dobsonians, small telescopes, and binoculars. Obviously some objects are better seen in a large telescope with lots of light-gathering power, and so they'll rate a higher number of Dobsonians than binoculars. But some objects are too spread out to fit into a Dobsonian field of view, such as the Beehive Cluster; these will look better in binoculars.

A few of these objects can be utterly breathtaking on a clear, crisp, dark, moonless night. The Great Globular Cluster in Hercules, M13, is an example. And even if the sky is hazy, they're big enough or bright enough to be well worth seeing. Such an object, and in general any object that is the best example of its type, gets a *four telescope* rating. (The Orion Nebula and the Large Magellanic Cloud rate *five telescopes.* If they are visible in your sky, they are not to be missed on any night.)

Objects that are still quite impressive but which don't quite make "best of class" get a *three telescope* rating. An example is the globular cluster M3; it's quite a lovely object, but its charms are more subtle than M13's.

Below them in the ratings are *two telescope* objects. They may be harder to find than the three telescope objects; or they may be quite easy to find, but not necessarily as exciting to look at as the other examples of their type. For instance, the open clusters M46 and M47 are pleasant enough objects in a small scope, but they're located in an obscure part of the sky with few nearby stars to help guide you to their locations. The open clusters M6 and M7 are big and loose, and easy to find, rating high for binoculars, but they look less impressive in a bigger telescope: only two Dobs.

Finally, some objects are, quite frankly, not at all spectacular. As an example, the Crab Nebula (M1) is famous, being a young supernova remnant, but it's very faint and hard to see in a small telescope. You may have a hard time finding such objects if the night is not really dark or steady. They're for the completist, the "stamp collector" who wants to see everything at least once, and push the telescope to its limits. In their own way, of course, these objects often turn out to be the most fun to look for, simply because they are so challenging to find. But they might seem pretty boring to the neighbors who can't understand why you're not inside watching TV on a cold winter's night. These objects we rate as *one telescope* sights.

Matching the telescope to the weather

Sky conditions control just how good your observing will be on any given night, and so they will determine what you can plan on seeing. The ideal, of course, is to be alone on a mountain top, hundreds of miles from any city lights, on a still, cloudless, moonless night. But you really don't need such perfect conditions to have fun stargazing. We've seen most of the objects in this book even amid suburban lights.

Any night when the stars are out, it's worth trying. True, the visibility of the fainter objects can be obscured by thin clouds, especially when they reflect the light of a nearly full Moon (or the glow of city light pollution); but that is not necessarily all bad news. Some things, like colorful double stars, actually look better with a little background sky brightness to make the colors stand out. And objects that require the highest magnification – double stars again, or planets – demand really steady skies which often occur when there is a thin cloud cover. (The clearest nights are often those associated with a cold front passing through, when the air tends to be turbulent… not to mention hard on the observer's hands and feet!)

On the other hand, the fainter objects on our list can stand no competition from other sources of light. They are listed as *dark sky* – wait for when the Moon is not up. To see them at their best, take your telescope along on your next camping trip. Not only will the dim ones become visible, but even the bright nebulae will look like brand new objects when the sky conditions are just right.

How to look, when to look: eyepieces, filters, best seen

As we mentioned above, telescopes have two different functions: to make small things look bigger, or to make faint things brighter. You control which of those roles it's doing by the eyepiece you choose. In **eyepiece power** you're trading off magnification against brightness and field of view. Your shorter focal length eyepiece will give higher magnification and the longer eyepiece will generally give a greater field of view. For more detail, see pages 18–21, *Know Your Telescope*.

Extended objects like open clusters need low power to fit in the field of view; dim objects like galaxies need low power to concentrate their light. Small, bright objects like planets and double stars may need a high-power eyepiece. Planetary nebulae are small but dim; since the lower-power eyepiece shows you more of the sky, a good technique is to find the object with low power, and then observe it at medium power; then try high power if it is bright enough or the skies are dark enough. Some objects, like the Orion Nebula, are interesting under both low and high power!

Low power makes faint objects appear brighter; unfortunately, it can also make a murky suburban sky brighter as well. Sometimes, once you've found a faint object in low power, you can increase its contrast against the sky by switching to a higher power eyepiece. A more elegant (but more expensive) way to increase this contrast

The ideal observing conditions are to be alone on a mountain top, hundreds of miles from city lights, on a cool, crisp, moonless night. Be sure to bring your telescope along when you go camping! You shouldn't wait for perfect conditions, however. Most of the objects we describe in this book can be seen from the suburbs; many of them, even within a city. The roof of an apartment building can make a fine place for an informal observatory.

is by using an appropriate **filter**. A specially colored filter that cuts out the yellow light put out by sodium street lamps can make a remarkable difference in what you can see through your telescope even when you're fighting light pollution. Even better, a filter that lets through only the greenish light emitted by some nebulae can really bring out detail in otherwise faint objects. But these filters only work for certain objects, mostly diffuse nebulae and planetary nebulae. We indicate when a nebula filter might be appropriate. (Of course, you don't want to use colored filters if you're trying to pick out the colors of a double star.)

The indication for *best seen* lists the specific months of the year when these objects can be found relatively high in the sky, from the end of twilight until about ten o'clock, at about 45° north – roughly, anywhere between Florida and Scotland. (For the southern-hemisphere chapter we assume you're at 35° south, roughly the latitude of Sydney, Cape Town, Buenos Aires, and Santiago.) Of course, as mentioned above, staying up later can gain you a few extra months to preview your favorites.

Where to look, what to look for

The first map for each object is a **naked-eye chart**. These charts generally show stars down to third magnitude. As we mentioned above, magnitude is the measure of a star's brightness, and smaller numbers mean brighter stars. On the best night, the human eye can see down to about sixth magnitude without a telescope; with the glow of city lights, however, seeing even a third-magnitude star can be a challenge.

The next chart shows the **finderscope view**. The little arrow pointing to the west indicates the way that stars appear to drift in the finderscope. (Since most finderscopes have very low power, this westward drift of the stars will seem very slow.)

Finderscopes come in all sizes and orientations. We have assumed that your finderscope turns your image upside down, but does *not* give a mirror image of the sky (see below); check to see how yours behaves, and use our charts accordingly. (Before sunset, aim your telescope at a distant street sign or car license plate: is it upside down? Is it mirror imaged?) Likewise, the field of view of finderscopes can vary, but they usually show about 5° to 6° of the sky. The circle we draw in our views assumes a 6° field, and the whole finderscope-view box is generally 12° square.

On the right-hand page we show the **telescope view:** what the object should look like through your eyepiece. Usually we give two views, one as seen in a small telescope through a star diagonal (see below) and the other as seen in a Dobsonian or other Newtonian telescope. Notice that one is the mirror image of the other, as we will explain shortly.

These pictures are based on our own observations with small telescopes. The idea is that if you can match up what you see with what's in our picture, then you'll know that you are actually looking at the same object we're talking about. What we have drawn is what a typical person is likely to notice, and we don't always show all of the fainter stars in the field of view. These pictures are not meant to serve as technical star charts – don't try to use them in celestial navigation!

Again, follow the links on our website **www.cambridge.org/turnleft** to find these charts in other orientations.

Note that we treat double stars somewhat differently. These stars are relatively bright and usually easy enough to see (if not split) with the naked eye. For that reason, you don't need the same kind of elaborate directions to find them as you do for the fainter clusters and nebulae. Within each season we have gathered the best of the doubles into occasional double-star spreads with a more detailed finder chart pointing out the locations of the best examples in the area. Then each double is pictured in a close-up circle: a circle of two thin lines represents a view ten arc minutes across, while a thick/thin pair of circles denotes a closer, higher-powered view, only five arc minutes in diameter. These close-up views assume a Dobsonian orientation. Their purpose is merely to give you an idea of the relative distance and brightness of the companion, and if there are any other stars in the field of view. These pages also include a little table describing the individual double stars, their colors and brightness, and how close together (in *arc seconds*) they appear to be.

(What are arc minutes and arc seconds? They are the way we indicate the size of a very small angle. The arc of the sky from horizon to zenith is an angle of 90°. Each degree can be divided further into 60 minutes of arc, usually written 60' (the full Moon is about 30' across) and each minute can be further divided into 60 seconds of arc, or 60". Thus one arc second is 1/3,600 of a degree – as tiny a separation as you'll ever be able to make out in a small telescope.)

The low-power eyepiece view assumes a power of between 35× to 40×; this is what you'd use for large clusters of stars and a few big galaxies. The medium-power eyepiece drawing assumes roughly a 75× view. The high-power eyepiece drawing gives a view magnified about 150×. And because larger telescopes like Dobs can handle even higher power, in a few of the Dobsonian fields we include an inset with a very high-power view of 300×. To get this kind of magnification, you either need a very short focal length eyepiece or an additional lens called a Barlow. (We talk about Barlow lenses with other *Accessories* on page 244.)

Note again that we've included an arrow to show the direction that stars will drift. As the Earth spins, the stars will drift out of your field of view; the higher the power you are using, the faster the drift will appear. Though this drift can be annoying – you have to keep readjusting your telescope – it can also be useful, since it indicates which direction is west.

In the text, we describe where to look and how to recognize the object. And we comment about some of the things worth looking for in the telescope field – colors, problems that might crop up, nearby objects of interest. Finally, we describe briefly the present state of astronomical knowledge about each object, a guide to what you're looking at.

East is east, and west is west... except in a telescope

What about the orientations of these pictures? Why do we seem to confuse south and north, or east and west?

There are several things going on. First, we're all used to looking at road maps or geographical atlases: maps of things on the ground under our feet. There, traditionally, north is up and east is to the right. However, when we look at the sky our orientation is just backwards from looking at the ground. Instead of being outside the globe of the Earth, looking down, we're inside the globe of the sky, looking up. It's like looking at the barber's name painted on a window; from inside the barber shop, the lettering looks backwards. In the same way, a sky chart keeping north up must mirror east and west: west is to the right and east is to the left.

Next, many finderscopes are simple two-lens telescopes. This means that, among other things, they turn everything upside down. (Binoculars and opera glasses have to have extra lenses or prisms to correct for this

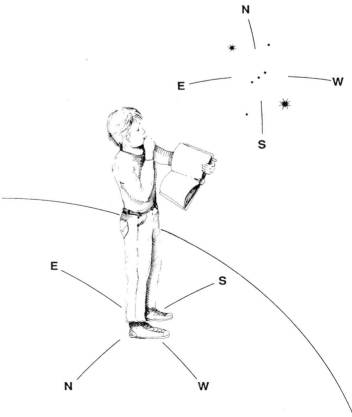

Standing on the Earth, looking out from inside the globe of the sky, the directions east and west appear to be reversed from what we're used to on ordinary maps.

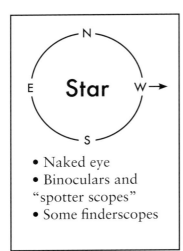

- Naked eye
- Binoculars and "spotter scopes"
- Some finderscopes

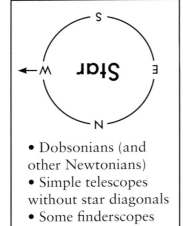

- Dobsonians (and other Newtonians)
- Simple telescopes without star diagonals
- Some finderscopes

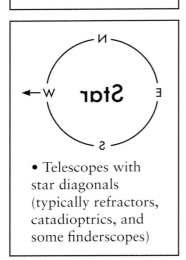

- Telescopes with star diagonals (typically refractors, catadioptrics, and some finderscopes)

A star field as it appears to the naked eye; upside down, as it appears in a Dobsonian; and mirror imaged, as it appears in most telescopes with a star diagonal. To determine your telescope's orientation, try reading the lettering on a distant sign.

The arrows show the directions that stars appear to drift, moving east to west, across the field of view.

effect.) So instead of north at the top, we'd see south at the top and north at the bottom, with east and west likewise reversed. But nowadays many finderscopes do have an extra lens or prism to turn the image rightside up again. Some evening before it gets dark, check out your finder by aiming it at a nearby street sign and see what kind of orientation it gives you.

We've chosen to orient the finder views with south at the top: this matches the view of what a typical simple two-lens finder will show to most of our readers (living in the northern hemisphere) when looking at most of the objects in this book as they arc across the southern sky. It also matches what southern hemisphere observers will see when looking south at all the glorious objects around the south pole. For our northern circumpolar objects, however, we twist the finder view to have "north" at the top. (Southern hemisphere observers looking at objects in the north can find rotated versions on our website; or just turn the book upside down!) And, of course, once you're pointed up at the sky, what's "up" is likely to be any direction: as the night progresses its apparent orientation will twist around as it moves across the sky.

Finally, most refractors and catadioptric telescopes sold nowadays have an attachment called a *star diagonal*, a little prism or mirror that bends the light around a corner so you can look at objects up in the sky without breaking your neck. This means, however, that what you see in your telescope is a mirror image of your finderscope view. It's also a mirror image of almost any photograph you'll find in a magazine or book. The orientation as seen through a star diagonal is what we use in our "small telescope with a star diagonal" views.

But Dobsonian telescopes do not use star diagonals. (A Dobsonian is the most common example nowadays of what is classically called a Newtonian design. What we say here applies to all Newtonian telescopes.) So it does not have that extra mirror image affect. But it does show the object upside-down from what you expect, which means that to push an object into the center of the field of view can often mean moving the telescope in just the opposite direction of what you might expect. (One trick is to "drag the object;" in other words, push or pull the telescope in the direction you want to push or pull the object in order to center it in your field.) Since a Dobsonian usually has at least twice the aperture (and thus at least four times the light-gathering area) of a small scope, the Dob view will also include more stars, fainter stars, and larger areas of nebulosity. We try to show just how much more, in our drawings.

Notice, finally, that many spotter scopes sold for terrestrial use nowadays come with a 45° angle prism called an *erecting prism*. This does *not* give a mirror-imaged view. Don't confuse these devices with true star diagonals; they're handy for bird-watching, but they'll still give you a crick in your neck when you try to use them to look at stars overhead. If you can, see if your telescope dealer will replace this with a true 90° findercope. The view you get through an erecting prism will be better represented by the Dobsonian view, at least in terms of the orientation of the stars, though not of course in the detail you'll actually see ... unless you have a very large spotter scope indeed.

There's more to be found online

Throughout this book you'll see links to the accompanying web resources. This book can be used without ever referring to these internet resources, but we hope that you'll explore the website and use it to enhance your experience.

How do you do that? Start by going to the site (www.cambridge.org/turnleft). Along the left banner you'll see a link for the Home Page and one for How to Use This Website. Below them are about a dozen more links. Let's explore each of these in turn:

The Moon: On pages 26–37 we follow the Moon through the month and we show images of features to look at each night. Depending on the telescope, finderscope, or binoculars you're looking through and the hemisphere in which you live, there are four different orientations in which things can appear: rotated 180° (or not) and mirror-flipped (or not). In the book, the way in which we show objects on the Moon is how they appear using binoculars or the naked eye in the northern hemisphere (this is the same orientation as with a Dobsonian telescope in the southern hemisphere).

But what if you're using a Dobsonian in the northern hemisphere (or binoculars in the southern hemisphere)? On the Moon part of the website (www.cambridge.org/features/turnleft/the_moon.htm), that's the first of the three alternative orientations for each Moon phase. The other two options are for a refracting or catadioptric telescope, in either the northern or southern hemisphere. Whichever version of the charts you need, just bring them to the telescope by either printing them out or having them on your laptop or tablet.

The Planets: Keeping track of where to find the planets is tough to do in a book; simple tables like we used in earlier editions are imprecise and quickly go out of date. Faint objects, like Uranus and Neptune, really need detailed finder charts; fast moving objects like comets and asteroids are even harder to describe. But in fact, precisely this material is best found online. We give links to our favorite resources.

Seasonal Skies: These pages are organized as in the book, by the seasons of the year (or parts of the world) that are favorable for observing each object. Simply choose the part of the sky (say, April–June or Northern skies) and click on the link. There you'll find all the objects in that chapter of the book. As with the Moon, we provide alternative orientations for all the seasonal object charts in the book.

The book's naked-eye charts are shown as seen in the northern hemisphere; the webpage has a version that is flipped for southern hemisphere observers.

The finderscope charts in the book are rotated 180°, to match the northern hemisphere view in a simple finderscope without a star diagonal. The charts in the website cover the three other possibilities: with a star diagonal in either hemisphere or without a star diagonal in the southern hemisphere. As with the Moon charts, simply find the appropriate version for your finderscope and hemisphere and use it at your telescope.

There are two versions of the telescope view charts in the book; one is for a small (3" or 4") catadioptric or refractor and the other is for a moderately large (8") Dobsonian. The website includes the corresponding charts as seen in either a small Dob or a large Cat or refractor.

For each object, there are also additional links to things like white-on-black drawings, Astronomy Picture of the Day images, data sheets on the objects, and chart numbers where they can be found in some popular star atlases.

Where Do You Go From Here? Here we give suggestions for the next steps to take after you've seen most of what is in this book – websites, magazines, software, clubs and organizations, and stargazing equipment.

What's Up This Month? It's really frustrating to decide that you want to see something like Uranus or Neptune, a lunar eclipse, a Saturn ring-plane crossing, or a favorable apparition of Mercury only to find that you're a month too late and you'll have to wait a year (or a decade or more) for your next chance. Magazines and other websites are the best place to get detailed information on what's happening in the sky, but this link can give you warning of some key things not to miss.

Tables: Here you'll find enhanced versions of the tables at the back of the book, in .pdf and .txt formats; and for those who want to keep records of their own observations of objects in the book, in .xls (spreadsheet) format as well.

Know thyself (and thy telescope)

If you're just starting to observe, the next section of this chapter on knowing and using your telescope may be handy. We explain how your telescope works, and how that affects your choices of what to observe, what lenses and accessories to use, and why objects look different from one telescope to another.

At the end of the book, we talk a bit more about this new fifth edition, and Dan suggests some more advanced books you might like to look into to become a more knowledgeable observer. For those thinking to upgrade their 'scopes, we talk about telescope accessories and the relative advantages of Cats and Dobs. And we provide **tables** of all the objects we've talked about, with their coordinates and technical details. Check it out while waiting for the Sun to set.

But, except for these first and last chapters, the rest of this book is meant to be used outdoors. Get it dog-eared and dewy. After a year of observing, you'll be able to tell your favorite objects by the number of grass-stains on their pages!

At **www.cambridge.org/turnleft** *you can find flipped, mirrored, and inverted versions of all our star charts.*

Using your telescope

Find a place to observe

The easiest place to observe is your own back yard. Certainly, you'll have trouble seeing faint objects if there are bright streetlights nearby; and you'll have trouble seeing anything at all through tall trees! But don't sit around waiting for the perfect spot or the perfect night. Observe from some place that's comfortable for you... say, someplace with easy access to hot tea or coffee (and a bathroom). You shouldn't have to make a major trip every observing night. Save that for special occasions.

If you live in a city, the roof of a building can be a fine place to observe. You'll be above most of the streetlights, and if your building is tall enough you may be able to see more of the sky than your suburban tree-bound friends.

However, you do have to be outdoors. The glass in most windows has enough irregularities that, magnified through the telescope, it can make it impossible to get anything into really sharp focus. In addition, the slightest bit of light in your room will cause reflections in the glass that will be much brighter than most of the things you'll be looking at. You could get around these problems by looking through an open window, but – besides the obvious problem that your field of view would be very restricted – during most seasons you have the problem of warmer air from inside mixing with colder air from outside: light travelling through alternately hot and cold air gets bent and distorted, causing an unsteady, shimmering image.

Find your finderscope

One thing that's well worth your time doing while it's still daylight outside is lining up your finderscope.

The finderscope is a big key to making your telescope fun to use. If it is properly aligned, then you can find most objects in the sky pretty quickly. If it isn't, finding anything (even the Moon!) becomes painfully frustrating.

Your finderscope comes with a set of screws that can be adjusted, either by hand or with a small screwdriver, to point it in slightly different directions. Go outside, and set up your telescope where you can aim it at some distinctive object as far away as possible... a streetlight down the road, the top of a distant building, a tree on a far-off hilltop. Just keep looking through the telescope itself until you see something distinctive.

The time you spend aligning your finderscope will spare you hours of fruitless searching for objects later on.

Now, try to find it in the finderscope. (We assume you've chosen something big enough for you to see through your finderscope as well as in the telescope.) Twist the screws until that object is *exactly* lined up with the cross-hairs of the finder. "Close" is *not* good enough. The set screws always seem to move the finderscope in some totally unpredictable direction, but be patient and keep working until you've got it exactly right.

Then – here's the worst part – be sure all the set screws are as tight as possible, so that the finderscope won't move out of alignment as soon as you move the telescope back inside. Tightening up the screws inevitably goofs up the alignment, so you'll probably have to go though this procedure two or three times before you've got it right.

But it is time well spent. A telescope with a misaligned finderscope is a creation of the devil, designed to infuriate and humiliate and drive stargazers back indoors to watch re-runs. Don't let it happen to you.

Cool it

Temperature changes cause problems when you first take your telescope into the cold outdoors from a warm, heated house. Warm mirrors and lenses may warp out of shape as they cool down and contract. Worse yet, warm air inside the tube can set up distorting currents as it mixes with the cold outside air. This can seriously hurt your telescope's resolution.

For a small telescope, ten minutes' cooling time should prevent this problem. You don't have to stand around doing nothing during this time; just wait awhile before looking at objects that require your telescope's best resolution.

But a Dobsonian has a lot more mirror to cool, and that mirror is at the bottom of a long tube. It takes a lot longer to cool down. Set it up outside just after sunset (no sense letting the Sun's rays heat up the tube) and wait at least half an hour, during twilight, before doing any serious observing.

Evening dews and damps

If every star looks like a nebula to you, check for dew on your lenses!

Dew forms on your telescope because, ironically enough, it's possible for the telescope to cool down too much. On nights with a moderate amount of humidity (not enough to form clouds, but enough to make the grass damp) you'll find that solid objects taken outdoors can radiate away their heat faster than the moist air around them cools off. (Why? Because solid objects can radiate in many wavelengths, including wavelengths not absorbed by the air; but water-laden air absorbs in the same limited wavelengths it radiates, so it takes a longer time for its heat to escape.) This means that your telescope will soon be cooler than the air, and it will start beading up like a cold can of soda pop on a muggy summer day.

Whatever you do, resist the temptation to wipe the surfaces of your lenses. It runs the certain risk of damaging the lenses. And it doesn't do much good in the long run, as they'll just dew up again anyway!

One solution is to blow hot dry air across the lens – try a hair dryer, or hold the lens near the defroster vent in your car, but be sure not to overheat the lens. If you can, bring a dewed-up telescope (or at least your eyepieces and maybe your finderscope if it easily detaches from the main 'scope) indoors for fifteen minutes or so to dry out. There's no point using a lens if you can't see through it!

The simplest solution is to block the telescope lens from radiating its heat. A long black tube (called a *dew cap*) sticking out beyond the lens stops the radiation and keeps the lens warm. Lacking a dew cap, in a pinch you can just roll a piece of black paper around the lens; or point the telescope down towards the ground when you're not using it. Or, if you can afford it, electric powered bands to heat your lenses are available (see page 244). Dan swears by his.

Dobsonian and other Newtonian telescopes, of course, have their mirror at the bottom of just such a long tube, so that's less likely to dew up (though if you're out long enough, they will dew, too). However, their finderscopes and eyepieces can still turn milky with dew before you realize it.

On cold winter nights, moisture trapped inside a telescope with a closed tube (a refractor or catadioptric) can condense on the *inner* sides of the lenses when the telescope cools off outside. The best way to deal with this kind of condensation is to eliminate the warm, moist air altogether. When you take out your telescope into the cold, remove the star diagonal so that the inside of the tube is open to the air, and point the telescope down towards the ground so that the hot air inside the tube can rise out and be replaced by the cooler outdoor air. But on especially dewy nights, beware lest condensation drip down the eyepiece tube into the telescope! This can be a serious problem with the larger catadioptric designs.

Prepare yourself

The whole point of your telescope is to gather starlight and bring it to a good, sensitive detector: your eye. And, just as professional detectors are housed in elaborate containers to keep the environment around them perfect, so too you should keep your eye's container – you! – warm and safe and comfortable.

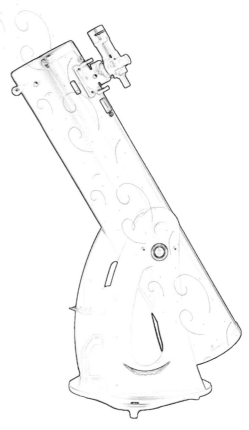

It takes a surprisingly long time to cool down a Dob. Set it up outside just after sunset and wait at least half an hour before serious observing.

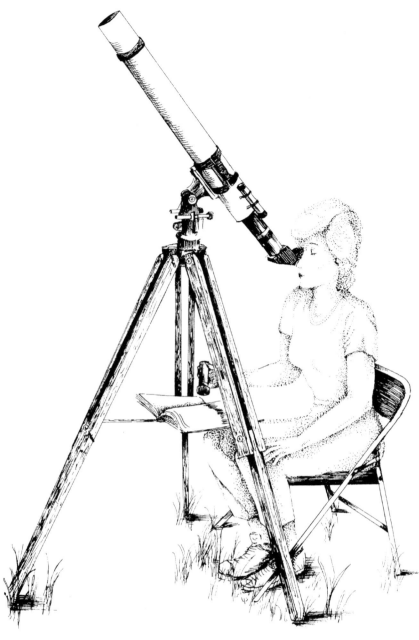

Be prepared for the cold. You'll be sitting still for a long time, so dress extra warmly. In the winter, an extra layer of everything is a good idea, and gloves are a necessity for handling the metal parts of a telescope. A thermos of coffee or hot chocolate can make all the difference in the world. In the summertime, keep your arms and legs covered as protection from the chill, and from bugs as well.

Get a chair or stool to sit on, or a blanket to kneel on, and a table or a second chair for your flashlight and book. These are the sorts of things you may think are too elaborate to bother with, but in fact they're just the sort of creature comforts that allow you to enjoy yourself while you're observing. It only takes a minute to set up a stool and a table, and it will keep you happy for hours. After all, the point is to have fun, not to torture yourself.

It takes time for your eyes to get used to the dark. At first the sky may seem good and black, but you may not see so many stars in it. Only after about fifteen minutes or so will you begin to see the dimmer stars … and the glare from the shopping center ten miles away. That's the point when you can try to observe faint nebulae.

And once you get dark-adapted, avoid lights. You can lose your dark adaptation in an instant, and it can take another ten minutes to get it back again.

You'll want a small red-light flashlight to read your book and star chart. Red fingernail polish on a flashlight lens is a stargazer's time-honored tradition; it makes it easier for your eyes to remain adapted to faint light.

Dobsonian users: align your optics

One of the prices you pay for the large size of a Dobsonian telescope is that the primary and secondary mirrors need to be adjusted almost every night to make sure that they are actually sending the starlight straight into your eyepiece. (Catadioptrics and refractors have their lenses and mirrors fixed at the factory, and usually don't need to be adjusted.) The more the tube gets shaken up, the more likely the mirrors are to get out of alignment. Since these telescopes tend to be rather heavy and bulky, carrying them regularly from your garage or closet to your back yard is just the sort of thing to shake up those mirrors.

How do you align your mirrors? Every telescope is different. The easiest way is to purchase a laser collimation device, but they can be expensive – a good fraction of the cost of the telescope. Still, if you find yourself using your Dob several times a month, they're worth the price.

Get a chair or stool to sit on, or a blanket to kneel on, and a table for your flashlight and book. Red fingernail polish on the flashlight lens is an astronomer's time-honored tradition. It cuts down the glare from the flashlight, so that your eyes can stay sensitive to the faint starlight.

Absent such a system, the simplest way to align your mirror is to aim the telescope at a very bright star and put it out of focus until you see a donut of light with a big black spot (the shadow of the secondary mirror). Adjust the primary mirror, twisting the set screws where it is mounted, until the spot sits right in the center of the donut.

When you're finished: storing the telescope

In big observatories, there's a regular procedure the astronomers go through every night after observing, to arrange the telescope so that the weight of the massive mirror won't cause it to warp out of shape. Your 2" telescope doesn't have that problem! Basically, anyplace that is reasonably clean and cool will serve well as a storage area. (The cooler the telescope is before you take it out, the less time you'll have to wait for it to adjust to the outdoor temperature.)

If you've got the room, you may want to consider keeping the telescope assembled on its mount, ready for use at a moment's whim. The easier it is to get at, the more likely you are to use it; and the more you use it, the easier (and more fun) it'll be to find and observe faint objects in the nighttime sky. As long as it's out of reach of puppies and toddlers, your telescope is better off set up and visible than stored disassembled in a closet where its pieces can get stepped on or lost.

The care and feeding of lenses

Dust is your lenses' constant enemy. Not only will it scatter light every which way, but as the dust moves along the lens it can scratch the delicate optical coatings, leaving behind a permanent scar.

Camera stores sell many different little gizmos that can blow dust off your lenses. Cans of compressed air are popular, for instance; but be careful in case the one you want to use ends up depositing something other than air. (Try it out first on a clean glass from your kitchen; any residue on that glass would also be deposited on your lens, bad news.) Above all, don't use your breath to blow on your eyepiece lens unless you want to mar it with whatever else (like moisture) that is in your mouth!

Fingerprints are a true horror. An imperfection on the lens of as little as a thousandth of a millimeter can degrade the efficiency of your lens; the layer of grease that comes off your finger every time you touch something is just such an imperfection. A year's worth of accumulated fingerprints and dust can turn an eyepiece into a milky nightmare. It's great to let children look through your telescope, but maybe have them use an older lens, lest sticky little fingers wander where you don't want them.

Prevention is the best solution. The only way to keep dust and prints off your lenses is to store them covered until you need them, so they won't get touched even by accident.

Lens caps are easily lost if you're not careful. It is well worth your while to become a fanatic about keeping track of them. Make it a habit always to keep them in the same place – a certain pocket, or a certain corner of the telescope case. And use your lens caps, even while you're observing, whenever you're not looking through the lens itself. It helps prevent you (or your helpful neighbor) from accidentally touching the glass. It can also help in limiting the amount of dew on the lenses.

How to clean your lenses

The default answer is always the same: Don't do it. The fact is, cleaning your lenses is always a risk. It's usually one not worth taking.

If you've been taking good care of your lenses, then they will be so clean that the tiniest imperfection may stand out like an eyesore. One fingerprint and a few dust grains don't really affect the performance of your telescope that much, certainly not enough to risk permanent damage to the optics. What may look glaring when looking at the Moon or a planet might have no discernible effect when looking at deep-sky objects: the cure (cleaning) may be much worse than the disease! This is doubly true of a reflector's primary mirror: it's amazing how much dust can accumulate there without causing problems.

But what if, in spite of all your efforts, there's a perfect thumbprint on your favorite eyepiece? Is it possible to clean it off without hurting the lens? In theory, yes. But the trouble is, every high-quality lens has a thin anti-reflection coating; it gives the lens its characteristic bluish color. Most obvious things around the house that you could think of to clean fingerprints off the lens will either leave spots on this coating or destroy it outright.

And anything you use to apply these cleaning solutions runs the risk of rubbing the grime across the coating, leaving behind permanent scratches far worse than the original fingerprint. What's more, any liquid used in cleaning can settle around the edges of the lens, carrying the grunge there where it can't easily be blown off.

That said, there are good commercial lens cleaning kits available nowadays, and the coating on modern lenses is more durable than in years past. Still, check your lens specifications: some cleaners might be safe on one surface, but damage another. For example, some cleaners use isopropyl alcohol which can stain certain eyepiece coatings. (Distilled water is usually safe.)

A number of websites have useful instructions on the care and cleaning of lenses and mirrors; our website directs you to some of them.

The important things to remember are that you should only use a high quality lens cleaning kit; and that cleaning large surfaces (like the primary mirror of a Dob or, worse, the front corrector lens to a Cat) is far more complicated than cleaning an eyepiece. Above all, remember that the default regarding dirty optics should always be to prevent rather than to clean.

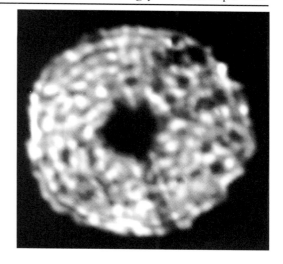

Dobsonian telescope mirrors need to be collimated regularly. The simplest way to align your mirror is to aim the telescope at a very bright star, and put it out of focus so that you can see a big donut of light, looking like this. The black spot is the shadow of the secondary mirror; the black lines, the shadows of the "spider" struts that hold the secondary in place. Adjust your primary mirror (usually there are small screws at the back of the mirror) until the black spot sits right in the center of the donut. Note: this image is not actually from a Dob, but from the 1.8-meter Vatican telescope – all large reflectors need collimation!

Know your telescope

Galileo discovered the four major moons of Jupiter (forever after called the Galilean satellites *in his honor); he was the first to see the phases of Venus and the rings of Saturn; he saw nebulae and clusters through a telescope for the first time. In fact, a careful checking of his observations indicates that he even observed, and recorded, the position of Neptune almost 200 years before anyone realized it was a planet. He did all this with a 1" aperture telescope.*

Charles Messier, who found the hundred deep-sky objects in the catalog that bears his name, started out with a 7" reflector with metal mirrors so poor that, according to one account, it was not much better than a modern 3" telescope. His later instruments were, in fact, 3" refractors.

The point is this: there are no bad telescopes. No matter how inexpensive or unimpressive your instrument is, it is almost certainly better than what Galileo had to work with. It should be treated well. Don't belittle it; don't apologize for it; don't think it doesn't deserve a decent amount of care.

Get to know your optics

An astronomical telescope has two very different jobs. It must make dim objects look brighter; and it must make small objects look bigger. A telescope accomplishes these jobs in two stages. Every telescope starts with a big lens or mirror called the *objective*. This lens or mirror is designed to catch as much light as possible, the same way a bucket set out in the rain catches rainwater. (Some astronomers refer to their telescopes as "light-buckets.") Obviously, the wider this lens or mirror is, the more light it can catch; and the more light it catches, the brighter it can make dim objects appear. Thus, the first important measurement you should know about your telescope is the diameter of the objective. That's called the *aperture*.

If your telescope uses a lens to collect light, it's called a *refractor*; if it uses a mirror, it's called a *reflector*. In a refractor, the light is refracted, or bent, by a large lens called the objective lens. In a reflector, the light is reflected from the primary mirror, sometimes called the objective mirror, to a smaller mirror sitting in front of the objective, called the secondary mirror. In both cases, the light bent by the objective is further bent by the eyepiece lens, to make an image that can be seen by your eye.

A reflector where the light is sent back through a hole in the main mirror is a *Cassegrain* reflector. *Catadioptric* reflectors have, in addition, a lens in front of the primary mirror that allows the telescope tube to be much shorter. A specific catadioptric design that works well for amateurs is one called the *Maksutov*

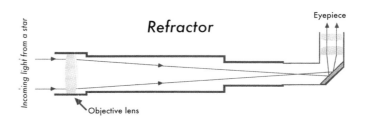

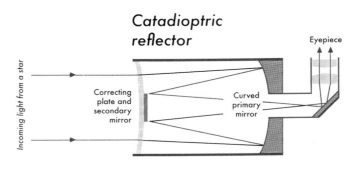

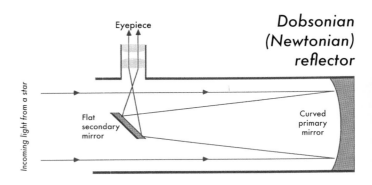

reflector; bigger Cats often use a slightly different design, called a *Schmidt*.

A reflecting telescope in which the secondary mirror bounces the light sideways through a hole near the top of the telescope tube is a *Newtonian* reflector. The most popular amateur version of the Newtonian design nowadays is the *Dobsonian*, a Newtonian with a simple but elegant alt-azimuth mount invented by John Dobson.

The primary mirror (or, in a refractor, the objective lens) bends the light to concentrate it down to a small bright image at a point called the *focal point*. The light has to travel a certain distance from the objective until it is fully concentrated at this point; this distance is called the *focal length*.

The small, bright image made by the objective seems to float in space at the focal point. Put a sheet of paper at that spot (or piece of film, or a photographic CCD chip, or a slice of ground glass) and you can actually see the little image that the objective makes. This is what's called the *prime focus* of the telescope. If you attach a camera body there, with the camera lens removed, you can take a picture. The telescope is then just a large telephoto lens for the camera.

The second stage of the telescope is the eyepiece. One way to describe how the eyepiece works is to think of it as acting like a magnifying glass, enlarging the tiny image that the objective lens makes at the focal point. Different eyepieces give you different magnifications. The shorter the eyepiece focal length, the higher the magnifying power – but with a fainter image, and a smaller field of view. You'll find yourself using high power less often than you might think.

Get to know your tripod

Small telescopes often come on a tripod similar to a camera tripod, which lets you tilt the scope up and down or turn it left and right. The up–down direction is the *altitude*; swiveling left and right moves you through the *azimuth*. Such a mount is called an *alt-azimuth mount*. For a small telescope, this is a perfectly reasonable sort of mount. This type of mount is lightweight, requires no special alignment, and it's easy to use since all you have to do is point the telescope wherever you want to look.

A popular, inexpensive variant of the alt-azimuth mount is the *Dobsonian*. Instead of a tripod supporting the center of the telescope tube, a Dobsonian is mounted near the bottom, where the mirror sits. Two design features keep the telescope from tipping over: the base is made especially heavy (and the heaviest part of the telescope itself is the mirror, which is already down at that end in a Newtonian design) and the tube is made of some lightweight material. Also adding to the low cost and simplicity of use, the two axes that the telescope moves about to point at stars have Teflon friction pads, which (when they're tightened just right) let you move the telescope from position to position, but hold it in place when you let go. (For more about Dobs, see page 245.)

Once you're focused in on an object in the sky, you'll discover that the stars move slowly out of your field of view. Using an alt-azimuth mount, you have to constantly correct in both directions; as the object you're looking at goes from east to west, it also moves higher or lower in the sky. With a little thought it's not hard to understand why. The stars are rising in the east and setting in the west, and so they're slowly moving across the sky. What's really happening, of course, is that the Earth is spinning, carrying us from one set of stars to another.

To correct for this motion, a fancier type of mount can be found (usually on bigger telescopes), called an *equatorial mount*. This can be thought of as an alt-azimuth mount, only tipped over. The axis that used to be pointing straight up now points towards the celestial north pole. (It is tilted from the vertical at an angle of 90° minus the latitude where you're observing.) It's called the *equatorial axis*. With this sort of mount, you can just turn the telescope about this tipped axis in the direction opposite to the Earth's spin, and so keep the object you're observing centered in the telescope.

You can even attach a motor to turn the telescope for you. The motor is called a *clock drive*; it turns the telescope about the tipped axis at half the speed of the hour hand on a clock. Thus, it makes one rotation every day (actually, every 23 hours 56 minutes, since the stars rise four minutes earlier every night).

The extra tip of the equatorial axis can make it awkward at first to find what you're looking for, however. And, depending on the design of the mount, it may shove the telescope off to one side of the tripod. The telescope's weight then has to be balanced out by heavy counterweights. That makes this sort of mount quite burdensome to lug around.

If your telescope does have an equatorial mount, the first thing you must do when you go observing is to make sure you set up the tripod in the right direction; the equatorial axis must be lined up with the spin axis of the Earth. Conveniently for folks in the northern hemisphere, the Earth's axis has pointed (for the past thousand years or so) at a spot very close to a bright star called Polaris. So all you have to do is make sure that the equatorial axis of your tripod is pointing straight at Polaris. But remember, every time you set up the telescope you've got to line it up again!

(Southern-hemisphere observers are out of luck; there is no south polar star.)

If you are planning to use your telescope with a motorized drive to take time-exposure astrophotographs, you'll need more careful alignment. But for casual observing, lining up on Polaris is quite good enough.

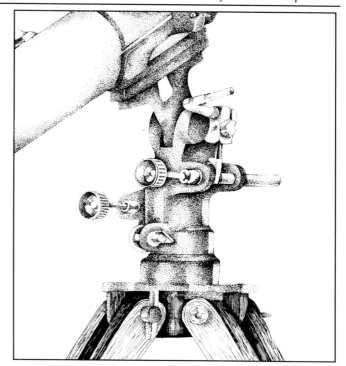

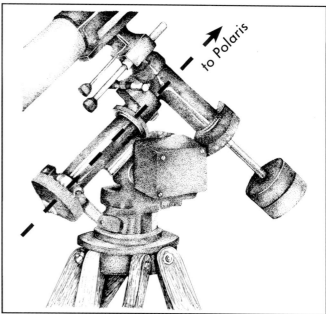

There are two basic types of mount, the alt-azimuth (top) and the equatorial (bottom). The equatorial mount is effectively an alt-azimuth mount which has been tilted so that one axis turns with the Earth.

Basic telescope math

Learning to calculate the magnification, resolution, or other properties of your telescope is a great way to get to know in detail what it is capable of doing. But be sure you use consistent measurement units: divide millimeters by millimeters, or inches by inches; don't mix them up!

A = **aperture** of telescope (mm): the diameter of the primary mirror of a reflector or the objective lens of a refractor

L_T = **focal length** of telescope (mm): the distance from the main mirror or lens of a telescope to where light from a star is focused to a point

L_E = **focal length** of eyepiece (mm): the distance from the eyepiece lens to where distant light is focused to a point

f = **focal ratio** of telescope = L_T/A

M = **magnification** = L_T/L_E

R = **resolution**, the smallest angle (in arc seconds) the telescope can see

V_A = **apparent field of view**: the angular size (in degrees) of the circle of light you can see when you hold an eyepiece to your eye

We measure the sizes of astronomical objects in arc seconds, arc minutes, and degrees. One degree is sixty arc minutes, written 60'; and each arc minute is 60 arc seconds, or 60". The full Moon is half a degree (30', 30 arc minutes) in size. The planetary nebulae in this book are typically about one arc minute across. An easy double star like Albireo has a separation of about half an arc minute, or 30" (30 arc seconds).

Resolution

$$R \approx 120/A \quad (A \text{ in mm})$$
$$\approx 4.5/A \quad (A \text{ in inches})$$

Resolution is the measure of how much detail your telescope can make out. It determines how far apart two members of a double star must be before you can see them as individual stars, and limits how much detail you can see on the surface of a planet.

Resolution is measured in terms of the smallest angular distance (in arc seconds) between two points that can just barely be seen as individual spots in the telescope image – for example, the separation of a close "cat's-eyes" pair of double stars.

There is a theoretical limit (resulting from the wave nature of light) to how much detail any telescope can resolve, which is approximated by the formula given above. Assuming good conditions, a telescope with a 60 mm wide objective mirror should be able to resolve a double star with a separation of 2 arc seconds. In practice, of course, you'd need very steady skies to do so well.

By this formula, an 8" Dobsonian should have a resolution limit of just over half an arc second. But in reality, that never happens. The general unsteadiness of the sky, even on the best of nights, means an amateur observer generally can't expect to do better than a 1 arc second resolution.

In some cases, the human eye is clever enough to get around this resolution limit. The eye can pick out an object that is narrower than the resolution limit in one direction, but longer than that limit in another direction, especially if there's a strong brightness contrast. The Cassini Division in the rings of Saturn is an example.

You can identify double stars that are a bit closer than the resolution limit if the two stars are of similar brightness; the double will look like an elongated blob of light. On the other hand, if one star is very much brighter than the other, you may need considerably more than the theoretical separation before your eye notices the fainter star. Experience helps.

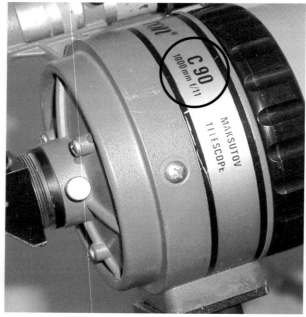

The easiest way to determine a focal length is simply to find the numbers written on the side of your telescope or eyepiece. The eyepiece to the left has a focal length of 25 mm; the lens/mirror combination of the Maksutov telescope on the right has a combined effective focal length of 1000 mm. It's important in any calculations to use the same unit of length – usually millimeters – for all lengths.

Note that this telescope also tells you its aperture – 90 millimeters, hence the name "C 90" – and the f ratio, f/11, which is (roughly) 1000 divided by 90. Using the 25 mm eyepiece in this telescope would give you a magnification of 40× (i.e. 1000 ÷ 25).

Magnification

$$M = L_{telescope}/L_{eyepiece}$$

To find the magnification you get with any of your eyepieces, take the focal length of your objective lens, and divide it by the focal length of the eyepiece (usually written on its side).

Be sure both numbers are in the same units. Nowadays, most eyepieces list their focal length in millimeters, so you must also find the focal length of your objective lens in millimeters.

For instance, with a telescope whose objective has a focal length of 1 meter (that's 100 cm, or 1000 mm), an eyepiece with a focal length of 20 mm gives a power of 1000 ÷ 20, or 50× (50 power). A 10 mm eyepiece would give this telescope a magnification of 100×. When you're making this calculation, don't confuse focal length with aperture!

$$M_{max} \approx 2.5 \times A \quad (A \text{ in mm})$$
$$\approx 60 \times A \quad (A \text{ in inches})$$

Maximum useful magnification is a consequence of your telescope's *resolution*. The image formed by a telescope's primary lens or mirror is never perfect, so there is a limit to how big you can magnify that image and see anything new. Looking through a telescope at extremely high magnification won't help any more than looking at a photograph in a newspaper with a magnifying glass lets you see any more detail. (A newspaper photo is made of little dots; a magnifying glass just shows the same dots looking bigger.) Once you've reached the limit of resolution in the original image, further magnification won't give you any more detail. Of course, even for a bigger telescope, the sky's unsteadiness limits your useful magnification to about 400×.

$$L_{eyepiece} < 7 \times f$$

Longest useful eyepiece: Low power eyepieces (the longer ones) gather light from a wider area of the sky. The lower the power, the wider the circle of light that comes out of the eyepiece: the "exit pupil." But the width of your own eye's pupil is about 7 mm (it gets smaller as you get older). An exit pupil wider than your eye's pupil is a waste, if your goal is to take advantage of your telescope's full aperture. Thus the longest useful eyepiece is just your eye's pupil diameter (i.e. 7mm) times the focal ratio (f) of the telescope.

Of course, there can be reasons to go to an even lower power, for instance to bring more of a large nebula into your field of view. However, in a Dob (or any reflector) if the power is too low, the shadow of the secondary mirror becomes visible as a dim spot in the middle of your view.

Focal ratio *f*

$$f = L_{telescope}/A$$

Focal ratios for refractors are typically large, f/12 to f/16. Catadioptric reflectors typically have *f*-ratios of f/8 to f/12. Newtonians, including Dobsonians, commonly have low *f*-ratios of f/4 to f/7.

The low *f*-ratio of Dobsonians is generally a good thing; it allows for large aperture in a relatively compact package, and it makes it easier to get beautiful views of diffuse deep-sky objects. But telescopes with low *f*-ratios have much greater problems with a field distortion called *coma*, which turns stars near the edge of the field from sharp points into blurry "v"shapes. Also, when *f* is small you need a more expensive eyepiece with a shorter focal length to get high magnifications to observe planets or double stars. And there is a limit to the longest useful eyepiece (minimum useful magnification) for small-*f* telescopes.

Field of view

$$V = V_A / M$$

The *field of view* V tells you how wide an area of the sky you can see in your eyepiece. When you hold a typical inexpensive eyepiece up to your eye, the field of view appears to be about 50° (the *apparent field*). Since the view through a telescope is magnified, the part of sky you can actually see using this eyepiece is equal to the eyepiece apparent field, divided by the magnification. Thus, a typical low-power eyepiece, about 35× or 40×, shows you roughly 80' of the sky; a similar medium power eyepiece (75×) gives you a 40' view, while a high power eyepiece (150×) should show you 20'. These are the values we assumed for our circles of low-power, medium-power, and high-power telescope views in this book.

Nowadays, most eyepieces that come with telescopes have apparent fields of view of about 50°. But even some modestly-priced eyepieces can range up to 70° apparent field, and special (expensive) designs give apparent views of up to 100°.

Thirty years ago, apparent fields of about 40° were more common. Thus, if you have an older eyepiece, you will see less of the sky than what we indicate here.

To estimate the field of view of an eyepiece, take it out of the telescope, hold it to your eye, and look through it at a bright lamp in a dim room. Move your head so that the lamp (looking like a blurry bright spot) appears at the edge of your field of view, then turn until it sweeps across to the opposite side of the eyepiece view. The angle you turn through is the eyepiece's apparent field of view.

The Moon

You don't need a book to tell you to look at the Moon with your telescope. It is certainly the easiest thing in the nighttime sky to find, and it is probably the richest to explore. But it can be even more rewarding if you have a few ideas of what to look for.

Getting oriented

Setting the stage

The Moon in a small telescope is rich and complex; under high power in a Dob, you can get lost in a jumble of craters and all the mare regions seem to meld together. So the first thing to do is to get oriented.

The round edge of the Moon is called the **limb**. Since the Moon always keeps the same side facing towards the Earth, craters near the limb always stay near the limb.

The Moon goes through **phases**, as different sides take turns being illuminated by the Sun. The whole sequence takes about 29 and a half days, the origin of our concept of "month" (think: "moonth"). This means that, except at full Moon, the round disk we see will always have one part in sunlight, one part in shadow. The boundary between the sunlit part and the shadow part is called the **terminator**.

The terminator marks the edge between day and night on the Moon. An astronaut standing on the terminator would see the Sun rising over the lunar horizon (if the Moon is waxing; or setting, if it's waning). Because the Sun is so low on the horizon at this time, even the lowest hills will cast long, dramatic shadows. We say that such hills are being illuminated at "low Sun angle." Thus, the terminator is usually a rough, ragged line – unlike the limb, which is quite circular and smooth. These long shadows tend to exaggerate the roughness of the surface.

This means that the terminator is the place to look with your telescope to see dramatic features. A small telescope has a hard time making out features on the Moon smaller than about 5 km (or roughly 3 miles) across; but along the terminator, hills only a few hundred meters high can cast shadows many kilometers long, making them quite easily visible in our telescopes.

Basic geography (lunagraphy?)

We see two major types of terrain on the surface of the Moon:

Around the limb of the Moon, and in much of the south, is a rough, mountainous terrain. The rocks in this region appear very light, almost white, in color. This terrain is called **highlands**.

In contrast are the dark, very flat, low-lying areas; they are called the **mare**. The term mare, Latin for "sea," describes how the first telescope observers interpreted these flat regions. One large region of the Moon filled with such dark mare material is called an ocean, Oceanus Procellarum.

Throughout the Moon, especially in the highlands but occasionally in mare regions as well, are numerous round bowl-shaped features called **craters**. Indeed, several of the mare regions themselves appear to be very round, as if they were originally very large craters; these large round depressions are referred to as **basins**.

Lunar phases

As the Moon proceeds in its phases, the area lit by the Sun first grows in size, or **waxes**, from the thinnest of crescents until it reaches *full Moon*; then it shrinks back down, or **wanes**, to a thin crescent again. The point where the Moon is positioned between the Earth and the Sun, so that only the dark, unlit side is facing us, is called *new Moon*.

As it proceeds in its orbit past new Moon, the Moon will be visible in the evening sky, setting in the west soon after the Sun sets. By the half Moon phase (also called *first quarter*) it will be at its highest, and due south, at sunset, and it will set at about midnight. The waxing Moon is shaped like the letter D; think of it as "Daring." The full Moon, its earthward face fully illuminated by the Sun, must rise as the Sun sets, and set at sunrise. The waning Moon will not rise until late in the evening, but stays in the sky after sunrise, well into the morning. It's shaped like the letter C; think "Cautious." (In the southern hemisphere, the D and C shapes are reversed, of course. You could say that a Moon visible in the evening is a Corker; one that won't rise until after midnight is Dill.)

We emphasize the waxing Moon, visible in the evening when most of us observe, in the following pages.

Nights of the day

It takes 29.5 Earth days (and nights) to complete one cycle of phases, so that means the terminator moves across the surface of the Moon at a rate of about 12.2° per day, or half a degree per hour. Not surprisingly, then, the appearance of the Moon changes radically from night to night. Indeed, once you start looking closely in a telescope you can easily notice that change occurring over a matter of a few hours, right before your eyes, if you watch a particular spot on the Moon's surface.

Thus, to plan what to look for, it is useful to know what night of the "day" you're looking at. That's how we will describe where to look, and what to look for, in the following pages. But we'll concentrate on things to see at the terminator, where the shadows are most dramatic and subtle features stand out the most… and where the changes from hour to hour are the most obvious.

East is east, and west is west… except on the Moon

Traditionally, Earthbound astronomers have referred to the limb of the Moon that faces our north to be the northern limb of the Moon, and likewise for south; and the limb of the Moon facing our west they call the west side of the Moon, and likewise for east.

This is perfectly logical if you're observing the Moon from Earth. It gets very confusing if you're an astronaut on the Moon itself, though, because (as we saw in the section "How to use this book") these directions are just backwards from what we're used to on terrestrial maps. Thus, with the dawn of the Space Age, NASA created a new terminology for their astronauts who would be exploring the Moon. Now we've got two conflicting ways to refer to directions on the Moon. What's referred to as "west" in one book will be "east" in another.

This book is designed for casual Earthbound observers, not astronauts. But looking at the Moon's surface does feel like looking at a map. So which convention should we use?

The two authors of this book took a vote. It was a tie.

And so, while north and south are unambiguous, for the other directions we'll merely direct you to look "towards the limb" or "towards the terminator." And remember, telescopes with star diagonals give you a mirror image of our photos!

Special events on the Moon

Eclipses

Roughly twice a year the full Moon passes into the shadow of the Earth, and we have a ***lunar eclipse***. The eclipses last for a few hours, not the whole night; and of course they can only be seen when the Moon is up, which is always at night during a full Moon. Whether or not you can see a particular eclipse depends on whether or not it's nighttime in your part of the world when the eclipse occurs. To find these times, check the links on our website **www.cambridge.org/features/turnleft/the_moon.htm**.

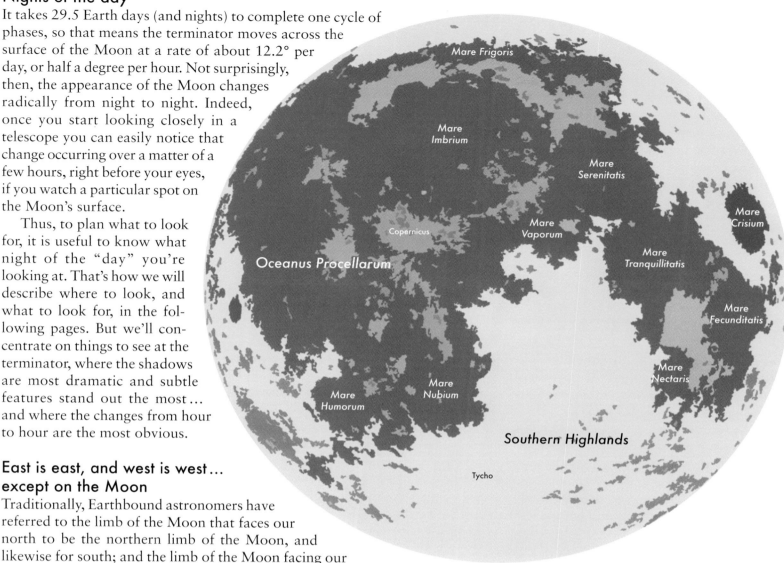

Mare Frigoris

Mare Imbrium

Mare Serenitatis

Mare Vaporum

Mare Crisium

Copernicus

Mare Tranquillitatis

Oceanus Procellarum

Mare Fecunditatis

Mare Nubium

Mare Nectaris

Mare Humorum

Southern Highlands

Tycho

The Moon is the one object in the sky where photos give an accurate idea of what you can see in a small telescope. The full lunar disks (pp. 26–36) are by permission of Robert Reeves (www.robertreeves.com), who assembled these from digital images he took with Celestron 8" and Meade 10" catadioptric telescopes. Most of the close-up images on pages 25–37 are by Rik Hill, who used telescopes from a 3.5" Questar to a 14" C-14 in his Tucson backyard. The rest are from Eric Douglass' digital version of the Consolidated Lunar Atlas *compiled in 1967 by Gerard P. Kuiper, Ewen A. Whitaker, Robert G. Strom, John W. Fountain, and Stephen M. Larson at the Lunar and Planetary Laboratory, University of Arizona.*

Observe in the Earthshine during the first days after new Moon!

Look for:
1. *Oceanus Procellarum (dark, irregular)*
2. *Mare basins (dark, round)*
   a. *Mare Humorum*
   b. *Mare Nubium*
   c. *Mare Imbrium*
   d. *Mare Serenitatis*
   e. *Mare Tranquillitatis*
3. *Grimaldi (dark basin at dark limb)*
4. *Crater Aristarchus (bright spot)*
5. *Large craters (pale circles)*
   a. *Copernicus*
   b. *Kepler*
6. *Crater Tycho (pale circle in south surrounded by a white ring, then a darker ring; then, radiating out from these rings, white rays.)*
   Look for Tycho's rays:
   a. *between Humorum and Nubium*
   b. *south toward the dark limb*
   c. *toward Nectaris.*

Recall how the Sun arcs high overhead during the summer, but seems to scoot along the horizon during the winter. Now realize that the daytime sky of the summer is the nighttime sky of the winter. (When a total eclipse in July blocks out the Sun's light, the brighter stars of Orion and other December constellations can become visible!) So at night during the summer, the Moon (and the planets) seem to scoot low along the horizon; while during the winter, the Moon and planets travel a path high overhead, following the same path the Sun will be on six months later.

New Moon

There is something pleasant, beautiful, and reassuring about the return of the new Moon in the evening sky. The best months for spotting a very new Moon are around your hemisphere's spring equinox, when the ecliptic makes a steep angle to the horizon, so the Moon is relatively far above the just-set Sun. Check the time of new Moon for good opportunities – a hair-thin crescent less than about 36 hours after new Moon can be a challenging but thrilling sight. It will be hard to see, so make sure to be prepared to know where to look and to have a horizon clear of obstacles, cloud, or haze. Look just after sunset, preferably with binoculars. The Moon may look beady near the ends of the crescent, due to roughness in its topography.

During the first nights of the new Moon, look into the dark side illuminated by *Earthshine*, sunlight reflected off Earth onto the Moon, for a sneak peek at things you won't get to see again for another couple of weeks.

Occultations

The Moon drifts slowly eastward through the stars of the zodiac, at a rate of about its own diameter per hour. The side that's dark during the waxing phases of the Moon is the leading edge.

If you find a star just beyond this dark limb, watch it in the telescope for about ten or fifteen minutes. As the unlit leading edge of the Moon passes in front of it, the star will suddenly appear to blink out. It is an eerie, surprising, and exciting sight even when you're expecting it. Somehow, one's subconscious is shocked at such a sudden and violent change occurring with no sound at all.

Occultations are easier to observe when the Moon is in its crescent phase, because then in the telescope you can see the dark limb slightly illuminated by Earthshine. This makes it easier to judge how long you need to wait for the occultation.

Some occultations seem to go in double steps, with the star's light blinking from bright, to dim, to off. This happens when close double stars become occulted. Then, when the star reappears on the other side of the Moon, it will likewise seem to "turn on" in stages. Both brightness and color can change in steps like this. For example, when the Moon occults Antares, its greenish seventh-magnitude companion reappears several seconds before the brighter red star.

If the edge of the Moon just grazes the star, then the star can blink several times as it passes behind lunar mountains and valleys. This is called a *grazing occultation*.

Planets get occulted, too! In fact, since both the Moon and the planets follow roughly the same path across the sky (the path of the ecliptic through the zodiac constellations), such occultations are rather common. It is astounding and delightful to see the edge of one of Saturn's rings, or one of the horns of the crescent Venus, peeking out from behind a lunar valley.

Librations

As everyone knows, the Moon always keeps the same face to the Earth. The time it takes for the Moon to spin once on its axis is the same as the time it takes to go once around the Earth; as seen from Earth, the two motions cancel. That's why we only get to see half of the surface of the Moon from Earth.

But as everyone also knows, what "everyone knows" isn't always exactly right.

For one thing, direction of the Moon's pole is slightly tilted (by 6.7°) from the plane of its orbit, so at times we can sneak a peek "under" or "over" the poles of the Moon. But in addition, the Moon's orbit around Earth isn't exactly a circle and so it doesn't move at exactly the same rate as it spins; sometimes it's slower, sometimes it's faster, and so at different times in its orbit we see first the east, then the west sides turn slightly towards us. As a result, we can actually see nearly 60% of the Moon's surface. It's fun to keep an eye out for the extremes of these *librations* when otherwise hidden bits of the Moon creep into our field of view.

This also means that the arrival of features that we identify with a certain "night" of the lunar day can vary significantly from month to month. Objects near the poles, which get tilted into (or out of) the shadows, can appear days before or after their average appearance date. If what you are seeing isn't what we're describing, check our descriptions of the night before (or after) your particular night.

Advanced lunagraphy

The Moon is much like a small planet orbiting the Earth. Indeed, one could argue that the Earth and Moon make up a double planet system, which do a dance around each other as they both orbit the Sun.

The current preferred theory supposes that the material forming our Moon came from a protoplanet, perhaps as big as Mars, impacting the Earth while it was forming four and a half billion years ago. Its debris formed a swarm of meteoroids that fell together into a ring around the Earth and eventually snowballed into the Moon.

The **craters** we see all over the surface of the Moon are the scars of meteorite impacts from the final stages of the Solar System's formation, the blast holes formed as meteoroids exploded upon impacting the Moon. The material blasted out from these explosions eventually had to fall back to the surface of the Moon. Most of the material is turned up out of the crater, forming a hummocky rim; other pieces, blasted farther out of the crater, made their own craters. Thus, around the bigger craters you'll often see lots of smaller craters in a jumbled strip of land around the crater at least as wide as the crater itself. The region including both the hummocky rim and the jumbled ring of secondary craters is called the **ejecta blanket**.

Some of this flying debris travelled hundreds of kilometers away from its crater. The churned-up surface where it landed is lighter in color than the surrounding, untouched rock, and so we see these areas as bright **rays**, radiating away from the crater. These rays are most easily seen during full Moon.

The highlands, with the largest number of craters visible, are clearly the oldest regions. After this crust was well formed, a few final large meteoroids crashed into the surface, making the round **basins**. The mare regions were formed over the next billion years as molten lava from deep inside the Moon erupted and filled the lowest and deepest of the basins.

Relatively few craters are seen in these regions, indicating that most of the accretion of impactors had already finished before these basaltic lavas arrived. The lava flowed in tubes beneath the mare surfaces; once the flow stopped, some of these tubes collapsed to form long cracks, called **rilles**. As the pools of lava cooled, they shrank in volume, causing the surface to buckle and form **dorsa**, or wrinkle ridges.

The Moon today... and tomorrow

Since the time of mare volcanism, which ended about three billion years ago, the Moon's surface has been essentially unchanged. An occasional meteoroid still strikes, forming fresh-looking craters such as Tycho. But more commonly, the surface gets peppered with innumerable micrometeorites, the tiny bits of dust that make shooting stars when they hit the Earth. On the Moon, with no air to stop them, they have pulverized and eroded the rocks, the hills, and the mountains into the soft, powdery surface seen by the Apollo astronauts.

By seeing how fresh a crater looks, you can get an idea of its relative age. Likewise, by counting how densely packed the craters are in a given region, you can tell (relatively) how old it is. The Apollo samples, taken from a variety of regions and including pieces of the ejecta from fresh craters such as Tycho, allow us to peg some real numbers (determined by decay of radioactive elements in the minerals) to these relative ages. From them we learn that the oldest highlands are around four and a half billion years old. Giant basins were formed by impacts about four billion years ago, then flooded with mare basalts over the next billion years. Since then, nothing much except the occasional impact has occurred. Tycho, for instance, was formed by a meteorite strike about 100 million years ago.

More recent impacts of comets and water-rich meteorites onto the Moon have added one final ingredient to the lunar surface. In 1998, the Lunar Prospector satellite orbiting the Moon found evidence that ice is mixed into the soil in the cold regions around the lunar poles. In 2009, a probe smashing into one of these always-shadowed polar craters confirmed this result, splashing water vapor out of the crater to where it could be detected. This ice could provide the water needed to make human settlements possible on the Moon within the not-too-distant future.

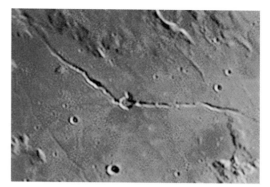

Rilles: *Hyginus Rille looks like a pair of cracks radiating from Hyginus Crater*

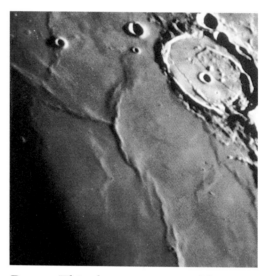

Dorsa: *This dorsum, or wrinkle ridge, is found at the limbward edge of Mare Serenitatis*

The Apollo astronauts themselves have already left their mark on the Moon. There are new craters now where spent rocket stages crashed into its surface. And at the landing sites themselves are the footprints of the astronauts, undisturbed by air or weather, which could outlast any trace of humankind on Earth.

Crescent Moon: Nights 3–5

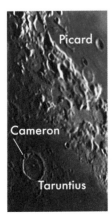

Nights 0–2

The Moon is too close to the Sun for easy observing during the first two nights after new Moon. But try observing the full disk in Earthshine, as described on page 24.

Night 3

• A bright crescent! This may be the best night for observing Earthshine, as tonight the Moon is up longer in reasonably dark skies: see page 24 for details.

• **Mare Crisium** is a flattened oval north of the equator, near the bright limb. Its surface is broken only by one north–south *dorsum* and three small prominent craters: **Peirce**, 20 km wide, to the north, with small (10 km) **Swift** just north of Peirce; and **Picard**, also 20 km, to the south.

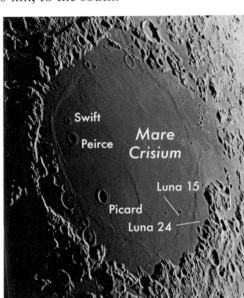

• 50 km from the southern shore of Crisium is where Luna 15 (an unmanned attempt to up-stage Apollo 11 by landing softly and returning samples to the Earth) crashed while Armstrong and Aldrin were on the surface 1200 km away. About 60 km from there, near the limbward shore of Crisium is the site of the 1976 unmanned Soviet probe Luna 24 which succeeded in returning about 170 g of Moon rock to the Earth. It would be 37 years before another craft would made a soft lunar landing: the Chinese Chang'e 3 probe, which landed a rover onto Mare Imbrium in 2013.

• Between Mare Tranquillitatis and Mare Fecunditatis, the 55 km crater **Taruntius** shows a prominent central peak and fresh little (10 km) crater **Cameron** on its northern rim.

• Look between Crisium and the limb. With a favorable libration, you should see an irregularly shaped patch of mare material, strung out very close to the limb. It is called, very appropriately, **Mare Marginis**. South of it on the limb is **Mare Smythii**.

• **Mare Fecunditatis**, south of Crisium, is just appearing along the equator. The low Sun angle nicely shows a complex system of rilles and wrinkle ridges (dorsa). In the north and terminator side of the mare, note an interesting pair of craters, **Messier** and **Messier A**. Note that unlike most craters, Messier is not circular but quite elongated; and it shows two rays radiating only towards the terminator direction. We'll visit these again, to better effect, on Night 12.

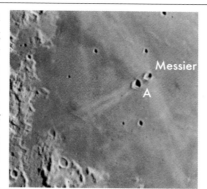

• The very large (130 km) crater **Langrenus**, with its easily visible central peak poking out of the shadow of the crater's sunward rim, is on the south, limbward shore of Fecunditatis.

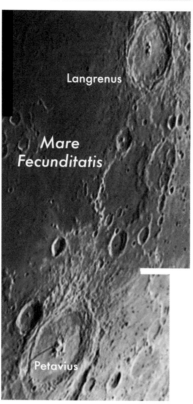

• Nearby, a little bay of mare lava forms a southern extension of Fecunditatis. Just south and limbward of that bay, the huge, 175 km crater **Petavius** also shows a prominent central peak.

Night 4

• In the north, the attractive crater pair of **Atlas** (87 km) and **Hercules** (70 km) are just appearing tonight. Hercules, which is closer to the terminator, has a floor with dark patches and a prominent 20 km crater, Hercules G.

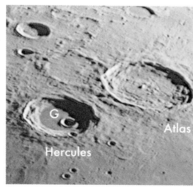

• The first part of **Mare Tranquillitatis** is now visible along the terminator. Halfway up the limbward shore of that mare, look for look for the sharp little (12 km) crater **Cauchy**. This crater will become quite bright closer to full Moon. Flanking Cauchy on either side is a pair of prominent scars across the face of the mare – **Rima Cauchy** close to the shoreline of Tranquillitatis, and **Rupes Cauchy** on the side toward the terminator. Rima Cauchy is a 200 km-long rille – a collapse feature associated with

ancient volcanism. Rupes Cauchy is more complicated – part of it is a rille, but part appears as a fault with the limbward side having moved upwards. As a result, shortly after sunrise in this part of the Moon, the fault casts a long shadow, clearly visible across the nearby mare.

• Look again at Mare Fecunditatis; craters Taruntius and Cameron; Langrenus; and Petavius (see Night 3).

Night 5

• Note the fractured floor of 95 km crater **Posidonius**, with 12 km crater Posidonius A and rilles. South of Posidonius, the 60 km crater **le Monnier** appears as a distinct, rounded extension of dark mare material into the bordering highland material. In 1973, the Soviet Moon rover Lunokhod 2 landed in the southern part of le Monnier and drove more than 36 km across the lunar surface. In the mare, **Dorsa Smirnov** stand out under low Sun angle.

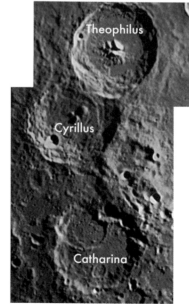

• South of le Monnier find a mountainous promontory between Serenitatis and Tranquillitatis: **Mons Argaeus**. Limbward of it in a small valley surrounded by mountains is the Taurus-Littrow valley, the landing site of Apollo 17 – the last humans to set foot on the Moon.

• At the southern end of Tranquillitatis, away from Serenitatis, the mare protrudes southward into a small bay (about 180 km across), **Sinus Asperitatis**. South of Asperitatis is **Mare Nectaris** (about twice its size, 350 km). Toward the terminator and just south of Asperitatis, three great (100 km) craters **Theophilus**, **Cyrillus**, and **Catharina** emerge tonight into daylight. Once the shadows of the rims leave the crater floors, these craters become complex and fascinatingly different areas to explore at high power. South of Catharina, the long escarpment of **Rupes Altai**, which runs hundreds of kilometers north–south, may be catching the light of the morning Sun.

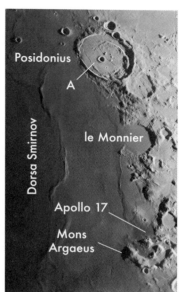

• South of Nectaris, the **Southern Highlands** are starting to get interesting, with a chaotic jumble of craters just starting to emerge. The best, though, is yet to come…

Also see **www.cambridge.org/features/turnleft/the_moon.htm**

Approaching half: Night 6

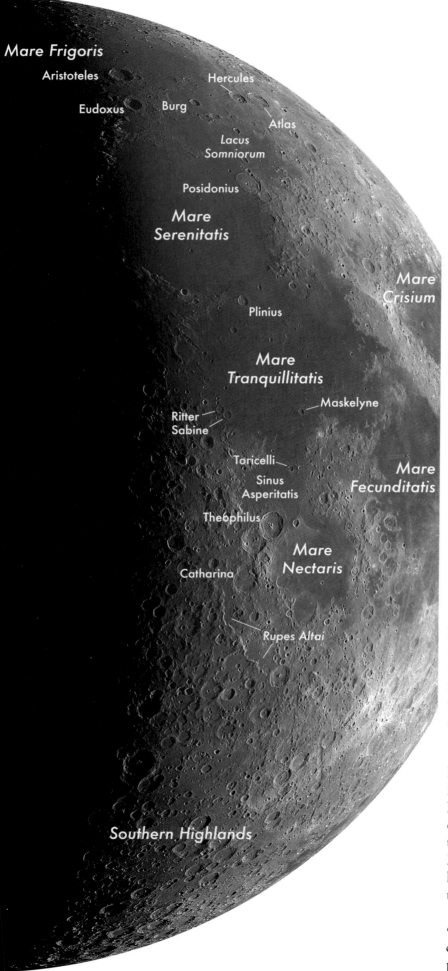

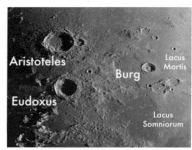

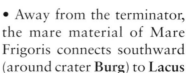

• The patch of mare appearing near the terminator far to the north is the start of **Mare Frigoris**. Cutting off its southern half, near the terminator, the 85 km crater **Aristoteles** should just be making its appearance. The complex, terraced look of its rim is the result of the several fault scarps that are part of the rim structure. The 65 km crater **Eudoxus** to its south shows similar morphology.

• Away from the terminator, the mare material of Mare Frigoris connects southward (around crater **Burg**) to **Lacus Mortis** and, farther south, **Lacus Somniorum**: the eerily named lakes of death and dreams, respectively. Just limbward of Lacus Mortis is **Hercules** (70 km) and **Atlas** (87 km). If you saw them when they were nearer the terminator a night or two ago, you will see the difference Sun angle makes now to the sharpness of their features and the length of their shadows.

• Southward of Lacus Somniorum and angling toward the limb are three large mare in succession: **Serenitatis**, **Tranquillitatis**, and **Fecunditatis**, each between 600 and 750 km across. The smaller oval-looking mare, limbward of Tranquillitatis, is **Mare Crisium**. Making a triangle southward from Tranquillitatis and Fecunditatis is **Mare Nectaris**.

• As described in Night 5, see the spectacular 95 km crater **Posidonius**, on the northern part of the limbward shore of Serenitatis, and **Dorsa Smirnov**, on the plains of Serenitatis.

• A promontory of highland material cuts off part of the connection between Mare Serenitatis and Mare Tranquillitatis. The 45 km crater **Plinius** is at the limbward end of the promontory; 18 km **Dawes** sits farther limbward and to its north, in the middle of the gap between the two mare. A prominent dark band, 20 to 25 km wide, arcs along the southern shore of Serenitatis and then just north of Plinius and Dawes.

Dark material like this, sampled by Apollo 11 and Apollo 17 astronauts, is rich in both titanium and small glass beads. It's the result of the last bits of mare lava freezing into glass instead of forming mineral crystals. A series of rilles, **Rimae Plinius**, run within the dark band along the southern shore of Serenitatis. They are presumably the collapsed lava tubes that brought this last bit of dark material to the surface of the Moon. Then, when the fresh impact of crater Dawes occurred, lighter colored rock from beneath this dark layer was splashed up on top of it, making the lighter apron around the crater.

• Across the northern part of Tranquillitatis runs a string of craters nearly buried under a series of volcanic flows, which produced a startling array of dorsa visible with the low Sun angle at this time. Look limbward and south from Plinius for **Jansen** and its companions, and follow the dorsa south to the partially buried crater **Lamont**. These impacts must have occurred after formation of the mare basin, but before the lava filled it. By contrast, craters **Ross** and **Arago** were made after the lava filled the basin.

Note how this image, at low Sun angle, shows the complex dorsa near Arago; they disappear as the Sun gets higher (as in the image below.)

• Fresh-looking 25 km crater **Maskelyne** is easily the most prominent crater in southern Tranquillitatis. It is located north and limbward of the mare passageway leading from Tranquillitatis into **Sinus Asperitatis**.

• Draw a line from Maseklyne, past the little (6 km) crater **Maskelyne G** (small craters are named for the nearest large crater, with letters assigned in a haphazard order), to the slightly worn-looking 30 km craters **Ritter** and its twin **Sabine**, just south and limbward of Ritter. North of Ritter, on the south-terminator corner of Tranquillitatis, you'll find a smaller pair of identical twins – sharp little (12 km) **Ritter B** and **C** (running north–south).

Step southward from Ritter, to Sabine, to a series of rilles; starting near Sabine, they run away from the terminator for 180 km, all the way to the start of the mare passageway into Sinus Asperitatis. Crater **Moltke** sits just north of their far end (away from Sabine and Ritter).

Now move back to the north. Between Moltke and Maskelyne G, look in high power for a string of tiny (4 km) craters. The first and easiest to find is **Armstrong**; the other two in the line are **Collins** in the middle (Surveyor 5 landed near here), and **Aldrin**, towards Sabine. Step from Maskelyne G to Armstrong. One step farther brings you to *Statio Tranquillitatis*,

as it is named on the official IAU Moon map: **Tranquility Base**, the landing site of Apollo 11. (Another way to find the site is to step from the crater **Manners** to **Arago B**; three steps farther is the Apollo 11 site.)

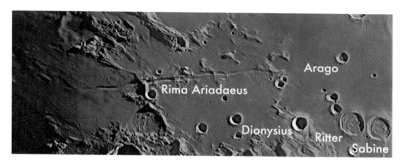

• Step northwards from Sabine to Ritter; four more steps brings you to a large rille, **Rima Ariadaeus**. It runs for 220 km in a nearly straight line from the shoreline of Tranquillitatis towards the terminator.

• The region just north of Rima Ariadaeus looks like it was scraped by a giant rake. These grooves in the craters and mountains all point back towards the center of Mare Serenitatis, and were probably formed by material thrown out of the Moon when the impact that formed Serenitatis occurred.

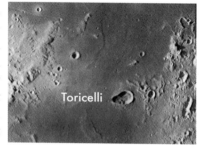

• Sinus Asperitatis, mentioned above, is a 200 km bay between Tranquillitatis and Mare Nectaris. In the north of Sinus Asperitatis, look for the 20 km crater **Toricelli**. It looks misshapen because of a 10 km crater that overlaps it on its nightward flank.

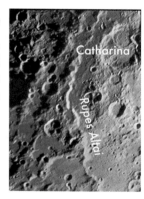

• South of Toricelli is the Theophilus/Cyrillus/Catharina trio (see Night 5). South of Catharina, see the **Rupes Altai** curve almost 500 km south and then limbward, with its marked mountain front catching the morning Sun.

• Now's the time to start exploring in the **Southern Highlands**. Take some time to pick out features there at the start of your observing session, and return sometime later, but before the Moon gets too low in the sky. You'll likely see some significant changes as more of the rough topography is exposed to the rising Sun.

Also see **www.cambridge.org/features/turnleft/the_moon.htm**

Half Moon: Nights 7–8

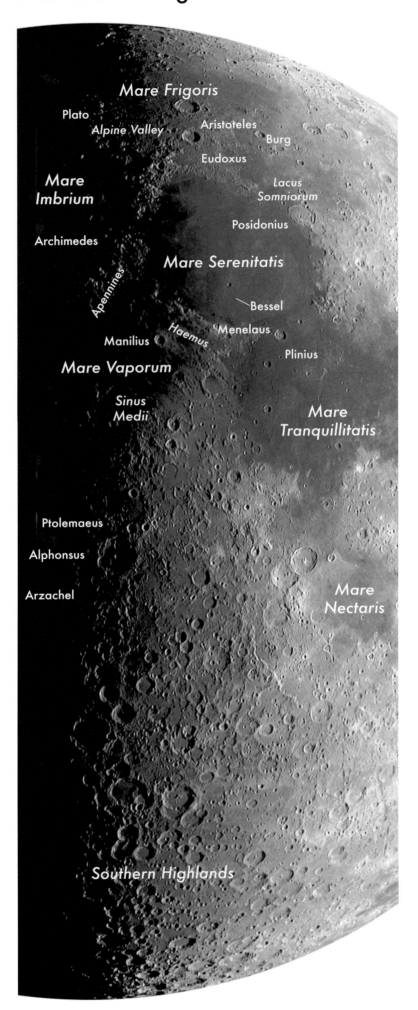

Night 7

• During these nights, the terminator moves very quickly across the lunar surface, changing what you see hour by hour. Keep looking, and compare your views as the shadows change!

• In the north, a fair bit of Mare Frigoris is now visible (depending on librations; see Night 6) paralleling the northern limb. Note the 85 km crater **Aristoteles** and 65 km **Eudoxus**. Southward are the irregularly shaped patches of mare: **Lacus Mortis** (surrounding crater **Burg**) and **Lacus Somniorum**.

• South of Frigoris, **Mare Serenitatis** shows only a few, small craters; the most prominent of them is **Bessel** (15 km), a bit south of the center. Note the faint, light-colored band passing through it, toward Lacus Somniorum, past the floor-fractured crater **Posidonius** (95 km) on the north-limbward shore. This is thought to be an ejecta ray from the crater Tycho, over 2,000 km away; it will become very prominent in the next week.

• The 45 km crater **Plinius** is in the narrow connection between Serenitatis and Tranquillitatis, south of a conspicuous band of darker mare lava (described in Night 6). From now on, subtle differences in mare darkness and color start to become more evident, evidence of variations in the composition of the mare basalts due to when and from how deep within the Moon they erupted.

• Running from Plinius toward the terminator, along the southern shore of Serenitatis, see the **Montes Haemus**, which were upthrust as part of the Serenitatis impact event. Along its spine is the 27 km crater **Menelaus**. From there, look toward the terminator for the 40 km crater **Manilius**, with its central peak emerging from the shadow of its rim.

• At this point of the month the **Southern Highlands** are spectacular, saturated with a nearly impenetrable mass of craters. Tonight and for the next few evenings, wandering through it brings much the same delight as wandering with a telescope through the richest parts of the Milky Way – pleasant sights seem to trip over each other.

The sights of these jumbled Southern Highlands can change noticeably in the space of just an hour. Find a crater right on the terminator and sketch it (or, if you're a poor artist, try to memorize the view). An hour or two (sometimes just minutes!) later, the view will be altered by the emergence of a new bright point of light where a mountaintop has caught the light of the rising Sun.

Night 8

• South of Manilius and toward the terminator is **Mare Vaporum**; the southern extension of that mare is **Sinus Medii**, a small sinus (bay) of mare that is just emerging these nights. It gets the name "Medii" because it is located at the very middle of the visible lunar disk. At the norther edge of the sinus are the **Rima Hyginus**, two parts separated by the small (11 km) crater **Hyginus**. When the Sun is low (Night 7) the terrain is seen to be quite hummocky; as the Sun rises (shown here) the relief is less visible but the darker mare stands out more against the lighter colored highlands.

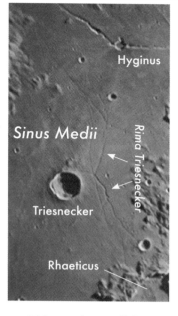

• South of the crater Hyginus is **Rima Triesnecker**, a series of intertwined rilles running 200 km southward to the degraded 45 km crater **Rhaeticus**. Halfway along the rilles is the odd 25 km crater **Triesnecker**. Unlike Rhaeticus, Triesnecker still has a well-preserved central peak and intact walls.

• In the far north, **Mare Frigoris** with craters **Aristoteles** and **Eudoxus** is now completely seen, a long and narrow band of mare running east–west. Separating Frigoris and Mare Imbrium to its south are the **Alps**. About 200 km toward the terminator from Aristoteles is the start of **Valles Alpes** (the **Alpine Valley**), a remarkable fault-bounded valley cutting southward and toward the terminator and **Mare Imbrium** from the southern shore of Mare Frigoris.

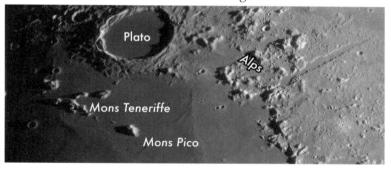

• At the northern end of the Alps near the terminator, within the bright material separating Frigoris from Imbrium, is the dark-floored plain of the 100 km crater **Plato**, very flat and almost completely devoid of craters. Just offshore from Plato, a short distance into Mare Imbrium, the morning light is casting long shadows from the irregularly shaped, 100 km-long **Mons Teneriffe** and, a bit farther from the terminator, the roughly 20 km wide **Mons Pico**. These mountains stand close to two and a half kilometers above the mare plain.

• The **Apennine Mountains** run from near Mare Serenitatis 600 km south and toward the terminator along the southern boundary of Imbrium. With peaks over 5 km high, this is the most impressive mountain range visible on the Moon.

• Approaching the Apennines, across Imbrium from Plato and the Teneriffe Mountains, lie three very prominent and very different craters. The largest is 83 km **Archimedes**, which (like Plato) has a nearly flat lava-flooded plain and no central peak. Look for subtle bands of white running nearly east–

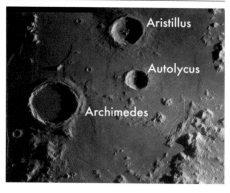

west across the plain of Archimedes, pointing toward the rough young 40 km crater **Autolycus**, only about 80 km away. They're probably ejecta from the impact that created Autolycus. Just north of Autolycus is the 55 km crater **Aristillus**, the youngest of the three. It has a well-developed, complex central peak. Look for the rough hummocky ejecta directly surrounding Aristillus. The rim of the crater is not perfectly round, but polygonal, probably due to pre-existing fractures in the crust.

• We noted Mare Vaporum and Sinus Medii on Night 7; by Night 8 these are coming well into view, looking quite different as the terminator crawls across them. Just south of Sinus Medii is a north–south string of three large craters; 150 km **Ptolemaeus** to the north, 120 km **Alphonsus** in the middle, and 95 km **Arzachel** to the south. *(Find these craters on the next page in the Moon image for Nights 9–10.)*

By the roughness of its features, it is easy to see that Arzachel is the youngest of the three. Ptolemaeus has no central peak, but its floor is curiously lumpy looking. Alphonsus, in the middle, is slightly polygonal and the interior of Alphonsus includes a triangular 1,500 m high central peak and several small rilles and fractures. A north–south ridge running across the center of the crater may be ejecta from a later impact. At high magnification and under favorable lighting, it is possible to see small circular dark areas, perhaps due to material emanating from extinct volcanic cinder cones.

• Toward the terminator from Arzachel, part of **Mare Nubium** has emerged into sunlight. Near its shoreline and just south of Arzachel is a prominent dark line, **Rupes Recta** (the **Straight Wall**). This is a 110 km long north–south fault zone with a total vertical offset of about a quarter of a kilometer. Now, shortly after sunrise, the high side is toward the Sun so it casts a shadow and looks dark. Around Night 20, when the Sun is about to set and the low side of the escarpment is toward the Sun, it looks like a bright line.

• Don't forget to explore the **Southern Highlands**. It's easy to get lost in there; but it's a fun place to get lost.

Also see **www.cambridge.org/features/turnleft/the_moon.htm**

Gibbous Moon: Nights 9–10

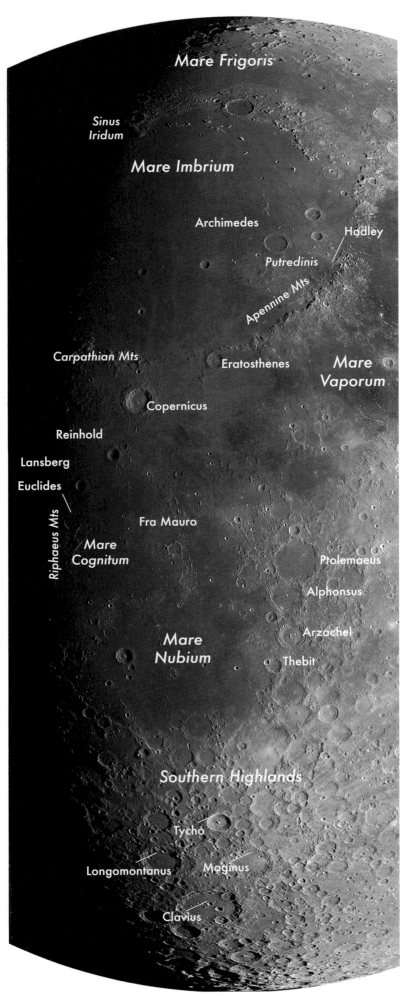

Night 9

• The large mare along the terminator to the north is **Mare Imbrium**. On its north-limbward edge, look for the arc of highlands around **Sinus Iridum**. Depending on the lunar libration, you may see all of Iridum, or just the limbward edge.

• On the opposite side of Imbrium, the **Apennines** are still prominent and they still cast shadows, even if they're not quite as spectacular as a night or two ago. Near the Apennines, the large flat-floored crater **Archimedes** makes a nice contrast with craters like **Clavius**, which we'll see to the south. You probably won't see anything inside the floor of Archimedes tonight.

• Follow the Apennines to the north (near where they are closest to Archimedes) for an area of mare (**Palus Putredinis** = marsh of decay) that bridges the gap between Archimedes and the mountains.

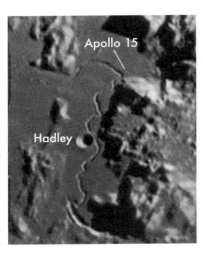

• At the far end of that marsh, very close to the mountains, a little crater (**Hadley**, 6 km) can be seen immediately up against the mountain front. Running past it, right along the mountain front, is **Hadley Rille**. Follow it a short distance (four Hadley crater diameters) to a gap in the mountains where it makes the first of two sharp left turns. The Apollo 15 landing site is just to the limbward of that first (more southerly) turn – not easy but worth the effort!

• The spectacular 95 km crater **Copernicus** is just past sunrise and extremely conspicuous. Watch how a low spot in the sunward side of the rim allows light into the bowl of the crater, and see the central peak begin to appear, first as a point, then as a series of points of light, and finally as a merged object that casts its own shadow. Many secondary craters (made by chunks thrown up by the Copernicus impact) appear on the sunward plain next to Copernicus.

• Follow the Apennines down to their southern end, where you will encounter the prominent 58 km crater **Eratosthenes**, between the Apennines and Copernicus. Right now it's hard to believe that in a few days, when the Sun no longer casts shadows, this crater will become nearly invisible against the background brightness. Check back later in the week and you'll see – barely. Between Copernicus and Eratosthenes, the nearly flooded ring of the rim of **Stadius** stands out because of the low Sun angle (it won't, in a day or so).

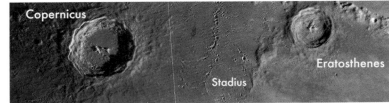

• South of Mare Nectarus is **Tycho,** looking very fresh, but much less prominent than it will look in a few days. Note, though, that over toward Mare Nectaris you can already see some of its rays.

• The rest of the **Southern Highlands** are fun to explore. Along the terminator there, and to the north, you will always see a few peaks visible, sticking up out of the dark side of the Moon.

Night 10

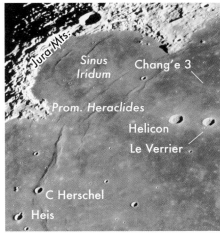

• The large mare along the terminator to the north is **Mare Imbrium.** On its north-limbward edge, look for **Sinus Iridum**, with its rim, the **Jura Mountains,** catching the light of the rising Sun. Limbward of Iridium are the 20 km craters Helicon and Le Verrier; to their north is the site of the 2013 Chinese Chang'e 3 lander. The southern end the rim is called **Promontorum Heraclides**; from there, a prominent north–south ridge heads southward toward the sharp little 13 km craters **Caroline Herschel** and **Heis.**

• The southern end of Mare Imbrium is marked by the rough east–west **Carpathian Mountains** and the great 95 km crater **Copernicus**, with its floor illuminated as it wasn't last night.

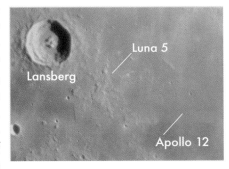

• Moving south and towards the terminator from Copernicus we encounter the 50 km crater **Reinhold** and then 40 km **Lansberg,** both of which are quite sharply defined. Just limbward of Lansberg is the landing site of the Soviet Luna 5 probe; south and limbward is the site of Surveyor 3 and Apollo 12.

• South of Lansberg we encounter the southern half-rim of the old, flooded 70 km crater **Euclides,** directly south of which are the **Riphaeus Mountains,** which run 160 km south of the rim of Euclides.

• Beside the Riphaeus Mountains, away from the terminator, is a small (250 km) patch of mare called **Mare Cognitum,** with the sharp little (7 km) crater **Kuiper** in the middle of it (look for it in the whole-Moon image on the facing page; it sits just above the u in "Cognitum"). Mare Cognitum – "The Known Sea" – got its name in the 1960s because it was the impact site of Ranger 7, the first successful American loan probe to the Moon. Gerard Kuiper was principal scientist of the Ranger missions.

Speaking of the Ranger missions, those of us of a certain age can remember the amazing feeling, watching on TV on March 24, 1965, as Ranger 9 headed toward the crater Alphonsus (see Night 8). As successively closer pictures appeared on screen, the legend below read: "Live from the Moon." It felt like science fiction … or the dawning of a new age.

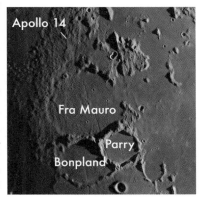

• On the limbward edge of Mare Cognitum, look for two middle-aged-looking largeish craters, first **Bonpland** (60 km) and then, touching it, 48 km **Parry.** Touching both of them on the north is the very worn-down looking larger (95 km) crater **Fra Mauro.** If your telescope has south up, the three craters look a bit like Mickey Mouse, face and ears, with the ears too close. Just north of Fra Mauro is the abortive Apollo 13, actual Apollo 14 landing site, among the rolling hills you see there looking like crinkly terrain.

• Heading down to the Southern Highlands, 85 km **Tycho** begins to stand out much better than last night – but nothing like it will a few days from now.

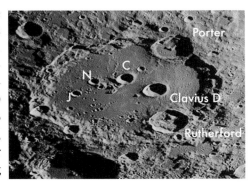

• Farther south, huge (225 km) **Clavius** is beautifully exposed. Look for the remarkable arc of six decreasing-sized craters: 50 km **Rutherford** (up against the south rim), then through the center of Clavius and curving to toward the terminator-side rim, **Clavius D** (30 km), **Clavius C** (22 km), **Clavius N** (16 km), and **Clavius J** (14 km). With high power and a steady night, an eight-inch telescope may catch as many as a dozen other craters in and around Clavius, including 50 km **Porter** (also known as Clavius B) on the northern rim and a sprinkling of others that, at 3–7 km in diameter, are just barely visible.

• Clavius and Tycho make a nice diamond pattern with two large but somewhat worn-down craters, 160 km **Maginus,** away from the terminator, and 145 km **Longomontanus** on the terminator side. In contrast to Clavius, note how much flatter the floor of Longomontanus is – the only thing interrupting the plain is one off-center mountain peak. See also how much younger and fresher-looking Tycho is than all of its three large neighbors.

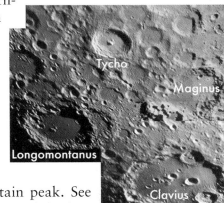

Also see **www.cambridge.org/features/turnleft/the_moon.htm**

Almost full: Nights 11–12

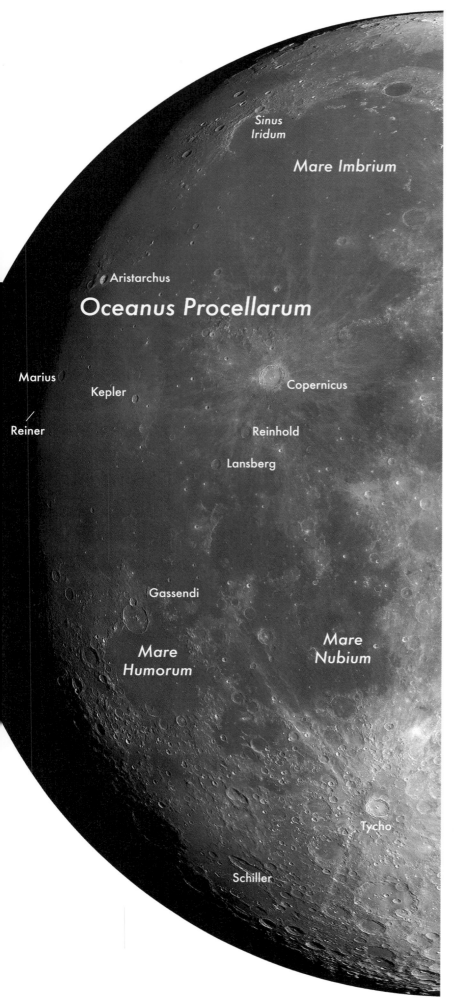

Night 11

The Moon is now dazzlingly bright, especially in a 6"–10" telescope like most Dobs. This is a good night to break out a neutral density Moon filter if you have one. Or even sunglasses!

• Near the terminator to the north is **Mare Imbrium**. By its southern edge is the crater **Copernicus**, now a bright ring surrounded by an irregularly shaped zone of ejecta and a system of radial rays. South of Copernicus are craters **Reinhold** and **Lansberg**. Making an almost right–angle triangle with Lansberg and Copernicus is the 32 km crater **Kepler**.

• Step from Copernicus to Kepler, and make a right angle northward to the 40 km crater **Aristarchus**, surrounded by very rough terrain to its north and toward the terminator. Now, at sunrise, Aristarchus shows prominently due to its long shadows. This crater is unusually bright and relatively young (less than half a billion years old). In addition, it is chemically unusual: its southern rim and the ejecta near it have been found to be especially rich in silica.

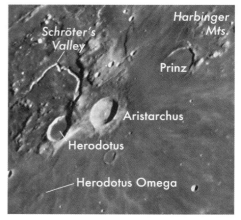

• About 6 hours after sunlight reaches Aristarchus, the 35 km crater **Herodotus** starts to appear on the terminator. Slightly misshapen and with its interior deeply flooded by lava, it makes quite a contrast with its neighbor Aristarchus. Just at this time, at very low Sun angle, look two crater-widths south from Herodotus for a small volcanic dome, **Herodotus Omega**, protruding from the surrounding plain.

• From Aristarchus move northward and away from the terminator to find the northern two thirds of a ring, all that remains of the 47 km crater **Prinz**, and scattered peaks of a small, mostly flooded mountain range, **Montes Harbinger**.

• As Aristarchus emerges into the light, look to the south, near a 40 km crater called **Marius**, for a grouping of small secondary craters, aligned roughly north–south. (Don't confuse Marius with the dark but rough-looking 30 km crater **Reiner**, four Marius-diameters to its south.) The largest of these is the 15 km **Marius A**, with 11 km **Marius C** at the northern end of the crater clump. A bit detached from the rest, 70 km north of Marius C, is the 12 km wide crater **Marius B**. If the sky is steady and the Sun angle is low, you may be able to make out **Marius Rille**, a delightful 250 km long but narrow (2 km wide) rille. Marius Rille starts in the south near Marius C, meanders northward to a point next to Marius B, and then makes a sharp turn toward the terminator. It's a fun challenge.

Once the crater Marius is in sunlight, it's time to start looking for the elusive but exciting volcanic domes of the **Marius Hills**, an area more than 100 km on a side that is absolutely peppered with a pattern of shadows looking like an egg carton.

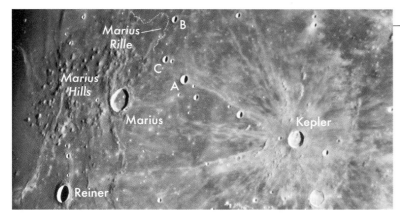

These are the shadows of numerous small volcanic domes. The volcanic hills causing those shadows are each only a few hundred meters tall, so they are visible only at a low Sun angle like tonight or last night (depending upon libration and the exact phase of the Moon). Because they are so small, you need the extremely low Sun angles of the first few hours after sunrise, when the shadows each extend several kilometers across the lunar surface. The area also includes several small rilles (collapsed lava channels). In 2009 the Japanese orbiter Kayuga found a partially collapsed roof, opening like a skylight into an otherwise still intact lava tube. Future lunar settlers may build their homes in these caves.

• Find **Mare Humorum** south of **Oceanus Procellarum**. On its northern shore, the 110 km crater **Gassendi** shows the effects volcanic uplift and subsequent collapse, as the lava flows that flooded its floor have become fractured by that flexing.

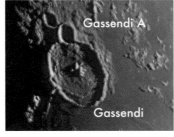

• Far to the south, one extraordinary crater stands out – the bizarrely elongated 70 by 180 km **Schiller**. Likely it is the result of two separate impacts, with the intervening rims somehow wiped out and the floor(s) partly flooded by lava.

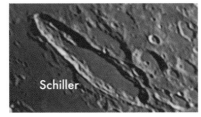

Night 12

By this point in the month, the Moon is mostly your enemy – filling the night sky with so much light that the majority of the deep-sky objects described in this book are out of reach. To make it worse, the Moon won't go away! It's already up at sunset, and it doesn't set until just before dawn.

If you can't beat the Moon, you might as well join it. Despite the near-total lack of shadows on the Moon tonight and its near-painful brightness, it's worth exploring. Instead of looking for rough topography, tonight is a good night to go hunting the mare for subtle coloration and brightness contrasts that reflect chemical differences in the basalt flows. Once the mare are old enough to have been space weathered, their color depends mostly on composition: the more iron and titanium a particular lava flow contained, the darker the mare surface.

The Moon is getting close to round now; it's hard to tell which side the terminator is on. So we'll refer to the side with

Oceanus Procellarum as the trailing side, and the other side (Tranquillitatis–Crisium) as the leading side: as you watch in the telescope it leads the way across your eyepiece field of view.

What really grabs your attention tonight are the great ray systems and the bright craters from which they emanate. Impacts in the highlands excavate bright, anorthosite-rich highland rock and toss it out across the face of the Moon; and even fresh, newly stirred-up mare basalts can look relatively bright because they have had less exposure to space weathering.

The greatest of these ray-producing impact craters is **Tycho**. In the south, Tycho stands out within the now shadowless **Southern Highlands**. Its rays stretch halfway across the visible face of the Moon. Tycho is young: Apollo 17 samples from within one of the rays suggest an age of just over 100 million years. That relatively recent date explains another remarkable feature – look carefully at Tycho and you'll easily see a dark ring of glassy melt deposits around the crater.

The next most obvious ray system belongs to **Copernicus**, surrounded by maria (**Imbrium** to its north and **Oceanus Procellarum** to its south). Copernicus, though less intense than Tycho, is still quite large and bright. It's surrounded by a large irregular area of bright grayish streaks of ejecta, extending hundreds of kilometers in all directions. At the same latitude as Copernicus, about halfway to the trailing edge, crater **Kepler** provides a scaled-down version of Copernicus and its rays.

The obvious ray crossing **Serenitatis** may be from Tycho – if so then it extends an astonishing 2000 km from the crater.

Perhaps the most intriguing rays on the Moon are much more modest in scale: the "Comet's Tail" rays of the Messier craters, as first described on Night 3 (see the image there). Look into **Mare Fecunditatis**, a bit off from the center of the mare away from the limb of the Moon. Find two little craters at the end of two nearly parallel rays. The crater closer to the center of the mare is called **Messier** (named after the comet-hunter who catalogued so many of the deep-sky objects discussed in this book). Messier is significantly elongated, about 9 by 11 km, visible even though our line of sight makes the view foreshortened – see Night 3. Next to it is the equally elongated (11 by 13 km) **Messier A**, though from our perspective the shape is complex, and hard to make out even at high magnification. Past Messier A, the two rays run for 150 km, to the edge of the mare. Apparently, this was the site of an extremely low-angle impact in which the impactor bounced after forming Messier, and then created Messier A!

• North of Kepler, find **Aristarchus** and the slightly smaller crater **Herodotus** (see Night 11). Just north of it is a 6 km crater, the "Cobra's Head," the head of a long, snake-like rille called **Schröter's Valley**. Viewed at high magnification, the topographic relief of the Aristarchus Plateau and Schröter's Valley seem to leap out at you – a fascinating view.

• Go from Aristarchus halfway to Kepler, and then turn at a right angle the same distance toward the terminator. The 40 km crater there is called **Marius** (don't confuse it with the dark but rough-looking 30 km crater **Reiner** four Marius-diameters to the south). Depending on the libration, this might be the night to look for the delightful **Marius Hills**. See Night 11.

Also see www.cambridge.org/features/turnleft/the_moon.htm

The full Moon

During full Moon, what you see are not the long dark shadows of peaks and crater rims. You can see no shadows now. Instead, what become visible are sharp contrasts between light and dark regions.

During most of the month, the Moon looks a sort of dull gray color. Moon rocks have very rough surfaces, since they've been pitted and powdered by eons of micrometeorites, and so any time the sunlight illuminating the Moon hits it from off to the side, you're going to see the effect of a lot of tiny shadows. Furthermore, as light gets reflected off this rough surface, it more than likely will hit some other bit of rock powder, and eventually a good deal of light gets absorbed before it can be scattered back off the Moon and into our telescopes.

But at full Moon, this all changes. There are no shadows. And the smooth surface of freshly broken rock is more likely to reflect light back where it came from, rather than absorbing or scattering it.

As a result, in the day or two before full Moon, the Moon starts to get much brighter. By full Moon, it can be painful to look at in telescopes of larger than 3" aperture. An inexpensive neutral density filter that fits in your eyepiece works wonders at increasing the contrast, making the full Moon far more pleasant to observe. Lacking that, you may prefer to observe with a higher magnification at this time; recall that increasing the magnification cuts down on the brightness. In a pinch, try wearing sunglasses – really! This is one situation in which wearing sunglasses at night is not just a fashion statement …

• As the Moon reflects the light of the Sun back to us more and more, the most recently disturbed areas of the surface, with the largest number of smooth surfaces and so the most places to produce glints and highlights, turn into bright spots. And of all these areas, the freshest and brightest spot you'll see is the crater **Tycho**.

• At the leading side of the Moon (to the west in the sky, to the east if you're standing on the Moon), **Mare Crisium** stands out as a dark oval all by itself near the limb. Near it, from north to south, are **Mare Serenitatis** (which is round), followed by the more irregularly shaped **Mare Tranquillitatis** and **Mare Fecunditatis**. **Mare Nectaris** is a bit south of Fecunditatis. In the north, on the trailing side of the Moon, **Mare Imbrium** opens out on its southern side into the irregularly shaped **Oceanus Procellarum**, with the bright crater **Copernicus** between them. In the south, **Mare Humorum** is near the limb and **Mare Nubium** is immediately adjacent to the **Southern Highlands**.

• Near the equator, right by the limb on the trailing (Procellarum) side of the Moon, the mare-filled oval basin of **Grimaldi** (220 km across) stands out clearly at full Moon.

• Bright rays emanate from several craters. The most conspicuous of these are Tycho (in the Southern Highlands), Copernicus (just south of Mare Imbrium) and the smaller but brilliant crater **Aristarchus** (between Copernicus and the limb). **Kepler** (making a triangle southward from Copernicus and Aristarchus) also has a nice system of rays. **Menelaus**, a 27 km crater on the southern shore of Mare Serenitatis that was relatively inconspicuous last week, is surprisingly bright tonight. The same is true of **Proclus**, located by the shore of Mare Crisium, on the side away from the limb.

• Take some time to wander around the surface of the Moon, comparing the mare. Subtle coloration and brightness contrasts in the mare reflect differences in the age and chemistry of the basalt flows that formed their surfaces.

The waning Moon

After its full phase, the Moon rises later and later every night. The full Moon doesn't rise until sunset; by the time of the last quarter phase, the Moon may not rise until after midnight, and not be high enough to be easily observed until the early morning twilight has begun. This means that most people, over most of the year, may never bother observing the Moon during this phase.

The easiest time of the year to observe the waning Moon is in autumn. That's when the Moon's orbit lies low along the southeastern horizon at sunset, and so the Moon scoots mostly northwards along the horizon from night to night without changing the time of moonrise very much. That's the "harvest Moon," when we have nearly a week of a bright Moon rising during twilight, helping to illuminate the fields during the months of September and October in the northern hemisphere (March and April in the southern hemisphere). At this time of year, the waning Moon six days past full – 20 days into the lunar month – rises at about 9:30 p.m. (That's at latitude 40°; farther from the equator, it rises even earlier.) It will be easily observed before midnight.

Observing the waning Moon: You will see the same features that we saw in the waxing Moon; the terminator reaches them the same number of days after full Moon as it did after new Moon. But many of these features won't look at all the same as they did during the waxing Moon, because now they are illuminated from the east instead of from the west. This will be especially true of any feature running north–south.

• A particularly dramatic example of this difference is the **Straight Wall** (Night 8), a fault scarp located near the crater **Birt**, at the western edge of **Mare Nubium**. This feature is the cliff between the eastern and western sides of Mare Nubium. The eastern side of the mare floor has dropped down some 250 m (800 feet) with respect to the uplifted western side. In the waxing Moon this cliff casts a broad, dark shadow; now, lit by the Sun from the east, it shows up as a narrow but bright line.

• Another feature that appears very different is the **Apennines**, the mountain range along the southern edge of **Mare Imbrium**. Because the shadows are now cast towards the west, it is easier to see features along the eastern flank of this mountain range, including **Hadley Rille**, the collapsed lava tube near the Apollo 15 landing site (see Night 9).

• There are a number of features in the various maria that also look quite different under this illumination. The basalts which give the maria their dark color were once flowing viscous lava. In many cases this molten lava froze before it reached the edge of the mare, leaving a cliff of rock around its edge called a *flow front*. These fronts are by their nature higher on one side than another; and so, like the Straight Wall, they will appear as lines of light where they before looked like dark shadows, or shadows where they were once lines of light. Look for them wherever the terminator crosses a mare plain.

• One special target, best visible at times of favorable libration when the trailing side is rotated toward us, is **Grimaldi**, a dark, mare-filled 220 km oval separate from Oceanus Procellarum near the trailing limb and just south of the equator.

• If the libration makes Grimaldi an easy target, then you can try a rare treat. Look along the extreme limb of the Moon, south from the Grimaldi basin, for as many as three thin little dark strings paralleling the limb of the Moon. These are parts of the great lunar farside **Orientale Basin** (the elusive third one is the central basin itself). Those dark bits of mare are separated by mountain ranges (the first, nearest us, is the **Cordillera** and the next is the **Rook Mountains**). At high magnification you may even be able to make out the roughness of those mountain ranges against the black of the sky. Mare Orientale is a sight worth seeking out. Make a habit of checking – just in case – whenever you see a waning Moon.

• One easy way to see the features of the waning Moon is to observe during the daytime! The Moon is still quite easy to find in the morning sky, and it can be surprisingly bright and clear in a small telescope. The morning air tends to be cool and steady, which helps. Of course, you don't want to try observing the Moon when it's close to the Sun – if it is within five days of new Moon, you're safer observing when the Sun is below the horizon.

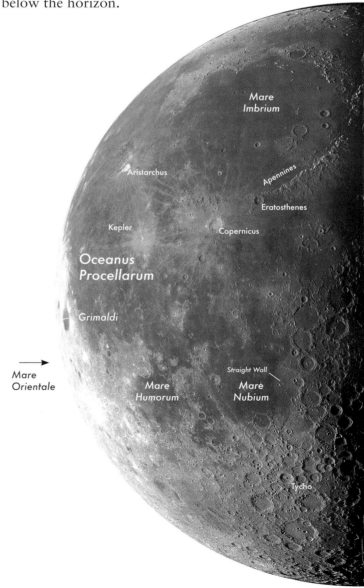

Mare Imbrium · Aristarchus · Apennines · Eratosthenes · Kepler · Copernicus · Oceanus Procellarum · Grimaldi · Mare Orientale · Straight Wall · Mare Humorum · Mare Nubium · Tycho

Also see **www.cambridge.org/features/turnleft/the_moon.htm**

Lunar eclipses

André Danjon invented a system (with values now called **Danjon numbers**) for describing the appearance of a lunar eclipse during the middle of totality.

According to his scale:
• **L = 4**: the Moon is bright orange or red in color, while the edge of the shadow will be bright blue. The color contrast is astonishing in such an eclipse.
• **L = 3**: a generally bright red Moon, with the shadow's rim being yellow.
• **L = 2**: the umbra is a deep rusty red, without a lot of color contrast, but it's much brighter towards the edge of the umbra.

For the L = 2, L = 3, and L = 4 eclipses, the contrast between mare and highland regions will be clearly visible; even a few bright craters can be seen.
• **L = 1**: the eclipse is so dark that the colorless gray–brown Moon shows virtually no detail, even in a telescope.
• **L = 0**: a very rare eclipse, so dark that you actually have trouble finding the Moon unless you know exactly where to look.

Standing on the Moon during a penumbral eclipse, an astronaut would see the Earth block off part of the Sun, but not all of it.

The Moon's orbit is tilted about 5° from the Sun–Earth plane, so usually it passes above or below Earth's shadow during full Moon. But twice a year the two planes intersect; thus in any given year, there are two opportunities for lunar eclipses.

If you're on the side of the Earth that can see the Moon while it is covered by the Earth's shadow, then you can see the eclipse itself. Thus, no matter where you are on the Earth, the odds are 50/50 that you'll get a good view of any given eclipse. In other words, you can expect to see one lunar eclipse a year... but some years you may be lucky and see both of them, other years might disappoint you completely.

On avarage, about half the eclipses will be total – the Moon will pass completely into the *umbra* of the Earth's shadow, the part where the Sun is totally blocked off by the Earth. The other eclipses can be classified as **umbral**, where at least part of the Moon passes into the umbra; and **penumbral**, where at least part of the Moon has some of the Sun's light blocked off by the Earth, but no part is in complete shadow.

You can easily see a lunar eclipse without a telescope. But binoculars or a small telescope lets you see the motion of the umbra across features on the Moon – you can really see the motion of the Moon in its orbit. And it lets you see the colors at their best. Observe the eclipse with your lowest power, so that you can see the whole Moon and not lose the colors.

Observing a lunar eclipse with a small telescope

The first thing to look for as the eclipse begins is a subtle darkening, a gradual dimming, at the edge of the Moon near Oceanus Procellarum. It takes about half an hour before the leading edge of the Moon becomes dark enough to notice.

The Moon moves at the rate of about its own diameter per hour. Thus, by the time you notice the darkening near the leading limb it's actually about half into the shadow and the center of Moon is just beginning to enter a partial eclipse. For this reason, eclipses that are less than 50% penumbral never get dark enough to notice.

In an **umbral lunar eclipse**, at least part of the Moon will see the Sun completely blocked by the Earth. The boundary between the umbra and penumbra, though not razor sharp, is easy to detect.

A **total lunar eclipse** means that the whole Moon passes into the umbra. At that time, the Moon does not merely get dark, but it may actually turn any of a variety of colors. There are two controlling factors that determine just how the Moon appears during totality. The first is how deep into umbra it goes – the deeper, the darker. The second factor is the weather on those places of the Earth where it is either sunrise or sunset while the Moon is being eclipsed.

This second factor is important because any light hitting the Moon once the Earth blocks off direct sunlight is light bent around Earth by its atmosphere. To some degree, the thin layer of transparent air around our spherical planet acts like a spherical lens, focusing light into the shadows where the Moon is sitting. When this air is clear – when the weather is fair along the edge of the Earth between the daytime and nighttime side – then the Moon can still be pretty bright even during an eclipse. If it is cloudy along the boundary, however, or if there's a lot of volcanic dust in air, then the Moon can look dark and colored during the eclipse.

Compare the brightness of the eclipsed Moon with that of familiar stars. Since the Moon's light is spread out while a star's light is concentrated in a point, it's tough to compare the two directly. One neat trick is to look backwards through your finderscope or binoculars at the Moon, to reduce the Moon to a pinpoint of light. Then compare it against a bright star, or a planet, also viewed backwards through your finderscope. From this you can estimate the magnitude of the Moon.

A normal full Moon is about magnitude –12.7; at mid eclipse, it's common for the Moon to be zero magnitude, like Vega or Arcturus. But during an especially dark eclipse it can get down to as faint as a fourth-magnitude star; and in extreme cases, invisible.

Finally, look for occultations. As we described on page 24, the Moon can pass in front of stars, "occulting" them, as it moves in its orbit about the Earth. An occultation during an eclipse is especially nice, since the Moon is so dark that you no longer have to worry about its glare wiping out fainter stars. Thus, you can see more stars being occulted.

The Sun

Observing the Sun needs special caution, but it can be worth it. The Sun is a constantly changing object, fascinating to observe. There are hours of fun to be had tracking sunspots, following Mercury move across the Sun's face during transits, and marvelling at solar eclipses. However, we cannot emphasize too strongly that the Sun is the one object that you should not try to observe without proper preparation.

Solar observing can be extremely dangerous not only to your eyes but also for your eyepiece and, in some cases, your telescope. Except with expensive, specially prepared filters that fit over the big objective lens (or mirror) of your telescope, observing the Sun directly is not worth the risk.

Projection of the Sun's image onto a screen, when done properly, can work well. Be aware, though, that it can still be quite dangerous to your eyepiece and in some cases it may damage light baffles inside your telescope. (And you must keep your eyes – and your clothing – out of the path of the projected light. It's said that Galileo set fire to his beard doing this!) But at least it won't lead to blindness if something goes wrong. Hold a sheet of white paper a foot or two away from the eyepiece, and let the light from the Sun project itself onto the paper. One useful trick is to put the paper inside a cardboard box: the sides of the box shade the image from direct sunlight. Aim the telescope by looking at its shadow – and keep the finderscope covered, since sunlight passing through it might damage it, or start a fire someplace! Even this method can ruin a good eyepiece if it gets too hot; still, better an eyepiece than an eye.

Above all, never use eyepiece solar filters, no matter how "safe" they are said to be. Dan did when he was thirteen years old and didn't know better. Fortunately he was looking away when the concentrated sunlight cracked the filter, or else his ability to see deep-sky objects – and everything else! – might have been impaired forever.

Eclipses: By a wonderful cosmic coincidence the apparent sizes of the Sun and Moon are, within a few percent, the same. When the Moon is near apogee its disk is too small to cover the Sun, so if the Moon crosses the face of the Sun we get an annular eclipse, in which a ring of the Sun's dangerously bright photosphere is still visible. With the proper filters, annular or partial solar eclipses can be fun to observe. But make no mistake: seeing a 99% partial eclipse isn't remotely as good as seeing totality. It's like driving 99 miles toward a great concert that is 100 miles away: you may have gone 99% of the way there but you'll totally miss out on the experience.

There are a few key dos and don'ts to remember if you're going to see a total solar eclipse. Obviously, ***don't*** stare at a partially eclipsed Sun. ***Do*** take advantage of every second of totality to look at it, not your equipment. Totality won't last for more than a few minutes, the fastest minutes of your life; if you plan to take pictures make it as simple as possible so you're not distracted from the view. Don't waste time adjusting telescopes. Having a nice medium-power eyepiece in your 'scope can be a delight, but don't let that keep you from looking with naked eyes or binoculars.

What, then, should you look for?

For a couple of minutes before and after totality, remember to glance downward to look for the spooky effect called shadow bands, subtle light fluctuations moving across the ground at speeds of just meters per second.

Know beforehand where the brightest stars and planets will be: Venus will appear long before totality. Just before totality look for the Moon's shadow racing across the sky from west to east and be ready to whip your solar filters off of binoculars or telescopes: practice doing this without having to look at them and waste precious seconds of viewing time. You don't want to miss seeing the diamond ring and Baily's beads, the lovely result of tiny bits of photosphere peeking through lunar valleys.

During totality, take every second you can to savor the delicate structure and lovely colors of the corona and the beautiful pink of prominences. Be prepared to be very deeply affected by the experience. And start saving your pennies; once you've seen totality you'll want to travel to wherever in the world you can see the next one!

Transits can be quite exciting to watch, if your telescope is equipped for observing the Sun. Unfortunately, the next transit of Venus won't occur until 2117. For Mercury, though, we don't have to wait so long. Parts of Mercury's passage across the Sun on November 11, 2019, will be visible in almost all of Europe, Africa, and the Americas – but South America and places near the Atlantic are favored. The next two transits, which occur on November 13, 2032 and November 7, 2039, will favor Europe and Africa. Consult your favorite astronomy magazine or websites (including ours) to learn where and when to look from your location.

Solar eclipses

The 2017 total eclipse seen in Montana; montage of long and short exposures by Claudio Costa.

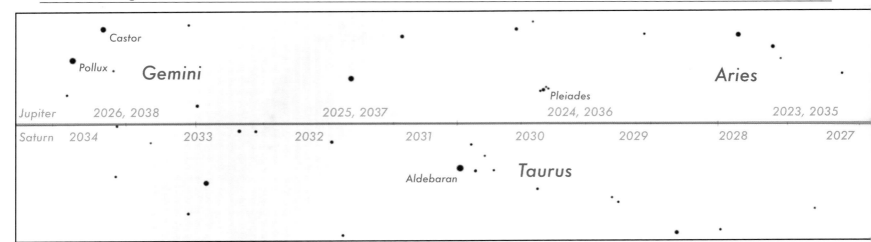

January through March evenings, look for planets among these stars (Jupiter and Saturn at the years indicated).

Observing Planets

*B*ecause the classical planets are bright, they can stand the highest-power magnification that the sky conditions and your telescope will allow. Because they're small, you'll want all that magnification.

The five classical planets – Mercury, Venus, Mars, Jupiter, and Saturn – are visible to the naked eye and thus were known even in the time of the Babylonians and Greeks. They are all at least first magnitude; all but Mercury can match or at times surpass even the brightest stars. But unlike the stars, they don't stay put in the sky from season to season. They wander; hence the term *planet*, which means wanderer.

They're not the only wanderers in the sky, however. Some comets are bright enough to dominate the sky, and their arrival can (with a few famous exceptions) be downright unpredictable. Furthermore, with a small telescope you can follow in the footsteps of the great astronomers at the turn of the 19th century and spot a number of wandering bodies – Uranus, Neptune, the brighter asteroids; even, if you're lucky with perfect conditions and a big enough telescope, Pluto.

The planets, the Moon, and the Sun follow the same narrow path through the sky, called the *ecliptic*. (Comets, asteroids, and dwarf planets like Pluto don't, however.) This path runs through the constellations of the zodiac, giving them their fame.

To find the ecliptic overhead, trace a line from where the Sun set, arching through where the Moon is (if it's up), and passing near the bright stars in these charts. There are only a handful of real stars as bright as the classical planets (Castor and Pollux, Aldebaran, Spica, Regulus, and Antares); once you rule them out, any other bright star you see that's not in one of these charts is probably a planet.

One other way to know you're looking at a planet, not a star? Planets don't **twinkle**.

Stars appear to twinkle because their light has to pass through Earth's atmosphere to get to our eyes. Think about what the bottom of a swimming pool looks like on a sunny day: it seems to be mottled with dancing lines of sunlight. That's caused by waves on the water; they bend the path of the sunlight back and forth as the waves move across the top of the pool. In the same way, irregularities in our atmosphere bend starlight ever so slightly as it passes through the air.

If you were sitting at the bottom of the swimming pool, you would see flashes of light when the streaks of sunlight passed over you. Likewise, at the bottom of our

July through September evenings, look for planets among these stars (Jupiter and Saturn at the years indicated).

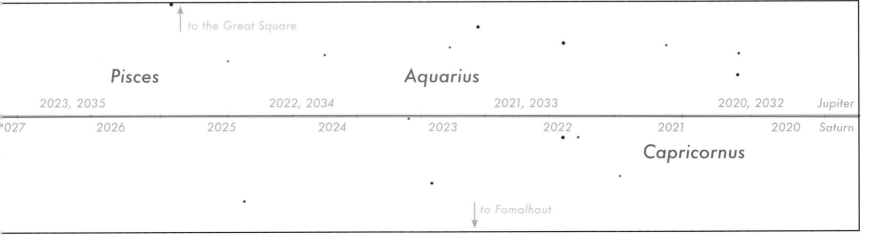

April through June evenings, look for planets among these stars (Jupiter and Saturn at the years indicated).

ocean of air, we see momentary flashes of starlight, especially if the air is turbulent (for instance when a cold front is passing through).

So why don't the planets twinkle, too? The difference is that planets have visible disks, and stars don't. Unlike planets, stars are so far away from us that they appear to be nothing but points of light, as you can see in your telescope. Even the brightest star is nothing but a very bright point, even under high power, while the planets are clearly extended circles of light.

When turbulent air momentarily bends a single beam of light away from our eyes, it appears to twinkle off. But when light from one part of the planet's disk gets bent away from our eyes, a beam of light from another part of its disk is just as likely getting bent into our eyes. When light from one point of a planet is twinkled off, another point gets twinkled on. As a result, we're always seeing some light from the planet. The light seems steady, and so the planet doesn't twinkle.

What you're looking at: Mercury, Venus, and **Mars** are all rocky planets. Mercury's surface is reminiscent of the Moon, lava flows peppered with craters. Venus is covered in a thick carbon dioxide atmosphere topped by clouds of sulfuric acid; landers sent by the USSR in the 1980s revealed a barren surface of volcanic rock. Mars is gray rock covered with a thin, shifting layer of red dust. Spacecraft have revealed a surface that once had a warm wet past. Today its atmosphere is thin and dry.

Jupiter and **Saturn** are balls of mostly hydrogen and helium, without anything like a planetary surface. From Earth, an observer sees distinct white zones (the ammonia clouds); where these clouds are absent, especially on Jupiter, belts of darker water and sulfur clouds are visible. Jupiter's **Great Red Spot** is as big as three planet Earths laid side to side. Its visibility in a small telescope depends on its color and contrast, which varies from year to year. Their **moons** are sizable places in their own right. The smallest of them, Europa, is nearly the size of Earth's Moon, and Io is a bit bigger; Ganymede and Callisto, and Saturn's moon Titan, are larger than the planet Mercury.

Planets are found among the stars shown in these four charts. The gray line is the ecliptic, *the average path of the planets. Saturn and Jupiter always lie close to the ecliptic and don't move much during a given year; the years marked here show where to look for them. But Mars, Venus, and Mercury can move rapidly through these stars during a year, and at times they might be found as much as several degrees above or below this line.*

W̱here can you find the planets tonight? Any planetarium software package can produce a finder chart for your time and place. Or check out our link to web resources at: www.cambridge.org/features/turnleft/planets.htm

October through December evenings, look for planets among these stars (Jupiter and Saturn at the years indicated).

Star maps courtesy Starry Night Education by Simulation Curriculum

The giant planets

Jupiter & Saturn

It takes 12 years for Jupiter to orbit the Sun, and nearly 30 years for Saturn. That means that Jupiter moves through the twelve zodiac constellations at the rate of one per year, while Saturn typically spends a couple of years in each zodiac constellation.

The illuminated face of Jupiter, and the rings of Saturn when seen at their maximum extent, show the largest disks of any of the planets: nearly 50 arc seconds in diameter. (As a result, they almost never twinkle.)

Nonetheless, they are small in a telescope – not much bigger than a double star like Alberio. Seeing detail is hard. Use your highest power.

Jupiter is yellowish and at magnitude –2.5, is almost three times as bright at the brightest star, Sirius. It is almost as bright as Venus; if it happens to lie near the western horizon, it can sometimes be mistaken for Venus.

Saturn looks like a first-magnitude star (it can brighten up to magnitude 0.6 when the rings are fully displayed). Though it does not stand out as the brightest "star" in the sky, nor does it have a distinctive color like Mars – it is a deeper yellow than Venus or Jupiter, and not nearly as bright – nonetheless it should be easy to spot.

Zones are white, and belts are dark. Within the belts, look for irregularities – festoons – that can be seen to move across the disk during the night as the planet spins. The famous Great Red Spot in the southern hemisphere is just barely visible in a small telescope nowadays.

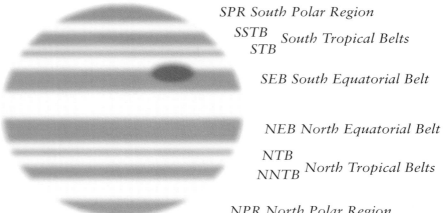

SPR South Polar Region
SSTB
STB *South Tropical Belts*
SEB South Equatorial Belt
NEB North Equatorial Belt
NTB
NNTB *North Tropical Belts*
NPR North Polar Region

What to look for: Because Jupiter spins so quickly, its disk is visibly flattened. It is wider at the equator and narrower at the poles.

Jupiter's clouds: Even a small telescope should be able to pick out some of Jupiter's clouds, alternating light and dark. **Zones** are white and **belts** are dark. With a larger telescope and very high power, look for irregularities, *festoons*, within the belts that can be seen to move across the disk during the night as the planet spins.

Jupiter's Red Spot: This spot in the southern hemisphere was described by observers with small telescopes in the 17th century, but it is just barely visible in a small telescope nowadays. (It was much darker through most of the 1960s and 1970s. It's also gotten smaller in the east–west direction through the 20th century.) Of course it is only visible when its side of Jupiter is rotated towards the Earth; and more than about an hour before or after its transit it's really hard to see. Also, the 'red' aspect of the spot varies a lot over time.

Saturn itself is featureless in a small telescope; the rings are the most spectacular thing to see at Saturn. They are utterly unmistakable. Alas, most binoculars are just not quite powerful enough to make them out.

Saturn and its rings are tilted relative to its orbit, and so as the years pass and Saturn moves around the Sun our view of the rings changes.

We see the rings at their widest twice during Saturn's orbit around the Sun, or about once every 15 years. When the rings are visible they stretch across 44 arc seconds at opposition. This is when ring shadows and the narrow Cassini division are best visible. That last occurred in 2017; the next good view is 2032.

A quarter of a Saturn orbit later (about seven years) we see the rings edge-on; for a brief period they become invisible. Actually that invisibility occurs more than once: when the Earth passes through the ring plane, and again when the Sun passes through the ring plane and thus does not illuminate the rings. One gets the eerie sensation then of seeing an entirely different planet; instead of the familiar ringed disk, all you see is just a small yellow ball about 20 arc seconds across. Look for that in 2025 (when, alas, Saturn will be very close to the Sun) and 2039.

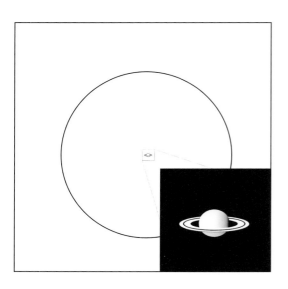

Jupiter's moons:

Io is bright, almost as bright as Ganymede, and slightly yellowish-orange.
***Europa** is more of a pure white, but dimmer.*
***Ganymede** is the brightest moon.*
***Callisto** is the darkest; often (though not always) it's seen the farthest from Jupiter.*

Moons and rings

 2018, 2047

 2019, 2046

 2020, 2045

 2021, 2044

 2022, 2043

 2023, 2042

At least three of **Jupiter's moons** are almost always visible at any given moment. On very rare occasions, there is a period of up to about two hours when none of the four Galilean moons are seen because they are either on a line with Jupiter or in its shadow. Look for this to occur on November 9, 2019 (visible from central Asia; duration 37 minutes); May 28, 2020 (from the eastern Pacific; duration 116 minutes); and August 15, 2021 (from the western Pacific; duration, only 4 minutes).

Because Jupiter's spin is only slightly tilted, and the orbits of the moons follow the spin, we always see the orbits of these moons more-or-less edge-on. From the eastern side of Jupiter, each crosses in front of the disk, continues to its maximum distance west of the planet, then moves back behind the disk to the eastern side again. The exception is Callisto: for a bit over half of its orbits it can be seen passing above or below Jupiter.

Transits – when a moon crosses in front of Jupiter so that its shadow, roughly the size of the moon itself, can fall on Jupiter – can be seen on a good crisp night even in a small telescope. Sometimes during transits it is still possible to see the transiting moon itself. Europa, which is white in color, stands out most easily against darker belts; dark Callisto can be seen against the brighter zones. It takes from two hours (for Io) to five hours (for Callisto) for a moon or its shadow to cross Jupiter's disk.

The location of the transits moves north to south of Jupiter's equator depending on what part of its 12 year orbit we're observing. Until 2021 they are north of the equator, then move to Jupiter's southern hemisphere until late 2026, then return to the northern hemisphere until early 2033… and the 12 year cycle repeats.

Mutual events, when the moons eclipse each other, occur over a period of a few months every six years. They only occur when we're in Jupiter's equatorial plane, but it's a delight to see one moon either occulted, transited, or (best of all) eclipsed by another. Look for them in 2021–22 and 2027–28. The one centered on early 2021 is best seen shortly after superior conjunction, so you'll need to catch the events with Jupiter in the eastern dawn sky.

Eclipses can cause very rapid brightness changes. Most are subtle, but others can cause a moon's magnitude to drop by a lot in just minutes – particularly if Ganymede, the largest moon, is eclipsing Io or, even more so, Europa (which can drop by more than 4 magnitudes in just 2 minutes). Occultations and transits are fun, and close approaches give a nice opportunity to compare brightness and color. Contrast with another moon can make Io's yellowish tint look almost orange by contrast – an illusion similar to what happens with some double stars.

 2024, 2041

 2025, 2039

 2026, 2038

 2027, 2037

 2028, 2036

 2029, 2035

Saturn's rings:

Because Saturn's axis is tilted, we see its rings at varying angles while it orbits around the Sun. To the right we show roughly how the rings present themselves over the next few decades.

Be sure to look for…

• the ring's shadow on the ball of the planet

• the planet's shadow on the rings; look for this when Saturn is towards the south or west at sunset, not at opposition

• the Cassini division, a sharp black line about half an arc second wide that separates the outer third of the ring from the inner part

Saturn's moons:

Titan, bigger even than Callisto and farther from its planet, is usually visible as a dot of light near Saturn itself.

Look for Saturn's smaller moons (in particular Rhea and Tethys) especially when the rings are nearly edge-on. Online resources can help you identify the positions of the different moons on a given night.

 2030, 2034

 2031, 2033

 2032

find more at: **www.cambridge.org/features/turnleft/planets.htm**

The fast moving planets

Venus

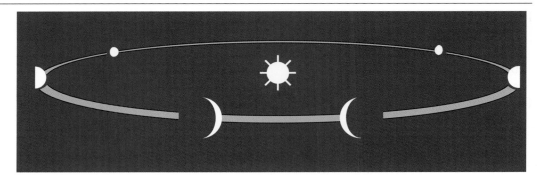

Because they orbit between us and the Sun, both Venus and Mercury show phases like the Moon's. These planets move fast (which is why Mercury is the "messenger" god). To know when and where to find them consult your favorite planetarium software or website, including ours: www.cambridge.org/features/turnleft/planets.htm

Venus is so absurdly bright that you can't miss it… The biggest issue with observing Venus is its glare. You won't see surface features, so colored filters are of limited usefulness in small telescopes. Sometimes, though, a simple neutral-density (Moon) filter can help with the glare and make it easier to make out the planet's phase.

The main thing to look for is the phase. For much of Venus' orbit, the phase doesn't change very quickly. As Venus approaches the time of its greatest evening elongation it's only half illuminated and its disk has slowly grown to about 25" (arc seconds) across. It is then at about magnitude −4.5.

Six weeks later it's only about 20% illuminated, but it has grown to almost 45" across, so it's slightly brighter than at greatest elongation. At this time, the angle it lies away from the Sun has only changed by about 10°, so it's still a good distance from the Sun and (depending on the time of year) still up well after sunset.

Over the next two weeks (6–8 weeks after greatest elongation) things change radically, almost night-to-night. By eight weeks after, it's low in the sky (but still as far from the Sun as Mercury is at all but the best elongations). It's now less than 10% illuminated, but so big (over 50") that it's lost only a couple of tenths of a magnitude in brightness.

At this point, you can actually see the crescent of Venus in either a good finderscope or binoculars that are well stabilized (if you don't have image-stabilized binocs, mount them on a tripod or hold then securely against a table or low wall). To reduce glare from Venus, this is easiest to do when the sky isn't totally dark.

On a really clear day near or in the weeks after eastern (evening) elongation, you can see Venus in the daytime with binoculars or (if you know exactly where to look) even with the naked eye. *But you must be sure to place yourself in a location where you won't look into the Sun by accident. Stand in the shade of a nearby building!*

One trick to catch Venus during the day is to spot it in the morning twilight, before sunrise, and then follow it as it and the Sun rises. But once you have seen Venus after sunrise, declare victory and go home. If you leave your telescope set up pointing where Venus used to be, eventually the Sun will rise into its field of view: bad news for your lenses (not to mention your eyes)!

The orbit of Venus, like Earth's orbit, is nearly circular, and it's also not inclined much to our orbit. Thus, every greatest elongation is at about the same distance (45° to 47°) from the Sun.

The best evening elongations of Venus are when the part of the ecliptic between the Sun and Venus makes the biggest angle to the horizon. That is near and a bit before the Spring Equinox (March in the northern hemisphere, September in the southern hemisphere). Conversely, the ecliptic is most favorable for morning elongations near the Autumnal equinox (September north, March south).

The time when Venus appears to be exactly at half phase is nearly a week sooner than you'd expect from its position in its orbit. (This is called the *Schröter effect*; Johann Schröter – also the namesake of a famous valley on the Moon – first pointed it out in 1793.) And when the crescent approaches its narrowest, the tips of the crescent, the *cusps*, can sometimes appear to extend a little bit beyond halfway around the planet. These effects are a result of the thick atmosphere, which bends the sunlight around the obscuring disk of the planet. Light entering the clouds of Venus gets scattered about by the atmosphere; a significant amount gets directed into the nighttime side of the planet, and then back out to space where eventually it can be seen by our telescopes. However, while it is being scattered, blue light is also absorbed by chemicals in the atmosphere. Thus, most of the light from the nighttime side that reaches us is red, not blue.

Mercury

The fun of observing Mercury is in the hunt, and knowing you've seen it. You must know when to look (within a week or so of elongation). You must know where to look (at what angle up from the setting Sun). A combination of naked eye and binoculars is best to tease Mercury out from the twilight glow. Count finding it as your victory!

The time windows for catching Mercury are short: just a few tens of minutes during twilight over just a week or two near greatest elongation. Don't let yourself get hung up by setting up a telescope if it's hard to find a good vantage point for it. In that case, just grab a pair of binoculars!

Obviously, Mercury is easiest to see when the Sun is out of the way. Thus Mercury's orbital ellipticity plays a role in the best time to look for it. Mercury is about 50% farther from the Sun at aphelion (farthest from the Sun, 70 million km) than at perihelion (nearest to the Sun, only 46 million km away).

Adding to the complication of finding Mercury, its orbit is inclined to ours by 7°, so from our perspective it can wander several degrees north or south of the ecliptic.

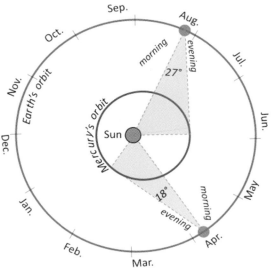

But what actually counts is how high above the horizon Mercury is during twilight. The high angle the ecliptic makes to the horizon in the early spring can make a springtime evening (eastern) elongation better than one in the autumn, even if Mercury is closer to the Sun itself at that time.

Unfortunately, August and September is a terrible time to see Mercury in the evening. Even though Mercury may be up to 27° or more from the Sun, that's 27° along the ecliptic; but the ecliptic makes a very low angle to the horizon in the summer.

This seasonal effect true for either hemisphere; it's the season, not the month, that affects the angle of the ecliptic. But southern hemisphere observers have the advantage that their winter (August) coincides with the larger elongation!

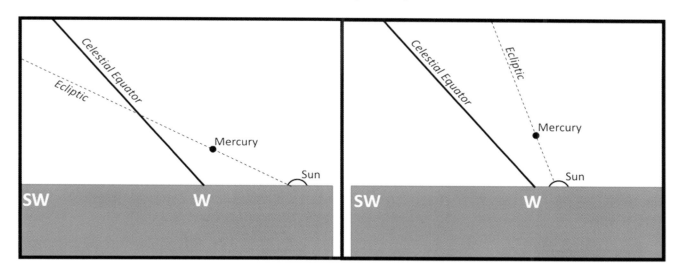

August: 27° elongation (northern hemisphere)   *April: 18° elongation (northern hemisphere)*

In your telescope, expect Mercury to look small – only about 8" – and since it's low in the sky you'll be looking through lots of atmospheric turbulence.

Look for its phase. The illuminated side should, of course, point to the Sun. Don't be confused by how the telescope flips the image! If in doubt, nudge the telescope in the direction of the illuminated side – you should find that you've nudged it toward the horizon.

Around greatest elongation Mercury will appear about half illuminated; from one night to the next it will increase very slightly in size and lose a couple of percent illumination as it starts to overtake the Earth in its orbit. Consider drawing what you see. Over the space of about a week the difference in phase should be noticeable.

Transits of Mercury are fun to observe, but there aren't many that are well placed over the next few decades. After Nov. 11, 2019 (best seen in South America but also visible in North America, Europe, and Africa), the next three are Nov. 13, 2032 and Nov. 7, 2039 (both visible in Africa, Middle East, and Europe) and May 7, 2049 (the Americas, Europe, and Africa).

Elongations around August are farther from the Sun than those seen in April. But the ecliptic rises more steeply from the horizon in the spring than in the autumn, and so Mercury is easier to see that season. Southern hemisphere observers get the best of both worlds… furthest elongation coincides with the steeper ecliptic angle!

find more at: www.cambridge.org/features/turnleft/planets.htm

Mars

Mars is usually not very interesting for a casual observer; most of the time it is too small to see much detail. But for that very reason, picking out its surface features has the same appeal as finding an elusive nebula or splitting a close double star. And during the best oppositions it should not be missed!

Mars looks like a bright red star to the naked eye. However, over a period of several months, you may notice that its brightness changes considerably. It can also move quite quickly through the constellations.

In the telescope, Mars is a bright but tiny orange disk. At opposition, the disk will be completely round; at a time far from opposition, it may appear slightly oval, like a gibbous Moon. On many nights, the view may be a bit disappointing. Be patient. There will come isolated moments of steadiness that will allow even very small telescopes to provide enticingly good glimpses of the red planet. Observe on a reasonably steady night, when the stars aren't twinkling. Then, be willing to keep looking. In half an hour you may get only a half dozen moments of perfect seeing, and each of them may last only a few seconds; but they'll be worth the wait.

The most prominent feature you'll be likely to see are the white polar caps, especially the south pole, which is generally larger and oriented towards Earth during the best oppositions. In addition, look for a large dark triangle, pointing to the north, called **Syrtis Major**; or a dark region toward the north, **Mare Acidalium**. Which one is visible depends, of course, on which side of Mars is facing Earth.

Mars has a thin atmosphere, with clouds and dust storms sometimes visible in a small telescope. Thin white clouds are most often seen in an area just north of the equator, a region of volcanoes known as **Tharsis**. Winds hitting the volcanoes force relatively moist air from the warmer Martian plains into the cold upper atmosphere; as the moisture freezes, the clouds form. Observing Mars with a blue filter helps make these clouds stand out.

Along with clouds, the Martian winds stir up planet-wide dust storms. The bright dust, similar to flakes of rust as fine as flour, can stay in the air for weeks, obscuring the darker volcanic rocks that form the dark surface markings on the planet. Lesser dust storms can also occur, covering smaller areas of the planet. A red filter can help you spot these storms.

Mars spins on its axis once every 24 hours and 37 minutes. Thus, if you observe at the same time each night, you'll be seeing pretty much the same side of Mars from evening to evening. It takes more than a month for all sides of Mars to be seen this way. On the other hand, you can detect the rotation of the planet in just an hour or two; compare sketches of the planet made at the beginning and at the end of an evening's stargazing session.

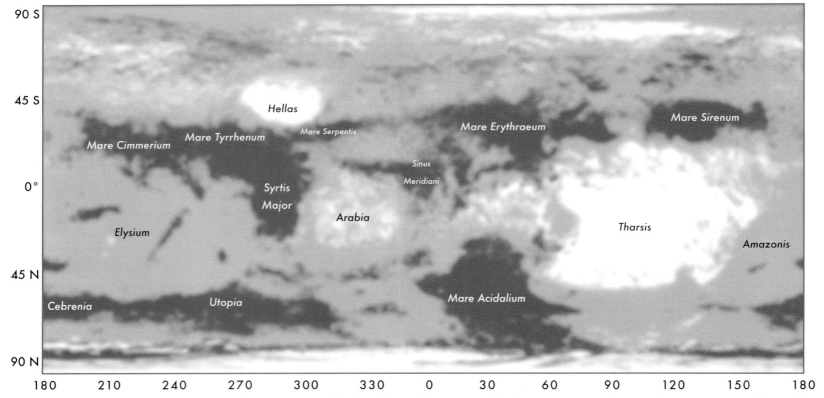

This shaded map indicated the main features most likely to be visible in a small telescope. Note that south is up, as is traditional. Don't be fooled by the detail; Mars is usually so small that you will be lucky to get a hint of Syrtis Major or Mare Acidalium.

As the Earth moves from the region of one season's constellations to the next in its yearly motion around the Sun, Mars follows along but takes nearly twice as long to complete an orbit. (Earth orbits closer to the Sun than Mars; we have the inside track, and we travel faster.) We are closest to Mars just as we are overtaking it, when we're right between Mars and the Sun. At that time, we see Mars rising in the east as the Sun sets, and at its highest at midnight. That is when Mars is said to be at *opposition*. Since we're closest to Mars then, that's when it's at its brightest.

In addition, both Mars and the Earth go around the Sun in somewhat eccentric orbits (more eccentric for Mars than for the Earth). Their paths are not simple concentric circles about the Sun, but ellipses which pass closer to each other at some places than at others. The place where their orbits come closest together coincides with the position of the Earth during the month of August. Those are the most favorable oppositions to see Mars close-up.

> *The best Mars oppositions occur about once every 15–17 years; the next good ones occur in 2018 and 2020, then not until 2033–35.*

During the best oppositions, Mars is an ominously brilliant (up to magnitude –2.9) blood-red star more than three times brighter than Sirius, rising in the east at sunset. That's also when the disk is its biggest, some 25 arc seconds across, and when a small telescope has its best chance to see detail on the surface. At this time, it is summer in Mars' southern hemisphere. Thus, the part of the planet tilted toward the Sun (and so most visible to us) will be the south pole of Mars; its ice cap is the most conspicuous feature of the planet. Less-favorable oppositions, those that occur in January or March, are only a fifth as bright, and the disk is less than 14 arc seconds across, barely half that of the best opposition.

With its red color, Mars can sometimes be confused with nearby red stars, such as Antares (whose name is, in fact, Greek for "not-Ares", i.e. not Mars). The easy way to tell them apart is that Antares twinkles quite a bit; Mars, like the other planets, does not twinkle.

> *During oppositions, Mars comes close enough for considerable detail to be seen on its surface, even with a small telescope. For more detail on what you might expect to see on any given night, go to www.cambridge.org/features/turnleft/planets.htm*

Mars oppositions

The best oppositions occur in July–September, when Mars is low in the southern sky for observers in northern climes. Southern-hemisphere observers, however, have a great view of Mars high in their winter sky then.

Mars oppositions

Date	Size	Mag.
July 2018	24"	–2.8
October 2020	23"	–2.6
December 2022	17"	–1.8
January 2025	15"	–1.4
February 2027	14"	–1.2
March 2029	14"	–1.2
May 2031	17"	–1.8
June 2033	22"	–2.6
September 2035	25"	–2.9
November 2037	19"	–2.0
January 2040	15"	–1.6
February 2042	14"	–1.2
March 2044	14"	–1.3
April 2046	16"	–1.6
June 2048	20"	–2.2
August 2050	25"	–2.9
October 2052	21"	–2.5

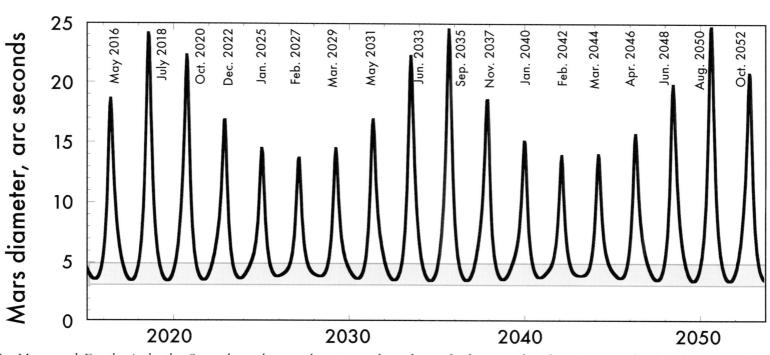

As Mars and Earth circle the Sun, they alternately approach and recede from each other. Because both orbits are elliptical (especially Mars' orbit) some approaches – those occurring in July through September – are especially close. Note that the size of Mars changes especially quickly during the time of closest approach; it doubles in size over the two months leading up to peak view, and shrinks just as fast afterwards. Mars should appear large enough to show surface features within a couple of months of the dates indicated above. (The shaded region above marks when Mars is on the opposite side of the Sun from us and thus difficult to see in any event.)

The far...

Uranus and Neptune

Uranus and Neptune are faint and small! They're tiny greenish disks of light; that's why "planetary nebulae," which look much the same, got that name. They are easiest to see at midnight around opposition, or around 10 p.m. a month after opposition. For the next 25 years, Uranus is at opposition in November–January, Neptune in September.

How do you find Uranus and Neptune? Start with a finder chart. The finderscope will show an oddly pale blue star. In the telescope, it looks like an greenish-blue star too "big" for its brightness or too dim for its size... more obviously for Uranus, less so for Neptune.

Uranus is surprisingly bright (magnitude 5.7 at opposition), and getting brighter as it approaches perihelion in 2050. This makes it easy in binoculars or your finderscope; in a really dark site, try looking for it with the naked eye! Its disk is small (about 3.7" at opposition, gradually growing to 4" at perihelion) and featureless even at high power.

Uranus moves slowly through the constellations; with a period of 84 years, it spends on average seven years in each. As of this writing (2018) it's in Aries – well placed in a November evening. Over the next few decades it moves through Taurus and Gemini – easy viewing in January; then to Cancer and then Leo, until by 2050 it will be approaching Virgo – well placed on an April evening.

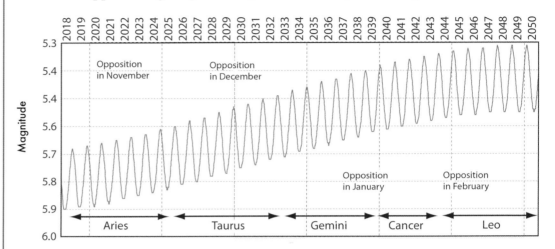

If you're in a dark place and have a telescope that is more than 10 inches in aperture, you just might be able to pick out the brighter moons, Titania and Oberon. Titania is the brightest (about magnitude 13.8), located as far as 32" from Uranus; Oberon is a bit dimmer (magnitude 14), but its orbit carries it ever farther from the glare of Uranus (up to 42" from Uranus), so it's also within reach for a 12 inch telescope or larger, even in suburban skies. However, Uranus is bright enough that it can drown out nearby moons even in a big telescope; that's why Ariel, which is about as bright as Oberon but three times closer to the planet, is much harder to see without an even bigger 'scope. Umbriel is dimmer and almost as close as Ariel, so even harder – you need a really big 'scope to spot it.

Of course, moons are most easily seen when they are placed as far as possible from the planet. But unlike the other giant planets, Uranus' spin axis is tilted nearly "on its side" (about 98° from the plane of its orbit) as seen from Earth. In 2028–29 we'll be looking down the north pole of Uranus, with its moons' orbits lying nearly face-on, so the moons might be anywhere, at any angle, around the planet – more opportunities to see them – each at its own distance. But in 2049–50 we'll be passing through the plane of its moons' orbits (as we did last in 2007); then, the moons' orbits will lie along a roughly north–south axis, and easier to see when at their furthest along one of those directions.

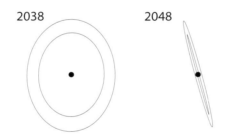

How do you know for sure if you've spotted Triton? At high power count how many planet diameters (about 2.4") from the center of Neptune your candidate lies. Is it preceding (ahead, or west) of Neptune as they move across your field of view or is it trailing (to the east)? Nudge the 'scope north and south, and see in which direction it is from the planet. Once you can say that your candidate is, say, north and east of Neptune, then go to your favorite planetarium software (or one of the websites we link to) and see if what you saw is where Triton actually is.

Neptune will look like a dim star in most finderscopes. In the telescope, it looks like an oddly pale blue star that, though its disk is very small, looks too big for its brightness or too dim for its size. High power will clearly show a (featureless) disk.

Neptune takes 165 years to orbit the Sun, thus spending on average fourteen years in each zodiac constellation. In 2022 it passes from Aquarius into Pisces, reaching Aries in 2039; by 2050 it will have just entered Taurus. Thus for the next few decades, the best evenings to look for it are in September.

Triton (magnitude 13.7) is about the same brightness as Uranus' brightest moons (and Pluto); but Neptune is dimmer than Uranus so in a large (more than 10 inch aperture) telescope in dark skies it's realistic to look for it. If you see a very dim star near the planet it's very likely Triton.

Asteroids and dwarf planets can be fun to track. They're just dots of light but they can move visibly from hour to hour. You'll need a planetarium program or an amateur astronomy website to find where to look for them.

There are nine large **asteroids** that can brighten up to better than 8th magnitude, making them easily visible in binoculars or finderscopes. The brightest of these, at times more than a full magnitude brighter than the others, is 4 Vesta.

...and the faint:

Asteroids, comets, and Pluto

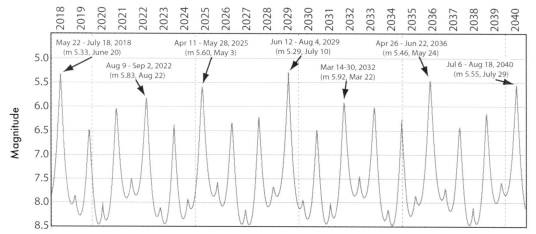

Every 16 or 17 months, Earth overtakes Vesta in its orbit. When that happens with Vesta near perihelion (every second or third close approach) Vesta reaches naked-eye visibility – as bright as magnitude 5.2.

The other asteroids visible in a small telescope, in order of maximum brightness, are 2 Pallas (mag. 6.5), 1 Ceres and 7 Iris (both mag. 6.7), 6 Hebe, 3 Juno, and 18 Melpomene (all mag. 7.5), 15 Eumonia and 8 Flora (mag. 7.9).

In addition, despite its small size (only 11 km around and 34 km long), the asteroid 433 Eros can be as bright as these larger asteroids during a close passage to Earth. But the next time that happens won't be until 2056!

On April 13, 2029, the tiny (320 m diameter) asteroid 99942 Apophis will have a near-miss with the Earth, passing by at a distance less than that of geosynchronous satellites. For those lucky enough to see its closest approach in darkness, it will get as bright as magnitude 3.4! Observers in Europe are especially well located to see it. At its closest it will appear 2 arc seconds across, just big enough to give a hint of a disk in an amateur telescope, and it will dash across the sky at up to ¾ of a degree (more than a full Moon's width) per minute. Mark the date!

Comets can be great fun to see in a small telescope. But with a few rare exceptions (Comet Halley being the most famous) the arrival of bright comets is not predictable. Most of them come to us from the outer reaches of the Solar System, too far away to be detected until they've nearly arrived on their once-in-a-million-years passage by the Sun. A bright comet is a rare enough event that when they do arrive, they are ballyhooed widely on amateur astronomy websites – check them to find out when and where to look.

Dwarf planet **Pluto** is a real challenge even in a good-sized Dob: at least 10 inches of aperture under perfectly dark skies, more realistically a 12 inch to 14 inch 'scope.

It is getting dimmer year by year as it moves away from the Sun in its very elliptical orbit: mag. 14.3 in 2018; mag. 14.7 by 2030; dimmer than mag. 15 after 2040.

Where to look? In 2018 it was in Sagittarius, at opposition in July and thus an evening object in September. From Sagittarius it spends sixteen years (2023–2039) in Capricornus; by 2050 it is well into Aquarius, an October evening object.

As for actually finding Pluto even in a large telescope... this is not for the impulsive!

First, prepare in advance a finder chart (using plantarium software, or an astronomy magazine, or a website) showing the area – say, a half degree – around the predicted location, but not showing exactly where Pluto will be. Point your telescope to the location covered by your chart; move to higher power. Mark on the chart where you see an extra dim "star". (Remember to use a red flashlight so you don't ruin your dark adaptation!) Afterwards, check to see if it's the right place; only then can you be confident that you saw it. If you get another good night, go out again and see if your first-observed "star" is no longer in that location but, instead, somewhere else. (And check again that it's in the right place!)

Whether it's Uranus or the asteroid Vesta, there's an almost illicit pleasure in seeing with the naked eye something that most people would reasonably think you're not supposed to be able to see. If you're near one of the dates indicated on the chart at left, consider looking for Vesta, first with binoculars and then naked eye. But beware! Some oppositions of Vesta come with extra challenges, such as full Moon or a far southerly declination. For more details, see our website **www.cambridge.org/features/turnleft/planets.htm**

In the late 1980s, Pluto was closer than Neptune and at mag. 13.8 visible with an 8 inch 'scope. No more!

The easiest way to find Uranus, Neptune, and the other fainter objects is to use a computer software program that plots planet positions for you. Likewise, almanacs and astronomy magazines such as Astronomy *or* Sky & Telescope *give detailed positions valid for a given month or year. The January edition of* Sky & Telescope *gives annual finder charts for Uranus Neptune, and Pluto We list links to sites with finder charts for them as well as Ceres and Vesta at* **www.cambridge.org/features/turnleft/planets.htm**

Seasonal skies: *January–March*

These months have some of the brightest stars and most interesting objects of the year, as the Milky Way runs straight overhead. Not only are there plenty of familiar bright constellations, like Orion and the Twins, but lots of wonderful telescope objects from the Orion Nebula to the faint but fascinating supernova remnant M1, The Crab Nebula, in Taurus.

It's cold at night in the northern hemisphere wintertime, even in California or Florida. Operating a telescope means standing still for a long period of time, which makes you feel even colder. And remember that you'll be adjusting knobs and levers on your telescope, which are often made of metal and get very cold indeed. Gloves, a hat, and several layers of underclothing are a necessity. A thermos of coffee or hot chocolate wouldn't hurt, either. Do what you can to make yourself comfortable.

The darkest and clearest nights are also the coldest, when you are looking through crisp air from the polar regions. Such nights are great for nebulae, but if a cold front has just passed

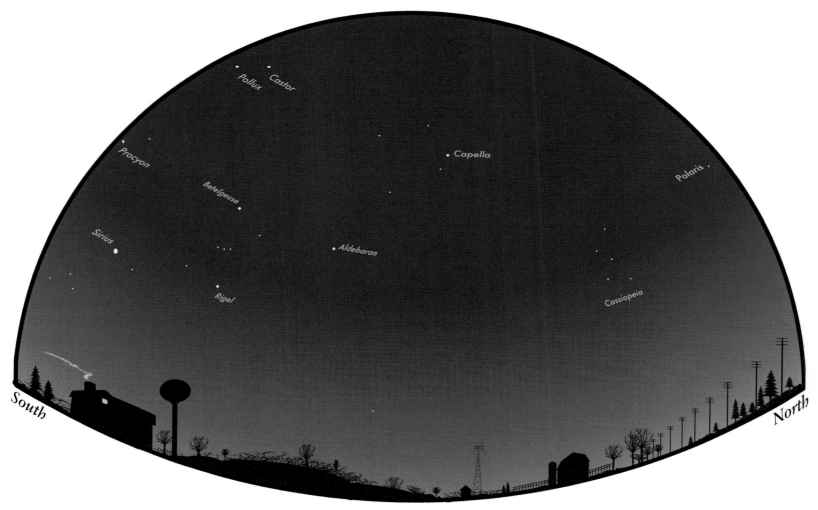

Looking west

Finding your way: January–March sky guideposts

The most prominent stars in the sky are to the south. First find *Orion*. Three bright stars make up his belt; two stars (including a very bright red one to the upper left) are his shoulders, and two more (including a brilliant blue star to the lower right) make his legs. The very bright red shoulder star of Orion is called **Betelgeuse** (pronounce it "beetle-juice" and you'll be close enough). The brilliant blue star in his leg is named **Rigel**. (It's pronounced "rye-jell".)

Above Orion and to the right is a very bright orange star called **Aldebaran**. It's the brightest star in the constellation *Taurus*, the Bull. This constellation is a member of the zodiac, the twelve constellations which the Moon and planets travel through. In a zodiac constellation, if you see a bright "star" that doesn't appear on the charts, there's a good chance it's a planet.

Above Taurus, north and east of Aldebaran and almost directly overhead this time of year, is a large lopsided pentagon of five stars. By far the brightest of these stars, the one to the north and east, is a brilliant star called **Capella**. These stars mark the location of the constellation *Auriga*.

High in the northern end of the western sky are five stars in the shape of a large W (or M, depending on your orientation). This is *Cassiopeia*. To its south are a number of the objects described in the October–December chapter; see page 162.

through, the air may be turbulent, making double stars very difficult to split and details on planets hard to see.

Southern hemisphere observers will be experiencing summer, but even so, be prepared for a chill in the air, dew on the grass (and on your lenses!), and insects hovering over your head. Don't underestimate the value of a good insect repellent. You also have fewer hours to observe; not only are the nights shorter, but also with daylight saving time it may not be dark enough to see faint stars until 9 p.m. or later, depending on where you live.

If you're in the southern hemisphere, be sure not to miss the Large Magellanic Cloud (pages 216–227), one of the richest fields for small telescopes to wander about.

To the west: The best of the October–December objects are still easily visible above the western horizon, especially in the northern hemisphere where the Sun sets early in the evening. So while you are observing, be sure not to overlook the following favorites:

Object	Constellation	Type	Page
Helix Nebula	Aquarius	Planetary nebula	172
NGC 247/253	Cetus/Sculptor	Galaxies	174
NGC 288	Sculptor	Globular	174
M31	Andromeda	Galaxy	176
M33	Triangulum	Galaxy	178
Mesarthim	Aries	Double star	180

Looking east

Looking east of Auriga, you'll find two bright stars close to each other. They're **Castor** and **Pollux**. Stretching out to the south and west from Castor and Pollux are the stars of *Gemini*, the Twins. With a little imagination you can "connect the dots" to make out the shape of two stick men, lying parallel to the horizon during this season. Gemini is also a zodiac constellation – watch for planets here.

To the southeast of the Twins, back towards Sirius, is the brilliant star **Procyon**.

Finally, the three stars in Orion's belt point down and to the left to a dazzling blue star, **Sirius**, rising in the southeast. This star is in fact the brightest star in the sky and belongs to the constellation *Canis Major*, Latin for "big dog." Hence it's often called the "Dog Star."

(If you are observing from south of 35° N, look south of Orion for the bright star **Canopus** – see page 208).

In all, the stars Sirius, Rigel, Aldebaran, Castor and Pollux, and Procyon make a ring around Betelgeuse. This is the part of the sky richest in stars of the first magnitude – and brighter!

To the north looking east, the *Big Dipper* is rising. Part of its handle may be obscured by objects on the horizon, depending on how far south you live. The two highest stars, at the end of the dipper's bowl, point to the north towards **Polaris**, the North Star. Face this star, and you'll be sure that you are facing due north.

South of the Big Dipper, rising in the east, is **Regulus** and the stars of Leo. As the night goes on, be sure to look ahead to the next set of seasonal charts starting on page 90.

In Orion: The *Orion Nebula*, M42 and M43

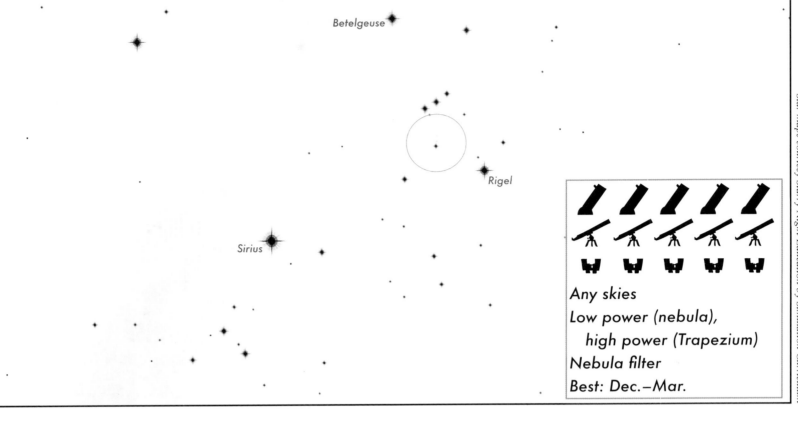

Star maps courtesy Starry Night Education by Simulation Curriculum

Any skies
Low power (nebula),
   high power (Trapezium)
Nebula filter
Best: Dec.–Mar.

- Spectacular nebula at all powers
- Trapezium: a mini-cluster (or a sextuple star!)
- Star-formation region

Where to look: Find the constellation Orion, high in the south. Three bright stars in a line make up the belt of Orion; dangling like a sword from this belt is a line of very faint stars. Aim your telescope at these faint stars.

In the finderscope: A star in the middle of this line (the sword) looks like a fuzzy patch of light rather than a sharp point. That's the nebula M42; center your cross-hairs there.

   Even in the finder (or a pair of binoculars) you should be able to see a rich starfield with lots of blue stars sprinkled in among the nebulosity.

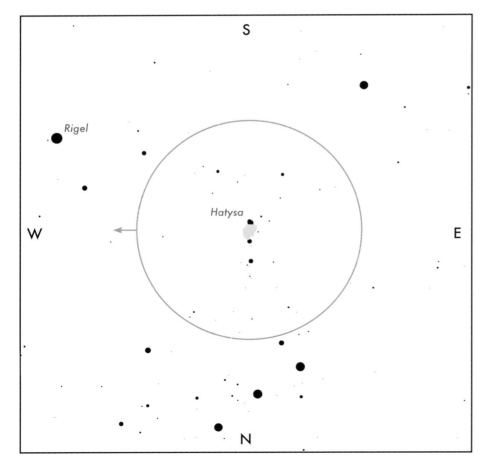

M42 in a star diagonal at low power

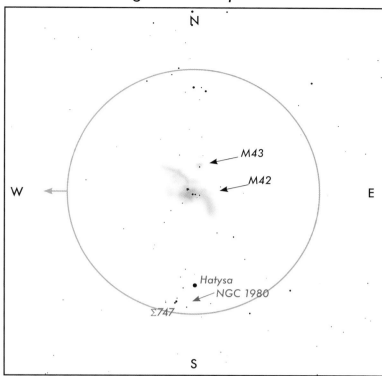

M42 in a Dobsonian at low power

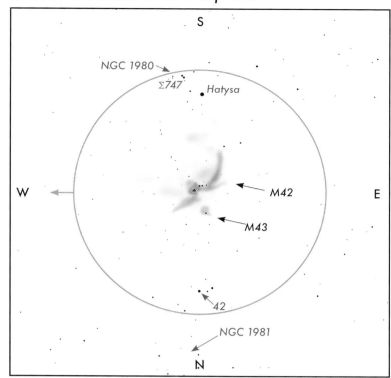

In a small telescope: Under low power, the nebula will appear to be a bright, irregular patch of light with a few tiny stars set like jewels in the center of the patch. Just to the north is an eighth-magnitude star; a patch of light envelops this star, and extends a bit north: M43, really part of the same nebula system as M42. At the southern edge of the low-power field of view, find the third-magnitude star Hatysa. It is a double (see next page), but hard to split in a small scope. However, just to its southeast is an easier double, Struve 747.

In a Dobsonian telescope: It's a bright, irregular patch of light with tiny stars set like jewels in its center. Enveloping the first reasonably bright (eighth-magnitude) star north of the Trapezium, M43 is really part of the same nebula system as M42. The associated open clusters NGC 1980 and NGC 1981 are indicated south and north of the nebula. Note the double stars Hatysa (see next page), a third-magnitude star at the southern edge of the low power field of view; and fourth-magnitude 42 Orionis, to the north, described on page 58.

The darker the sky, the more detail you can see in the cloud of light that is **M42/M43**, and the farther you can see the cloud spread out as it fades away into space. Under perfect conditions you might just be able to detect a slightly green hue in the nebula. With a large aperture and a nebula filter, the view is nothing short of spectacular.

The area around the **Trapezium** is a remarkable place to let your eye roam. Under low and medium power, you may make out several lanes of light, separated by darker gaps. Immediately around the Trapezium, there does not appear to be as much nebulosity. Part of this appearance is an optical illusion, as the brightness of the Trapezium stars makes our eye less able to pick up the fainter glow from the nebula; but part of this effect is real, too, as the light from the brighter stars apparently pushes away some of the gas, thinning out the nebula around these stars.

Looking at M42 and M43, one gets the impression that a tendril of light almost extends between these two nebulae. But there is a definite dark gap between these two, possibly the result of a dark cloud of dust obscuring that part of the nebula from our sight.

The Orion Nebula and the Trapezium are part of a large region of star formation, enveloping most of the stars we see as members of the constellation Orion. (Sigma Orionis is another member of this group – see page 56.)

The diffuse nebulae M42 and M43 are a region of active stellar formation within this system, where the gas is strongly illuminated by newly formed stars. The cloud visible in a small telescope is about 20 light years across from one side of the nebula to the other, while radio waves from this region indicate the presence of a cold, dark cloud of gas over 100 light years in diameter. There is enough matter in the visible nebulae to make up hundreds of Suns, while the surrounding dark clouds may hold many thousand solar masses. The nebulae lie some 1,350 light years away from us.

Of the four bright stars in the Trapezium, all but the brightest one (Star C) are in fact extremely close binary stars. None of them can be split in a telescope; but two of them are *eclipsing binaries*, stars that dim periodically as one member of the pair passes in front of the other.

Star B, also known as BM Orionis, is the star at the north end of the trapezoid. It consists of two massive stars; together, they have over a dozen times as much material as is in our Sun, and they shine with about one hundred times its brightness. An eclipse occurs once every six and a half days and lasts for just under 19 hours. During that time it fades by over half a magnitude.

Star A, the one to the west, is also called V1016 Orionis. Even though astronomers have been studying the Trapezium for hundreds of years, it wasn't until 1973 that anyone real-

find more at: www.cambridge.org/features/turnleft/seasonal_skies_january-march.htm

M42 and M43 in a star diagonal at high power

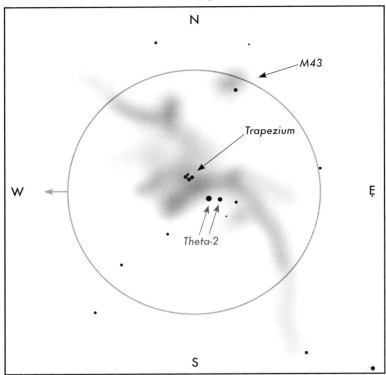

M42 and M43 in a Dobsonian at high power

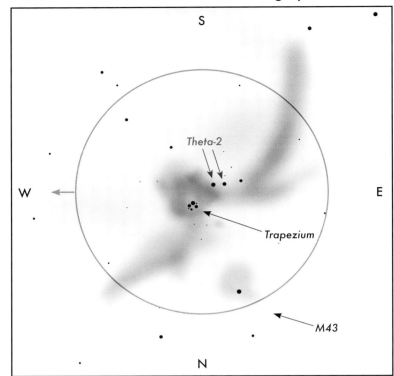

In a small telescope: Under high power and good conditions, pick out four closely spaced stars which form a diamond or trapezoid shape: the Trapezium.

In a Dobsonian telescope: Under high power and good conditions, pick out four closely spaced stars which form a diamond or trapezoid shape, the Trapezium, with two fainter eleventh-magnitude companions.

ized that this was a variable star. It goes into eclipse once every 65.4 days, fading by as much as a magnitude over 20 hours. Normally, this star is equal in brightness to Star D; if it appears dimmer, it's in eclipse.

In addition, there are at least four other fainter stars which have been discovered in and around the Trapezium. Two of these, at eleventh magnitude, may be visible with a Dobsonian.

The stars of the Trapezium are just the brightest of more than 300 stars that large telescopes can see in this nebula. The westernmost of the Trapezium stars is itself a double star. Two of this assemblage of young, newly formed stars near the Trapezium form the wide, bright double star Theta-2 Orionis.

These stars, of magnitudes 5 and 6.5, are located about two arc minutes to the southeast of the Trapezium.

Although Galileo first saw it with his telescope in 1610, this nebula was first studied intensely only in the late 1700s by, among others, Sir William Herschel. He was one of the great astronomers of his day; in fact, he discovered the planet Uranus. However, his drawings of the Trapezium region fail to show stars E and F, stars visible in a Dob that other observers (including his son John), starting in the 1820s, had no trouble finding. It may be possible that a cloud of dark material between those stars and us has been dissipating; or perhaps we're watching new stars being born.

*Also in the neighborhood: To the south of M42, just on the edge of the low-power field, is a multiple star, **Hatysa**. Three components are visible in a small telescope. Southeast of the third-magnitude primary is 7.5-magnitude star separated by 11 arc seconds, while a tenth-magnitude star is located 49 arc seconds east and a little south. This translates into quite a large system: stars A and C are separated by about a third of a light year!*

*Less than 9 arc minutes to the southwest of Iota is a wide, bright, easy double called **Struve 747**, comprised of a magnitude 5.5 primary star, with a magnitude 6.5 star located 36 arc seconds southwest of it. The two components of Struve 747 line up nicely with Hatysa.*

*Less than 5 arc minutes west of Struve 747 is a much dimmer pair, **Struve 745**. These are two equally faint stars (about magnitude 8.5), 29 arc seconds apart, aligned north–south. If they really do orbit each other, they would be separated by 2,000 AU.*

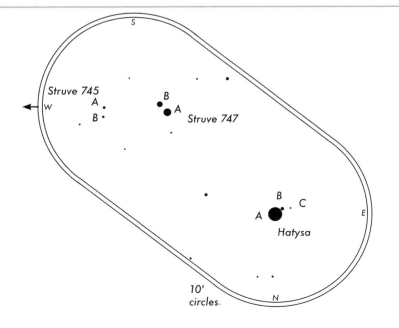

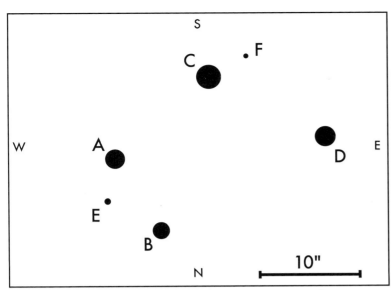

Six stars of the Trapezium (Theta-1 Orionis) can be seen in a typical Dob or a large Cat. This view represents a distance of 9000 AU across. Stars A and B are eclipsing variables; each eclipse lasts about 16 hours, but the period for B (6.5 days) is ten times shorter than A (65 days) – you're 10 times more likely to catch B (BM Orionis) dimming to magnitude 8.6 than you are to see A (V1016 Orionis) dimming to 7.6.

The Trapezium (Theta-1 Orionis)

Star	Magnitude	Color	Location
C (south)	5.1	White	Primary star
A (west)	6.7–7.6	White	13" NW from C
B (north)	7.9–8.6	White	17" NNW from C
D (east)	6.3	White	13" NE from C
E	11.1	White	5" N from A
F	11.5	White	5" E from C

Theta-2 Orionis

Star	Magnitude	Color	Location
A	5.0	White	Primary star
B	6.2	White	52" E from A

Hatysa (Iota Orionis)

Star	Magnitude	Color	Location
A	2.9	Blue	Primary star
B	7.0	White	11" SE from A
C	9.7	White	49" ESE from A

Struve 747

Star	Magnitude	Color	Location
A	4.7	White	Primary star
B	5.5	White	36" SW from A

Struve 745

Star	Magnitude	Color	Location
A	8.5	White	Primary star
B	8.5	White	29" N from A

In addition to these doubles, there are eight additional, more challenging double stars in Orion that can be split in a Dobsonian telescope. *See page 58.*

About diffuse nebulae

The Orion Nebula is a huge cloud of gas, mostly hydrogen and helium, with a little oxygen and other elements thrown into the mix.

What's going on in this gas cloud? Apparently, gravity is pulling clumps of gas to gather together until they're big enough to fuse hydrogen into helium at their centers, at which point they become stars. And so, we think, what we are looking at here are stars being born.

There are several arguments supporting this theory. For one thing, the spectra (the relative brightnesses of the colors in the starlight) of the stars within the nebula match the spectra which theory predicts should occur in young stars. As one moves away from the Trapezium region, progressively older stars are found. Tracing backwards the motions of the stars around the Trapezium, one can conclude that they all started out from more-or-less the same place within the last few hundred thousand years – these stars are mere infants in terms of the billions of years which most stars live. Star formation probably began in this nebula about 3 million years ago.

From these stars' orbits, one can also calculate that the whole mass of the Trapezium is about 1800 times the mass of our Sun. Star C is the biggest, at 45 solar masses; A and D are 15 to 20 solar masses, while B is 7 solar masses. In fact, it's been suggested that there may be an intermediate mass black hole (about 100 solar masses) within the Trapezium; that would help explain the observed velocities of the stars, which are higher than expected.

Many other examples of diffuse nebulae exist throughout our Galaxy, often in association with open clusters of stars. By looking at different nebulae and clusters, we can see different stages in the processes by which stars are born.

The gas of such a nebula is irradiated, and glows, by the energy of the young stars inside it. For example, the Trapezium and the other stars inside M42, and the eighth-magnitude star inside M43, provide more than just the light that shines through the gas. High-energy ultraviolet light from these stars also causes the electrons of the gas atoms, especially hydrogen and oxygen, to break apart from the atoms much like an electric current acts in a neon light. When the electrons recombine with the atoms, they emit particular shades of red (from the hydrogen) and green (from the oxygen).

Electronic CCD chips (and old-fashioned color film) tend to pick up the red light better than our eyes can, so color pictures of nebulae like this one tend to be reddish. But our eyes see it as green. Because we see only green light from these nebulae, we can use a nebula filter tuned to precisely that wavelength to cut back all the other light in the sky (due to the Moon, or streetlights, or even the other nearby stars) and emphasize the nebulosity itself.

Note that such a filter doesn't actually make the nebula brighter; it just makes everything else look dimmer, and thus pulls the nebula out from the ambient sky brightness, making it easier to see.

find more at: **www.cambridge.org/features/turnleft/seasonal_skies_january-march.htm**

In Orion: Two multiple star systems, Sigma Orionis and Struve 761

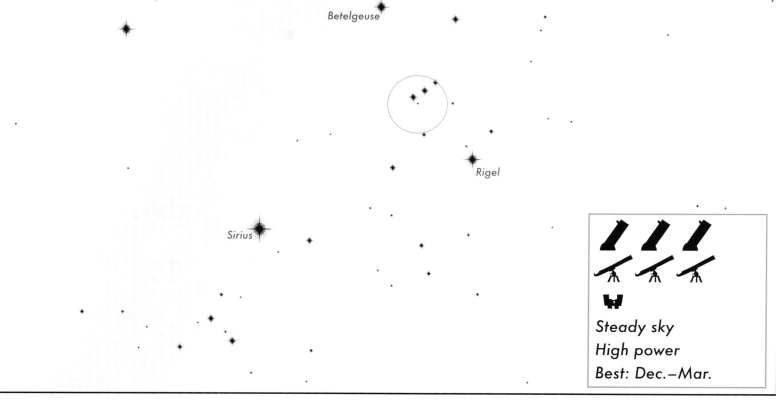

Star maps courtesy Starry Night Education by Simulation Curriculum

Steady sky
High power
Best: Dec.–Mar.

- Quadruple star and triple star
- Easy to find
- Easy to split

Where to look: Find Orion, and look to the three stars in his belt. From left to right, their names are Alnitak, Alnilam, and Mintaka. Just below Alnitak (the eastern one) is another, somewhat less-bright star. This is Sigma Orionis.

In the finderscope: The three stars of the belt will be visible in the finderscope. Sigma Orionis is only a little dimmer than these three bright stars, and should be very easy to find. Using Alnitak as the hub of a clock face, with south at 6 o'clock and the other belt stars lying in the 2 o'clock position, Sigma Orionis is positioned at about 5 o'clock.

Sigma Orionis and Σ761, star diagonal, high power

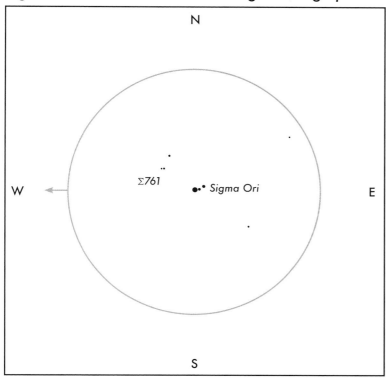

Sigma Orionis in a Dobsonian at high power

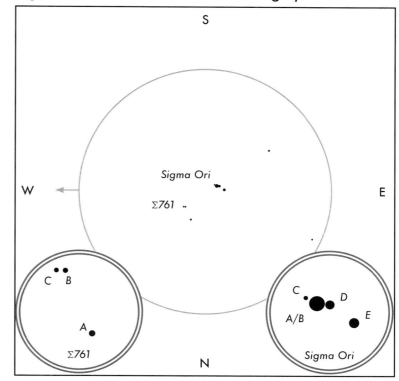

In a small telescope: The brightest star in the field is Sigma Orionis A/B/C, a triple system which is too close together for a small telescope to split. To the east of A/B/C is star D; northeast and about three times farther away is star E.

In the same field of view, to the northwest, is another triple-star system, called Struve 761. These three stars form a long, narrow triangle, pointing to the north.

In a Dobsonian telescope: The brightest star in the field is Sigma Orionis A and B, a double which is too close together for a small telescope to split. In a Dobsonian at high power, you can also see star C, a faint star southwest of A/B. To the east of A/B is star D, which shows a reddish tinge in a Dob. Northeast and about three times farther away is star E.

In the same field of view, to the northwest, is another triple star system, called Struve 761. These three stars form a long narrow triangle, pointing to the north.

The insets show a pair of 2' close-ups for each of the double stars.

With two complicated multiple stars so close together in the same field, it can be something of a challenge to keep track of which star is which.

The **Sigma Orionis** system is located about 1,200 light years from us. A and B are a pair of extremely bright and massive stars, only about 100 AU apart (too close to be separated in a small telescope). Both C and D are separated from A by great distances, at least 3,800 AU and 4,500 AU respectively, and E is almost a quarter of a light year from A.

Sigma Orionis, Struve 761, the belt stars, and the Orion nebula are all part of the same aggregation of young double stars and nebulae, travelling together through the Galaxy. Within our measurement uncertainty, all these stars are about the same distance away from us, and so are probably relatively close to each other.

We can infer from this that these stars were all formed from the same star-formation region, much like the Orion Nebula, which then became an open cluster of stars (see page 71) as the nebula gas was captured and incorporated into the young stars. Now these stars are slowly drifting apart as each follows its own slightly different orbit about the center of our Galaxy.

Some useful definitions when talking about double stars:
*An **astronomical unit** (AU) is the average distance from the Earth to the Sun – about 150 million kilometers, or 93 million miles.*

*A **light year** is the distance light travels in one year, about 9,500,000,000,000 kilometers, or six trillion miles. The nearest star, Proxima Centauri (see page 232), is 4.22 light years away from us. In this book we describe galaxies you can observe that are more than 20 million light years away.*

Sigma Orionis			
Star	**Magnitude**	**Color**	**Location**
A/B	3.7	White	Primary star
C	8.8	White	10.9" SW from A
D	6.6	Blue	12.7" E from A
E	6.3	Blue	41" ENE from A
Struve 761			
Star	**Magnitude**	**Color**	**Location**
A	7.9	White	Primary star
B	8.4	White	67" SSW from A
C	8.6	White	8.7" W from B

find more at: www.cambridge.org/features/turnleft/seasonal_skies_january-march.htm

In Orion: Orion's Dobsonian doubles

The Orion complex includes many newly formed stars in a region relatively close to us, and so not surprisingly there are many multiple-star systems here that can be observed with a backyard telescope.

We've already mentioned a number of multiple stars near the Orion Nebula (see page 54) and Sigma Orionis (page 56) that are well within the reach of even a small telescope. However, there are also quite a few other doubles in this region that need a Dobsonian (or a large refractor or catadioptric) to see.

In some cases, the stars are so close together that only the greater resolving power of a Dob can split them. In other cases, the contrast of brightness means you need a Dob's greater aperture to pull in the fainter star in the glare of its brighter primary.

Either way, these make for really fun challenges. They are especially good targets on less-than-perfect nights: the stars are generally bright enough to be seen even through thin clouds, while those clouds tend to keep the air still and steady.

Star	Magnitude	Color	Location
Meissa (Lambda Orionis)			
A	3.5	White	Primary star
B	5.4	White	4.1" NE from A
52 Orionis			
A	6.0	White	Primary star
B	6.0	White	1.1" SW from A
Alnitak (Zeta Orionis)			
A	1.9	Blue	Primary star
B	3.7	White	2.5" SSE from A
42 Orionis			
A	4.6	White	Primary star
B	7.5	White	1.1" SSW from A
32 Orionis			
A	4.4	Yellow	Primary star
B	5.7	Blue	1.2" NE from A
Mintaka (Delta Orionis)			
A	2.2	White	Primary star
B	6.8	Blue	53" N from A
Eta Orionis			
A	3.6	Yellow	Primary star
B	4.9	Blue	1.8" ENE from A
Rigel (Beta Orionis)			
A	0.3	Blue	Primary star
B	6.8	White	9.3" SSW from A

Meissa: Find Betelgeuse and Bellatrix, the shoulders of Orion. Meissa is the brightest of a cluster of stars where Orion's head should be. Aim the finderscope at Bellatrix, the right shoulder, and move northeast; you'll see a clump of at least four stars come into view at in the finderscope. Meissa is the brightest, and northernmost, of the bunch. It's a fairly easy split in a Dob.

This system is located more than 1,000 light years from us. Given that distance, the separation between the two stars comes to over 1,300 AU.

52 Orionis: Start at Betelgeuse, the bright red star at Orion's left shoulder. To the southwest in the finderscope will be a sixth-magnitude star; that's 52 Orionis. It's a nice cat's-eye pair, but very close; you'll want your highest power to split them.

52 Orionis is located about 500 light years from the Earth. Its companion orbits at a distance of at least 160 AU from the primary.

Alnitak, the easternmost star of Orion's belt, is a tough object to split in small telescopes, but should be easy in a Dob. (The primary, a lovely blue star, is itself a double but too close to split in an amateur telescope.)

Alnitak and its companions are located 800 light years away. The star we see is at least 600 AU from the primary star.

42 Orionis is located in the midst of the Orion Nebula field (see the Dob view on page 53). It's a very close double, with a significant brightness difference, so it's a challenge even in a Dob. It is the brighter of the pair of stars half a degree north of M42 and the Trapezium. Use your highest power to split them – and don't let the nebula distract you!

These stars are not actually in the Orion Nebula itself but just happen to lie about two-thirds of the way from us to the nebula. They are 800 light years away, and the separation between the two stars is about 250 AU.

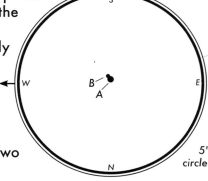

32 Orionis: Aim at Bellatrix, the right shoulder of Orion. Visible to its east in the finderscope is 32 Orionis. The colors are particularly notable for this pair, but they are quite close; you'll need your highest power to split them.

The 32 Orionis system lies nearly 300 light years away. The separation between the two stars is about 100 AU.

Mintaka is the westernmost star of Orion's belt, bright and trivial to find. The secondary star is well separated from the primary, but the brightness difference makes this a challenge in a small telescope since it can be overwhelmed by the glare of the primary. Note the secondary's blue color. (The primary is itself a double but too close to split in an amateur telescope.) This system is located about 900 light years from us; given that distance, the separation between the two stars comes to nearly 15,000 AU, or almost a quarter of a light year.

Eta Orionis is Orion's left knee. Start at Mintaka; Eta should be a reasonably bright star near the southwest corner of the finderscope view. At just under 2" separation, it's a nice challenge to split in a Dob at high power. Note the pleasant yellow–blue color.

The Eta Orionis system is about 900 light years from us. The separation between the two stars is about 500 AU.

Rigel is the brightest star (and right foot) of Orion and a wonderfully challenging double because the secondary is 6.5 magnitudes, or 400 times, dimmer than its primary. It can easily get lost in the diffraction spike made by the supports of the Dob secondary mirror. But it's great fun to find.

The distance to Rigel is just under 800 light years. The separation between it and its companion is over 1,500 AU.

Note that most of these stars lie at roughly the same distance from us, about 800 light years. They are probably related to each other, but not to the Orion Nebula itself; it's 500 light years farther away.

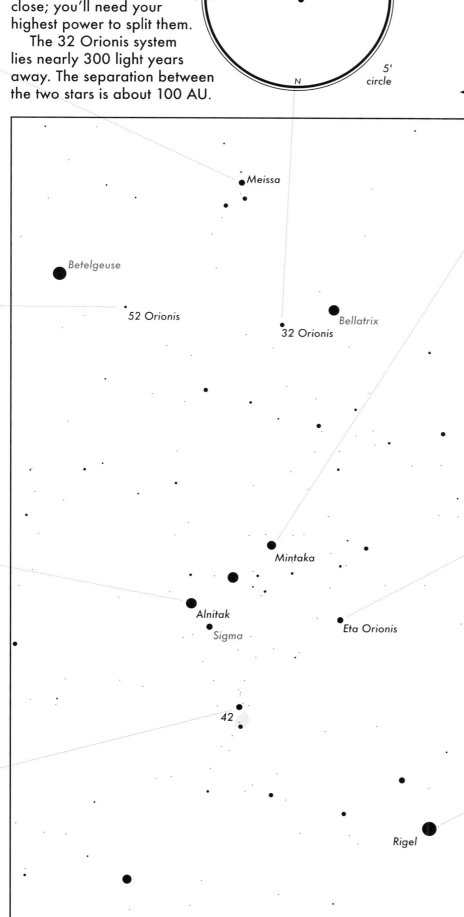

In Eridanus: A planetary nebula, NGC 1535, and *Keid*, a multiple star, Omicron-2 Eridani

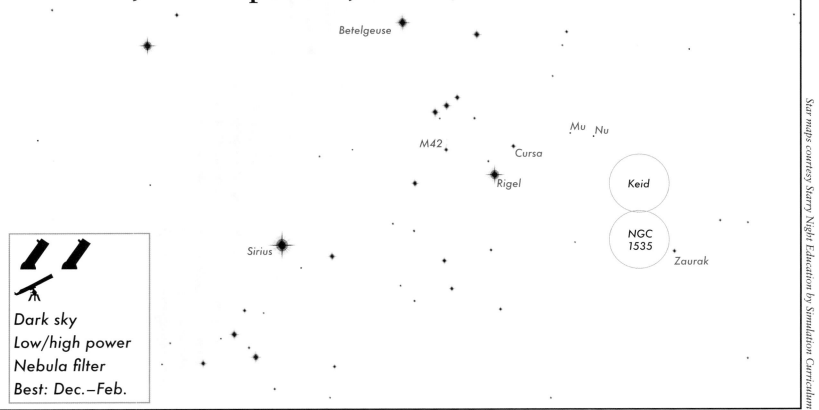

Star maps courtesy Starry Night Education by Simulation Curriculum

Dark sky
Low/high power
Nebula filter
Best: Dec.–Feb.

- NGC 1535: a fun challenge
- Keid B: brightest, easiest white dwarf companion
- Keid: home star of "Vulcan"

Where to look: The constellation Eridanus is named for a river; it is a string of faint stars in a dim part of the sky. The best way to find these objects is to follow the river. Find Orion high above the southern horizon and go to the bright star at its right foot, Rigel. Look northwest to a third-magnitude star, Cursa, and start the finder there.

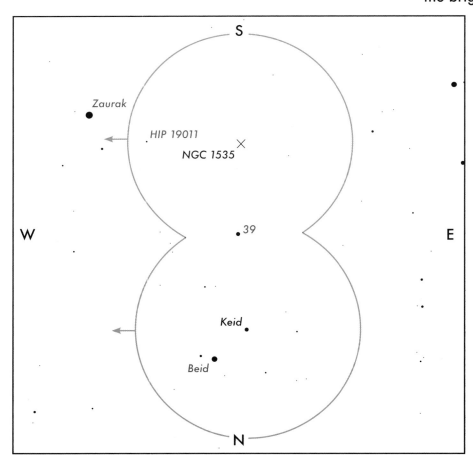

In the finderscope: In the finder, you'll know Cursa as the brightest of a tight triangle of stars. Move the finderscope to the west to a line of three stars; then nudge northwest to the pair of Mu and Nu Eridani. In the finder you'll also see stars northeast and northwest of Nu. From Nu, keep going west to a fourth-magnitude star by itself. The next step, southwest, takes you to Beid and Keid, a pair of stars in a southeast–northwest line. Beid, the northwest of the pair, has a companion star to its west; **Keid** is to the southeast. It is the southeast corner (and brightest member) of an equilateral triangle of stars.
NGC 1535: Take one step south (about half a finder field) from Keid to 39 Eridani (itself a challenging Dob double: a magnitude 5.0 star with a magnitude 8.5 companion, 6.3" to the southeast). NGC 1535 is one more step in the same direction. Aim for a point due south of 39 Eridani and due east of the faint star HIP 19011. (If you can't spot HIP 19011, look for the much brighter star Zaurak just to its west.)

NGC 1535 in a star diagonal at low power

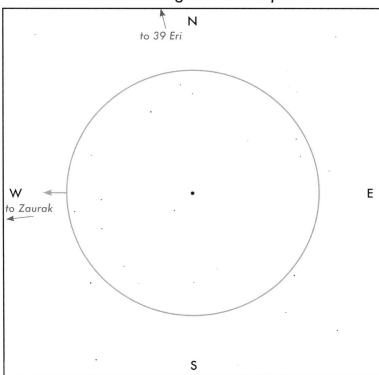

NGC 1535 in a Dobsonian at low/high power

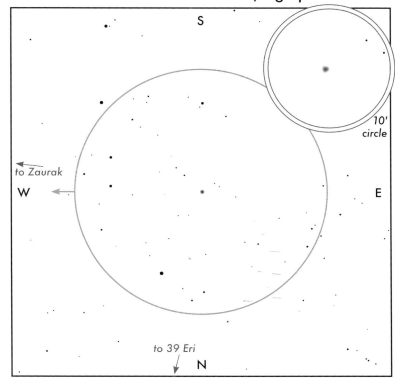

NGC 1535 in a small telescope: This is a real challenge, one for the thrill of the hunt more than the pleasure of the view. But a thrilling hunt it can be! We show the low-power view here to indicate which of the faint starlike objects is the one to aim for. Once you think you've found it, center on it and switch to high power to confirm that you see a fuzzy blue disk and not just another slightly out-of-focus star.

NGC 1535 in a Dobsonian telescope: At low power you can spot the planetary nebula as a small but fuzzy blue star. Once you find it, switch to your highest power to see it as a dreamy blue disk. Very high power and larger apertures reveal a lumpy, mottled surface; under excellent conditions you may see the 11th-magnitude central star.

NGC 1535 is a typical planetary nebula, the ball of gas erupted from a dying star (see page 99 for more about planetary nebulae). Best estimates for its distance put it about 7,000 light years from us, with a diameter of about a light year. The gaseous nebula surrounding the 11th-magnitude central star is rather thick, indicating that this is still a relatively young nebula. The mass of the nebula is about one tenth the mass of our Sun, and models suggest that this cloud has been growing for only a few thousand years.

Keid A has a pair of faint companions to its east; B is easy to see in a low-power Dob, but C is just enough fainter that seeing it requires high power and a good night. Star B can be hard to see (and C even harder) in a small telescope because it is so faint; use high power to increase its contrast with the sky.

Keid A is a K-type star, a bit fainter than our Sun, located only 16.5 light years from us. Its distant companions B and C orbit more than 400 AU from A, circling each other at an average distance of about 35 AU (about the Sun–Neptune distance).

Keid B is a white dwarf, with half the mass of our Sun packed into a volume no bigger than the Earth. In fact, Keid B is the easiest white dwarf to see; the other nearby white dwarfs, companions of Procyon and Sirius, are close to bright stars and so beyond the reach of a small telescope.

White dwarfs are the cooling remains of giant stars that have burned quickly through all their nuclear fuel and then, no longer hot enough to support their mass, collapse into a central core. They glow now only by the leftover heat of that collapse.

By contrast, Keid C is a small red M-type star. Though the smallest of these stars, it burns the slowest and so will continue to shine long after Keid A and Keid B have turned to embers.

The companions are far enough from Keid A that they should not perturb any planets near Keid A. Since Keid A is itself not all that different from our Sun and not too distant from us, it's an obvious candidate to search for extraterrestrial intelligence (SETI).

So far, no planet has been detected around this star. But in the fictional world of *Star Trek*, it is indeed inhabited by a most intelligent and very logical race. Though never mentioned specifically in any show, except for one offhand reference of traveling 16 light years to reach it, the show's producer Gene Roddenberry and three Harvard astronomers in a letter to *Sky & Telescope* (July, 1991) identified Keid as the location of Mr. Spock's home planet, Vulcan!

Keid (Omicron-2 Eridani, a.k.a. 40 Eridani)			
Star	**Magnitude**	**Color**	**Location**
A	4.5	Yellow	Primary star
B	10.2	White	83" ESE from A
C	11.5	White	7" N from B

find more at: www.cambridge.org/features/turnleft/seasonal_skies_january-march.htm

In Lepus: A globular cluster, M79

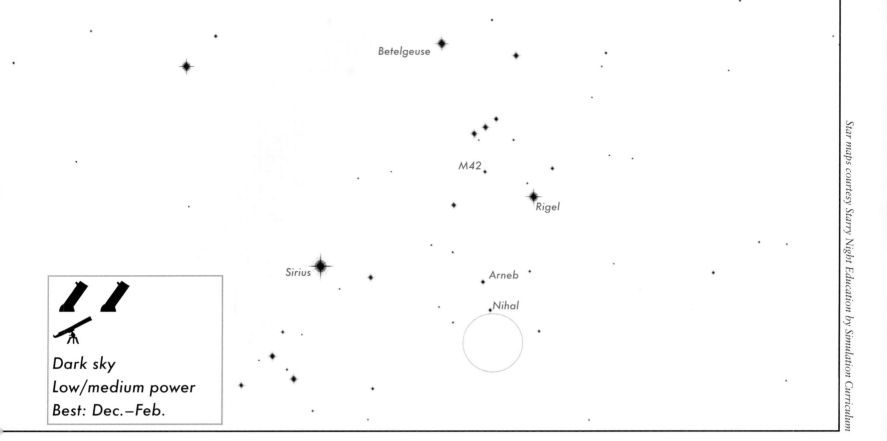

Star maps courtesy Starry Night Education by Simulation Curriculum

Dark sky
Low/medium power
Best: Dec.–Feb.

- The only easy globular cluster this season
- Nearby double star
- Interloper from another galaxy?

Where to look: Find Orion high above the southern horizon. Due south you will see two second-magnitude stars in a roughly north–south line, Arneb and Nihal. Step from Arneb, to Nihal, to one step farther south.

In the finderscope: There are a number of slightly fainter stars in this neighborhood, so it is important to be sure you've found the right ones. Arneb in the finderscope stands by itself; Nihal is in a cluster of slightly fainter stars. They should both fit in the finderscope field of view. Once you've identified Nihal, move south (away from Arneb) and look for a star called h 3752; it's the only star you're likely to see in the finderscope field of view. Aim for it. Once you're there, the globular cluster should be visible in the low-power telescope field of view, to the east and a bit north.

M79 in a star diagonal at low power

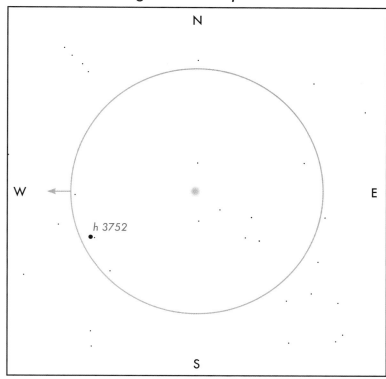

M79 in a Dobsonian at medium power

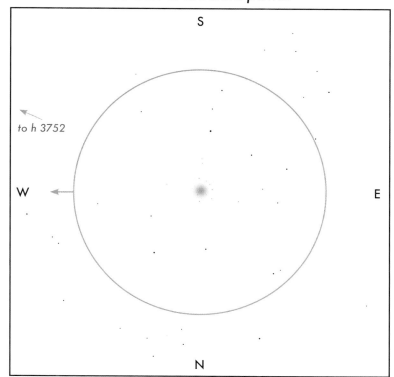

M79 in a small telescope: The globular cluster is subtle but sweet in a small telescope, a faint ball of light, easy to find between two brighter field stars. (Note a string of four faint stars to the northwest; if you see them, follow their line southeast to the cluster.) Once you have found the globular cluster, try it at higher power.

Note h 3752, in the southwest corner of your low-power field. It is a double, just splittable at high power: a magnitude 5.4 primary with a magnitude 6.6 companion 3.4" to the east, in the same direction as a ninth-magnitude field star an arc minute farther east.

M79 in a Dobsonian telescope: The cluster is a small ball of fuzz situated between two easy-to-see stars. At high power, the outer edges can be resolved into stars.

Note h 3752 to the southwest, just out of the field shown here but within the southwest corner of a low-power field of view. It is a double: a magnitude 5.4 primary with a magnitude 6.6 companion 3.4" to the east, an easy split in a Dob. The ninth-magnitude star an arc minute to the east is just a field star, not part of the system, but it makes a nice grouping with the double.

M79 is a globular cluster, 40,000 light years from us, but 60,000 light years from the center of the Galaxy. In fact, we lie between it and the center of the Milky Way.

Most globular clusters orbit the galactic center, which is in the direction of the constellation Sagittarius and thus hidden behind the Sun this time of year. Indeed, you'll notice that M79 is the only globular cluster visible during this season. What's it doing in a place like this?

One controversial theory has it that M79 is not from our Galaxy at all. Instead, it may be an interloper from a neighboring galaxy, a spheroidal dwarf galaxy known as the Canis Majoris Galaxy, which is currently undergoing a close encounter with the Milky Way.

Since we have seen other galaxies merging and colliding, it seems logical that such occurrences should be happening to our Galaxy as well. But from our vantage point inside our own Galaxy, such mergers are hard to make out.

The evidence for the existence of the Canis Majoris Galaxy, first proposed in 2003, is indirect. We can't see an obvious collection of stars that make up this galaxy. Instead, its existence is deduced from an excess of red giant stars in a part of the plane of the Milky Way towards Canis Major. Such stars, in the last stages of stellar evolution, ought to be distributed uniformly across the Galaxy. Even if stars are formed preferentially in

one spot, the relative speeds of their orbits around the galactic center should spread them out evenly throughout the disk by the time they turn into red giants. So their excess in this part of the sky suggests to some astronomers that another, smaller galaxy has recently been merged into ours at this point and contributed its red giants to our population.

If there is such a galaxy close to the Milky Way, the gravity of our much larger number of stars will eventually rip it apart. The distance between the stars in these galaxies is so great, however, that the odds of individual stars actually colliding is miniscule.

For more about globular clusters, see page 115.

find more at: www.cambridge.org/features/turnleft/seasonal_skies_january-march.htm

In Taurus: The *Pleiades,* an open cluster, M45

Star maps courtesy Starry Night Education by Simulation Curriculum

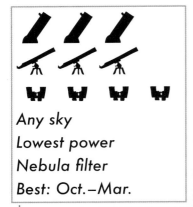

Any sky
Lowest power
Nebula filter
Best: Oct.–Mar.

Aldebaran

Betelgeuse

Bellatrix

- Easy to find
- Great in binoculars or finderscope
- In a Dobsonian, look for nebulosity

Where to look: Find Orion high above the southern horizon. Up and to the right is a V shape of dim stars, tilted towards the left, with a very bright orange–red star (Aldebaran) at the top left of the V. Step from the upper right shoulder of Orion, Bellatrix, to Aldebaran, to a small cluster of stars. That cluster is the Pleiades. This collection of stars is sometimes called the "Seven Sisters" though, in fact, only six stars in this cluster are easily visible to the naked eye. (Actually, as many as 18 stars may be seen by the naked eye if the sky is very dark and your eyes are exceptionally sharp.)

In the finderscope: In many ways the finderscope gives the best view of this large, bright, nearby open cluster. Along with the six brightest stars in a dipper-shaped cluster, another dozen or more may be visible.

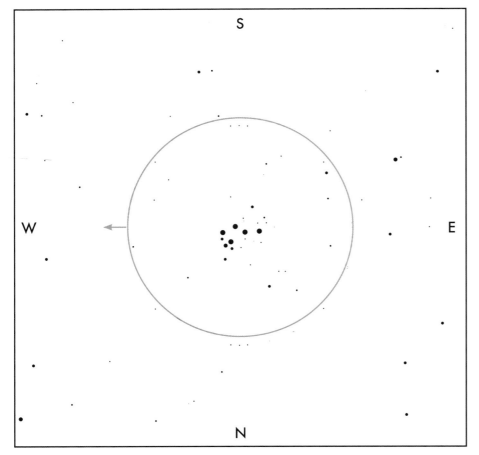

The Pleiades in a star diagonal at low power

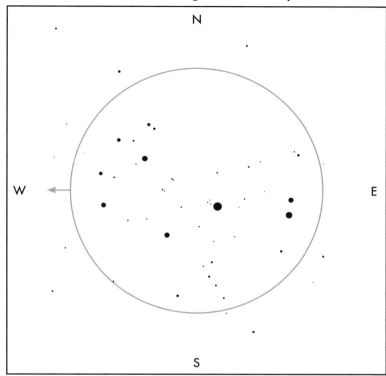

The Pleiades in a Dobsonian at low power

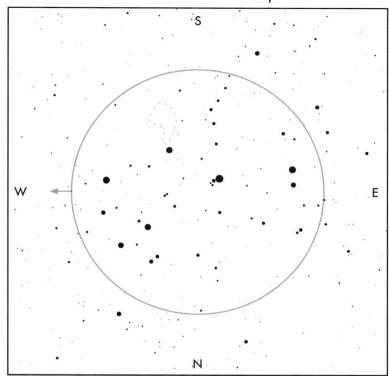

In a small telescope: Some 40 to 50 stars may be visible, but the group is so large that even at low power you probably won't be able to fit them all into the field of view at any one time.

In a Dobsonian telescope: Under very dark skies you can begin to see faint wisps of nebular gas near some of the stars. (However, if you see "nebulosity" around *all* the brightest stars, that just means you've got dew on your lenses!) Look for it especially south of the star named Merope, the southernmost of the four stars that make up the bowl shape, in the region denoted by the dotted line.

The **Pleiades** are easy to find, and there's something worth looking at in any sized telescope. Binoculars are best at getting across the richness of the cluster, while very dark skies and a much bigger telescope can allow you to see the nebulosity around the stars. That said, in some ways a two-inch or three-inch telescope is just the wrong size to really appreciate this cluster; nonetheless, seeing so many bright blue stars in one place, especially against the contrast of a very dark sky, can be quite impressive.

This star cluster, easily visible to the naked eye, has been known since antiquity. It is noted in Chinese annals dated at 2357 BC, and it is mentioned in the Bible. The brightest nine stars are named for seven sister nymphs and their parents (Atlas and Pleione).

If we were sitting in the center of the cluster, these nine stars would each be brilliant, at least as bright as Venus looks to us on Earth.

Over 200 stars have been counted in this cluster; it lies about 400 light years away from us. Most of the stars we can see with the naked eye fit within an area 7 light years across, but the outer reaches of the cluster extend across 30 light years.

The wisps of nebular gas seen around Merope and (more faintly) the other stars have long been used as evidence that the Pleiades is relatively young cluster, but recent theory suggests that we may just be seeing light from these stars shining fortuitously through a dust-rich region of interstellar gas that just happens to lie between those stars and us.

For more on open clusters, see page 71.

Asterope

Taygeta

Maia

Celaeno

Pleione   Alcyone

Electra

Atlas

Merope

In Taurus: The *Crab Nebula,* a supernova remnant, M1

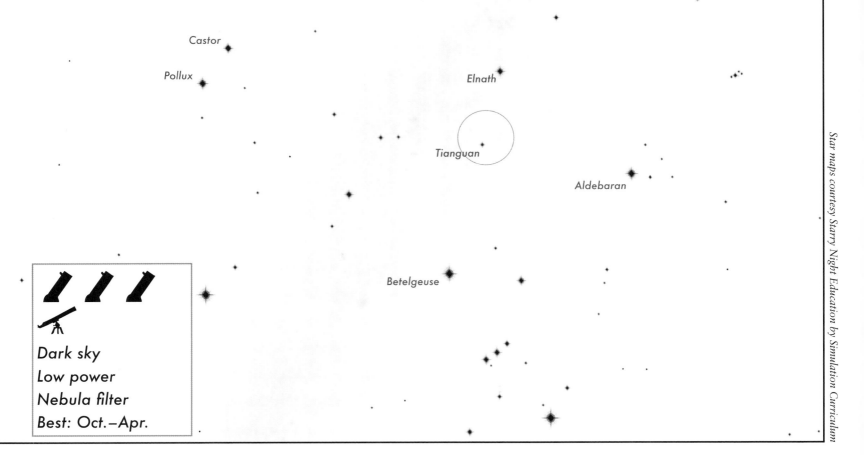

Star maps courtesy Starry Night Education by Simulation Curriculum

- Challenging in suburban skies
- Supernova remnant and neutron star
- Historical interest

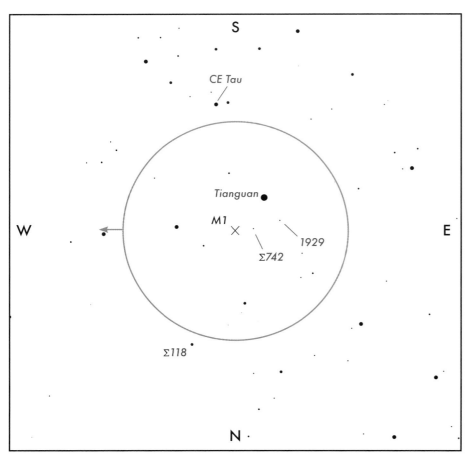

Where to look: Find Orion high above the southern horizon. Up and to the right is a V shape of dim stars, opening to the northeast. A very bright orange–red star, Aldebaran, sits at the top left of the V. Call the distance from the star at the point of the V to Aldebaran one step; step from the V point, to Aldebaran, to four steps farther to the left. At that point, just above the line (to the north) you'll find a third-magnitude star, Tianguan (Zeta Tauri). Don't confuse Tianguan with a brighter star to its north, Elnath. Aim the telescope at Tianguan.

In the finderscope: Start at Tianguan. Moving towards Elnath, look for two seventh-magnitude stars, HR 1929 and Struve 742. Aim just west of 742. If it's not dark enough to see 742 in the finderscope, you'll have no hope of seeing M1 in the telescope!

Also in the neighborhood: Struve 742 is a double star. The pair are magnitudes 7.1 and 7.5, in an east–west line separated by 4.1"; switch to high power to split it. The colors are subdued, yellow and blue.

Just outside the finder field is Struve 118. It's a yellow–white pair, a magnitude 5.8 primary with a 6.7 companion 4.4" to the south–southwest. In fact, it's a quadruple star: each star has a faint companion, but they are too close to be easily seen even in a Dob.

CE Tau is a distant supergiant star, 600 times as big as our Sun, emitting 40,000 times as much light.

M1 in a star diagonal at low power

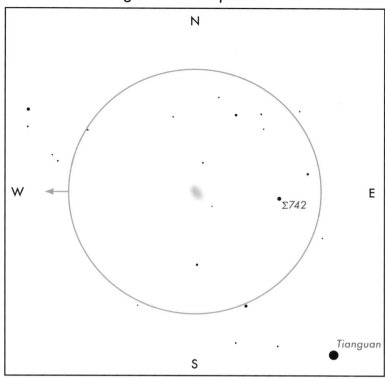

M1 in a Dobsonian at low power

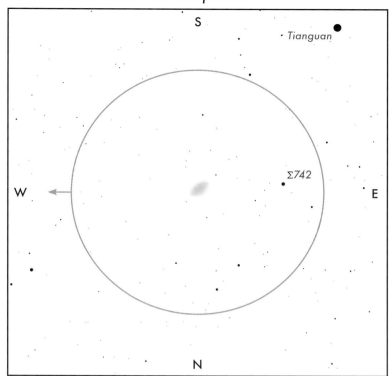

In a small telescope: You'll see two stars, one south and one east of the nebula. The object looks like a faint but reasonably large blob of light. It is distinctly elongated, not quite twice as long as it is wide. Averted vision helps you make out its shape. Note the double star Struve 742, a challenge to split in a small telescope.

In a Dobsonian telescope: The nebula is a faint, ghostly smudge in a starry field. If there is the slightest thin cloud cover to reflect the sky brightness back at you, it will be very hard to find; but on a moonless, cloudless night, in a Dob the nebula can jump out at you even in suburban skies. Note the double star Struve 742, a nice split at high power.

In a small telescope, **M1** is quite faint, an oval-shaped cloud of dim light. It's virtually impossible to find if the Moon's up, even with a good-sized telescope. As something to look at, it's interesting in a small telescope only for the challenge of finding and seeing it. Its real appeal, however, comes not from what it looks like, but what it is: a supernova remnant, a spinning neutron star wrapped inside a gas cloud.

In July of 1054, Chinese, Japanese, Korean, and Turkish stargazers (and possibly Native Americans) recorded the appearance of a bright star in the location where we now see the Crab Nebula. It was bright enough to be visible even in the daytime, and undoubtedly would have been a spectacular sight at night except that, in early July, that part of the sky rises and sets with the Sun.

We discuss under the Ghost of Jupiter (page 99) how a star that has exhausted its fuel can collapse, creating a cloud of gas which becomes a planetary nebula. When a very large star has completely exhausted its supply of nuclear fuel, the subsequent collapse results in a spectacularly violent explosion, which we call a supernova. The light of a supernova can, briefly, become as bright as the light from all the other stars in the Galaxy combined.

After the explosion, the gas expands into space while the core of the exploded star becomes a super-dense lump of degenerate matter called a neutron star. Gas in its very strong gravity and magnetic fields can become so accelerated that its energy will radiate away as radio waves. Supernova remnants, like this one, are strong radio sources. Indeed, soon after World

War II, when astronomers were first equipped with radio dishes and radio technology developed during the war, they found that this nebula was a strong source of radio waves. In 1968, the radio signals from this nebula were found to have a regular pulsing pattern, emitting two bursts of energy 30 times a second, one of the first such *pulsars* ever discovered.

The best estimates are that the original star which blew up was five to ten times as massive as our Sun. Today, most of that mass is packed into the dense star at the center (a 16th-magnitude star, visible only in large telescopes). The pulses occur as a result of that star spinning 30 times a second. Such a rapid spin rate is possible only if the star is very small, only a few kilometers in diameter. All that mass packed into such a small space makes this central star fantastically dense, and the force of gravity at its surface must be enormous.

This nebula is located about 5,000 light years away from us. It is already about 7 light years wide, and it continues to expand at a rate of over 1,000 km (600 miles) every second. From our perspective, this means that the nebula grows by about one arc second every five years.

As the nebula grows, it gets dimmer, since its light energy has to be spread out over a larger cloud. In the past, this nebula must have been brighter – perhaps twice as bright, 200 years ago. Dim as it is today, it is unlikely that Charles Messier would have noted it in his catalog.

find more at: www.cambridge.org/features/turnleft/seasonal_skies_january-march.htm

In Auriga: Three open clusters, M36, M37, M38

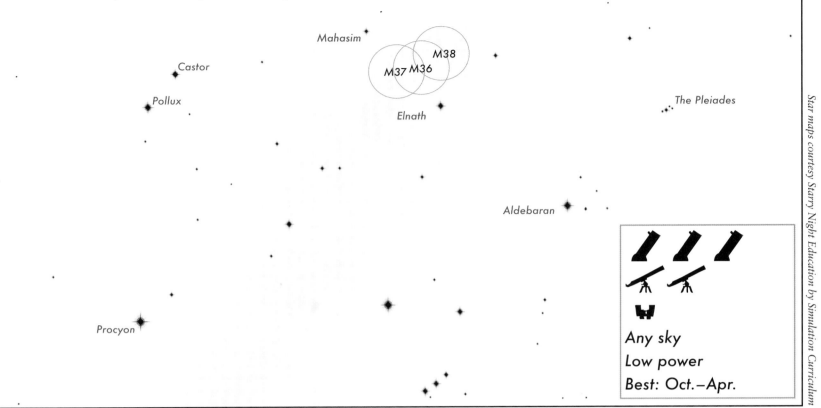

Star maps courtesy Starry Night Education by Simulation Curriculum

Any sky
Low power
Best: Oct.–Apr.

- Elegant views
- Range of types, from loose to grainy
- Easy (M36) to more challenging (M37) to find

Where to look: Find Auriga, north of Orion, and look for the bright star Capella. The constellation of Auriga looks like a ring of stars; Capella is at the upper right (calling north up); the star Mahasim is at the lower left; Elnath, the second-brightest of the stars (after Capella), is down and to the right from Mahasim, at the bottom of the ring. Start by aiming at Elnath.

In the finderscope: If the sky is clear and dark all three objects may be visible in the finderscope, though dim and small. **M36:** This is the brightest and easiest to find. From Elnath move your finderscope north, towards Capella, until you see a line of stars (a good finderscope will see a box at the northeast end), including 14, 16, and 19 Aurigae. Follow that line to the northeast until you see another cluster of stars off to the east; the brightest is Phi Aurigae. Move to Phi, and notice another star to the south and east, Chi Aurigae. Imagine an equilateral triangle with Phi, Chi, and the third corner pointing to the north and east; aim for the empty third point. **M38:** From Phi, move north and west, halfway to Sigma Aurigae; the cluster is just to the east of the halfway point. **M37:** Step from Elnath northeast to 26 Aurigae; from there, continue northeast to a right triangle of faint stars. Move northeast from the right-angle corner, mirroring the same distance away from the corner as the southwest corner star of that triangle. Look for M37 near there.

M36 in a star diagonal at low power

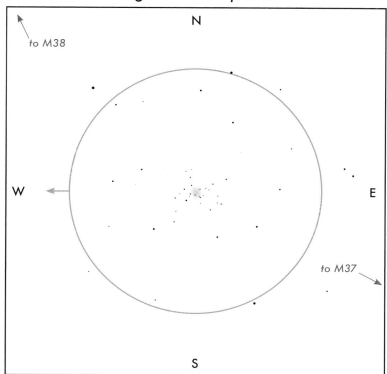

M36 in a Dobsonian at low power

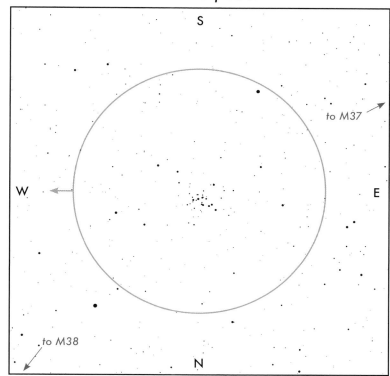

In a small telescope: You'll see a loose disk of stars, slightly brighter and denser at the center, with about 10 fairly bright stars including a nice close pair near the center of the cluster. In the area around this pair, you may see a bit of a grainy haze of light. Depending on how dark the skies are, and how big your telescope is, you may see clearly a dozen or so dimmer stars, with even more (perhaps 30 or so) visible with averted vision: let your eye wander around the field of view and try to catch the dimmer ones out of the corner of your eye. Most of them form a ring around the rest of the cluster.

In a Dobsonian telescope: The cluster is a loose disk of stars, slightly brighter and denser at the center. Notice a close pair of stars near the center of the cluster; in the area around this pair, you should see perhaps 30 stars visible, forming a ring around the rest of the cluster.

M36 is probably the finest of this trio of Auriga open clusters for telescopes of three inch or smaller aperture, since it is brighter (though smaller) than M38, with more bright stars than M37. On the other hand, M37 is the most impressive of the three in a Dobsonian; what appears just as grainy light in a small telescope can be resolved by a Dob into a gathering of individual stars. M38 is not as easily seen in the finderscope as the others, nor is it impressive in a telescope. Instead, the fun in observing M38 is in the hunt: there is a distinct feeling of satisfaction when you finally do catch it.

M36: This cluster of 60 young stars is about 15 light years in diameter, located 4,000 light years away from us. The brightest of the stars are several hundred times as bright as our Sun. They are mostly type-B stars – bright, blue, and hot. With its pretty knot of jewel-like blue stars against the background of hazy light from the dozens of stars too faint to be made out individually in a small telescope, this is an easy and very rewarding object. Even though it is a bright object, visible even in the finderscope, it is most impressive to look at under low power. That way you get both the individual bright stars and the grainy haze of unresolved stars around them.

M37: Larger than M36, this cluster is made up of several hundred stars, with 150 of them just bright enough (11th to 13th magnitude) to be picked out easily in a Dobsonian. In a smaller telescope, they will look like a faint cloud of light with only hints from the graininess of the light that the cloud is really made up of individual stars. The stars lie in a cloud some 25 light years across, a bit larger than M36; however, it is also a bit farther away, some 4,500 light years distant.

M38: This is a neighbor of the other two open clusters, located about 4,000 light years away from us. With about 100 stars in a cloud about 20 light years across, it is in between M36 and M37 in size. It is not as bright and compact as M36, nor do you get the feeling with a small telescope (like you do with M37) that this cloud of light is just on the verge of being resolved into individual stars.

M37 in a star diagonal at low power

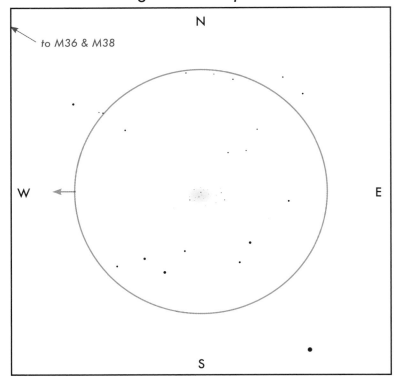

In a small telescope: You'll see an oblong, grainy cloud of light almost on the verge of resolution but with only a few individual stars, five or so, standing out. There is a distinctly orange star near the center. Other than these stars, it looks a bit like a globular cluster in a small telescope.

M37 in a Dobsonian at low power

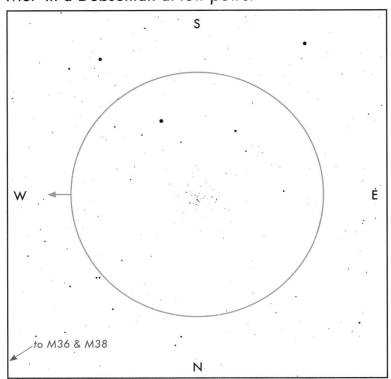

In a Dobsonian telescope: The best of the Auriga clusters in a Dob, you can see individual stars, not just grainy light, as faint as the sky allows; the eye picks out lots of lines of stars. Note the orange ninth-magnitude star in the center of the field of view. Compare its color with the much brighter field star (sixth magnitude) southwest of the cluster.

Also in the neighborhood: There are a number of nice doubles nearby. **14 Aurigae** *is actually a quadruple star; the main star A is a fifth-magnitude yellow star with a colorfully blue companion C apparently of magnitude 7.3 set 14" to the southwest. But in fact C is a double itself, a pair of eighth-magnitude stars only 2" apart – a challenge to split in a Dob. Meanwhile, star B is ninth magnitude, 10" to the north of A (and hard to see for being so much dimmer).*

The star **26 Aurigae**, *used to find M37, is actually a triple. Although the A/B pair are only 0.2" apart, too tight to split with an amateur telescope, A (with B) and C make a pleasant yellow–blue pair. The A/B pair has a combined magnitude of 5.5; the C star is magnitude 8.4 and it's found 12" to the west.*

Due west of Sigma Aurigae, visible at the edge of the finder scope view (page 68), is the double star **Omega Aurigae**. *The yellow fifth-magnitude primary has an orange eighth-magnitude star 4.8" due north.*

M38 in a star diagonal at low power

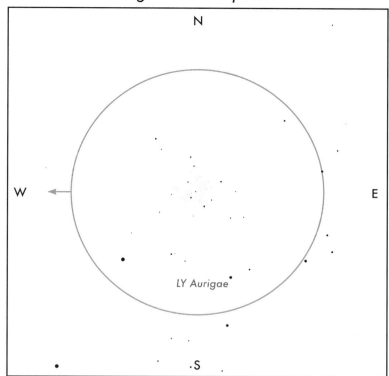

M38 in a Dobsonian at low power

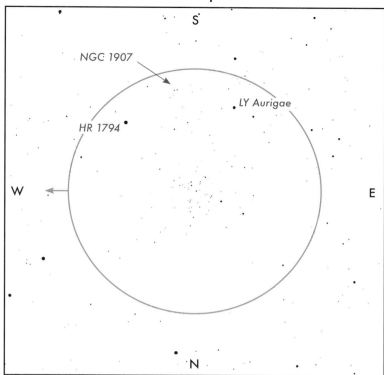

In a small telescope: The whole cluster, which is rather loose and large, is enmeshed in a subtle, irregular haze. Only about five stars are easily visible in a small telescope; another ten or so are possible, depending on how dark the sky is.

Look south of M38 for the eclipsing binary LY Aurigae (varying from magnitude 6.5 to 7.4 with a four-day period).

In a Dobsonian telescope: M38 looks like a loose collection of stars, not as bright and compact as M36. There is no single dominant star in this cluster. There is a hazy background of light behind these stars that is hard to resolve even in an eight-inch.

Look south of M38 for the eclipsing binary LY Aurigae (varying from magnitude 6.5 to 7.4 with a four-day period). To its west, halfway to the sixth-magnitude star HR 1794 and a bit south, is another open cluster, NGC 1907, looking like a small faint haze in a Dob.

About open clusters

Clusters of stars are born from the same large cloud of gas (like the Orion Nebula; see page 52). Unlike globular clusters (see page 115) the stars do not start out so close to each other nor in such large numbers that their own gravity holds the cluster together forever; instead, as they slowly move away from their stellar nursery, their initial velocities and the pull of the other stars in the Galaxy will serve to spread out these newly formed stars until eventually the whole cluster is dispersed. But while they are young, they will still be found grouped together as an open cluster.

Dating open clusters *can be done by looking at the types of stars present in the cluster, determined by their* **spectral class.**

For instance, most of the stars in M37 are of a type called spectral class A, *which means that they are bright, blue, and hot. By comparison, a star as bright as our Sun (a class-G star) in this cluster would be dimmer than 15th magnitude; you'd want a 20-inch telescope to see it.*

The brightest star in M37, however, is not bright and blue, but bright and red/orange. What's going on here?

The stars in an open cluster are in general very young, still spreading out from the region where they first were

formed. However, big bright stars, like O, B, and A stars (in that order), tend to use up their fuel and evolve faster than ordinary stars. They're so big that they can burn brighter than most stars, but because of their great brilliance they burn their fuel at an enormous rate. Thus, the biggest and brightest of the stars are the first to finish off their fuel and puff out into a dying **red-giant** *star.*

It's not unusual, therefore, to see one or two red stars in an open cluster, stars which may be very young in a cosmic scale yet have already lived out their normal lifespan. Ultimately such a star may become a nova or even a supernova, exploding and thereby returning material to interstellar space, where it goes into making new stars and solar systems.

The O and B stars in M37 have turned into red giants. That means that the brightest blue stars left in this cluster are spectral class A.

By estimating, from theory, how rapidly this process takes place we can put ages on these clusters. A cluster like M36, with many B stars, is young – about 30 million years old. M38 also turns out to be quite young, about 40 million years old. But because the O and B stars in M37 have had time to burn out and turn into red giants, it must be an older cluster; its age is estimated to be about 150 million years.

find more at: www.cambridge.org/features/turnleft/seasonal_skies_january-march.htm

In Gemini: An open cluster, M35

- Bright, colorful cluster
- Interesting in all apertures
- NGC 2158 also in same field of view

Where to look: Find the Twins; Castor is the very bright blue star to the northwest (up and to the right, facing south) and Pollux is the slightly brighter yellow star to the southeast. These two bright stars are the heads of two stick men. At the southwestern foot of the northwestern Twin (the stick man with Castor as its head) you'll see two stars of roughly equal brightness lying in an east–west line. The eastern star is Tejat, and the western star is Propus. Aim the finderscope towards these stars.

In the finderscope: Look for three bright stars in a crooked line. The two brighter ones, to the right (east) are Propus and Tejat, making the east–west line described above. With the finderscope centered on Propus, the third star, 1 Geminorum, will come into view just to the western edge and a bit to the north. The cluster and 1 Geminorum are both about the same distance from Propus. To find the cluster, start from Propus towards 1 Geminorum, but veer a bit left. You should be able to see the cluster as a faint lumpy patch of light.

Star maps courtesy Starry Night Education by Simulation Curriculum

M35 *in a star diagonal at low power*

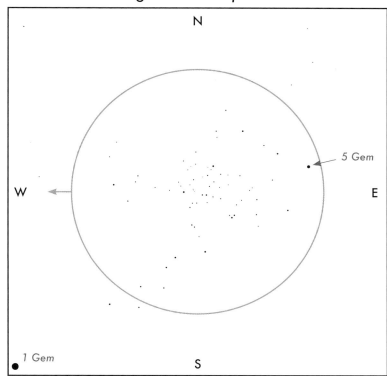

M35 *in a Dobsonian at low power*

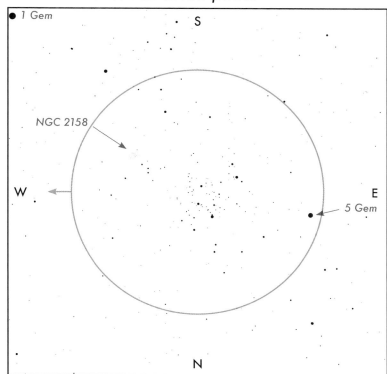

In a small telescope: At first glance, this cluster looks large but not particularly rich in stars. Your eye is drawn to about half a dozen brighter individual stars. But after a while you begin to notice many fainter stars in the background; everywhere you look there are more faint ones; as you let your eye relax and wander, with averted vision you will start to pick up more and more of the fainter stars. This gradual unveiling of a richer and richer star field is an effect most noticeable in binoculars or small telescopes (3" or less). A bigger telescope just shows all the stars at once, and as a result the final effect loses some of its charm.

In a Dobsonian telescope: You'll see a broad range of stars, from quite bright to just at the edge of visibility. Take your time to soak up the light, and trace out the various patterns that your eye can catch among them.

A little hazy patch to the southwest of M35 is the open cluster NGC 2158. Under high power (especially on a dark night, with averted vision) you can begin to see a graininess in its light.

M35 is quite a pretty open cluster. Most of the stars are hot, blue stars, but some of the brightest are yellow or orange in color, giant stars that have already evolved past the main sequence stage.

This is a particularly pretty region of the sky to scan about in, so rich in stars that you may at first mistake other stars of the Milky Way for the open cluster. Once you've found it, however, the cluster is unmistakable.

This open cluster consists of a few hundred young stars, clustered together in a region roughly 30 light years in diameter, located just under 3,000 light years from us. From the colors of the stars present, one can infer that this is a young open cluster, probably formed about 50 million years ago.

The contrast in size and brightness between M35 and NGC 2158 (if you can find it) is delightful. In fact, NGC 2158 is about seven times as far from us as M35, which accounts for the difference in size and brightness. That means that the intrinsic size and brightness of these two open clusters are really quite similar. One big difference, though, is that NGC 2158 is much older – more like 2 billion years old.

For more on open clusters, see page 71.

Also in the neighborhood: Propus is a double star, too close to split in a small telescope but a fun challenge with a Dobsonian at its highest power. The primary is an orange star, magnitude 3.5, while its companion is much dimmer (magnitude 6.2) and only 1.7 arc seconds (to the west–southwest). By contrast with the primary, the secondary's color looks almost green.

*Looking southeast of Tejat you'll come across Nu Geminorum; just before you reach it, you may see two faint stars in the finder field. The one farther from Nu is **15 Geminorum**, and it's also a nice double. This one is easy even in a small telescope: the two stars, magnitude 6.6 and 8.2, are separated by 25 arc seconds (look south–southwest). You can see the split even at low power, and with the extra light from low power you can appreciate their colors: the primary is orange, while its companion is blue.*

*An open cluster, **NGC 2129**, is located a degree west of 1 Geminorum. Look for a pair of stars, seventh and eighth magnitude; a small telescope sees a dim haze surrounding them, which a Dob resolves into a loose sprinkling of fainter stars in strings running east to west.*

find more at: **www.cambridge.org/features/turnleft/seasonal_skies_january-march.htm**

In Gemini: The *Clown Face*, a planetary nebula, NGC 2392; and *Castor,* a multiple star, Alpha Geminorum

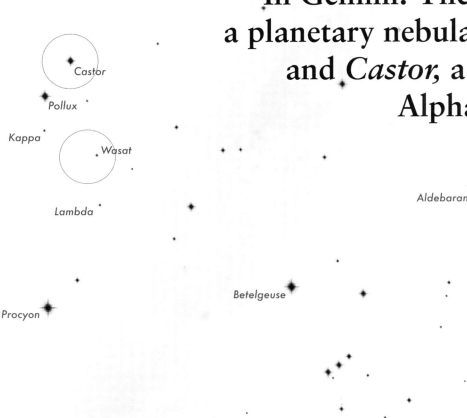

Star maps courtesy Starry Night Education by Simulation Curriculum

Dark/steady sky
Low/high power
Nebula filter
Best: Dec.–May

- Bright, easy to find planetary nebula
- Castor: easy to find, a fun challenge to split
- Several other doubles nearby

Where to look: Find the Twins, a pair of bright stars north and east of Orion. **Castor** is the blue one to the northwest (closer to Capella) and Pollux is the slightly brighter, yellow star to the southeast. If Castor and Pollux are the heads of two stick men, the waist of the stick man on the left (the second star down from Pollux) is Wasat. The nebula **NGC 2392** lies to the east of this star.

In the finderscope, NGC 2392: Aim first at Wasat. In the finderscope, this will be the brightest of three stars that make an equilateral triangle. The northeastern star in the triangle is 63 Geminorum; it's the one with two faint companions. Center first on this star, then move a bit more than a full-Moon diameter to the southeast.

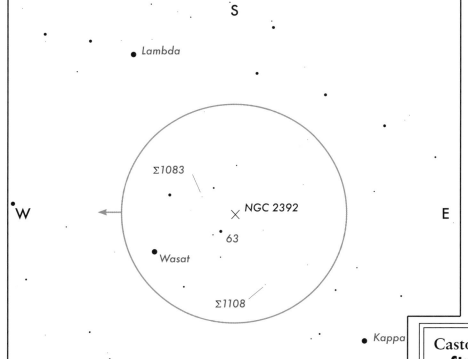

Castor (Alpha Geminorum)			
Star	**Magnitude**	**Color**	**Location**
A	1.9	White	Primary star
B	3.0	White	4.8" ENE from A
C	9.8	White	70" SSE from A

NGC 2392 *in a star diagonal at medium power*

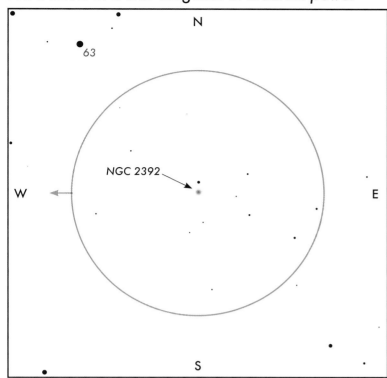

NGC 2392 *in a Dobsonian at low/high power*

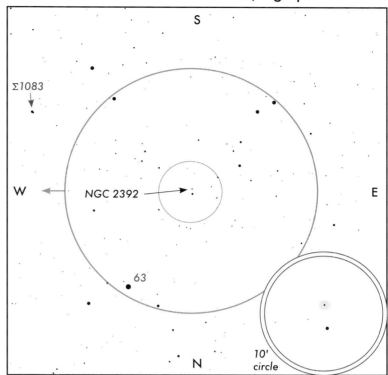

In a small telescope: The nebula looks like a blue–green out-of-focus star south of another star. Use a medium power to find it; the two may look like a double star of nearly equal brightness. High magnification shows the nebula as a round and fuzzy disk, while the nearby star remains a point of light.

In a Dobsonian telescope: As is the rule for planetary nebulae, find it with low power and then observe it in high power – with a nebula filter, if you have one. On a still night, look for the central star. The high power and filter will reveal a ghostly disk of green light with just a hint of the structure inside it that, in photographs, gives the nebula its face.

NGC 2392: You'd need to use a very large telescope to see the dark features that give this nebula its name **the Clown Face**. The nebula is located roughly 5,000 light years from us; the cloud itself is about 40,000 AU across (about two thirds of a light year), and judging from the motions of the gas it appears to be growing at more than 20 AU per year. From this, one can infer that this nebula may be less than 2,000 years old, making it one of the youngest of the planetary nebulae (see page 99).

Castor: This is a lovely triple star, a challenge in a small telescope but easily within reach of a Dobsonian. With good magnification and a clear night, you will see a definite separation into two fairly bright stars; then note a third, dimmer star with a slightly orange tint to the south-southeast.

In fact, Castor is at least a sextuple star system. A and B are each a pair of hot young A-type stars, roughly twice the size of our Sun and perhaps ten times as bright; star C is actually two dwarf K-type stars, about two thirds the size of our Sun and less than 3% as bright.

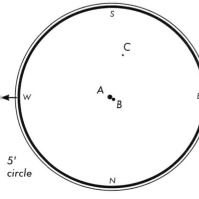

Each pair is separated by only a few million miles; the A pair orbits once every nine days, B once every three days, and C in less than a day. The B pair orbits the A pair with a period of about 450 years. At their closest, they are only 1.8 arc seconds apart; at present, their separation is about four arc seconds, and growing

slowly. Over the next hundred years star B will eventually move northwards, reaching a maximum separation of 6.5 arc seconds before circling back in its orbit towards A. By the middle of the twenty-first century, this will be an easy pair to split even in the smallest telescope.

The C pair must be more than 1,000 AU from the other stars. They take over 10,000 years to complete an orbit around A and B.

Also in the neighborhood: To the southwest, **Struve 1083** *is a nice yellow–blue pair, a magnitude 7.3 star with a magnitude 8.1 companion 6.8" to the northeast. Another yellow–blue pair nearby,* **Struve 1108** *is magnitudes 6.6 and 8.2 separated by 11.6" (the companion lying to the south); it's a challenge for a small scope, but easy for a Dob.*

Wasat itself is a double star. Because the secondary is so much fainter (magnitude 8.2, 5.6" southwest of the magnitude 3.6 primary) it's hard to see except in a Dob at high power. The primary is white, while the orange secondary is so faint it almost registers as purple to the eye.

Lambda Geminorum (Pollux's right knee) is a hugely unequal double, magnitudes 3.6 and 10.7 with a 9.7" separation north-northeast. The brighter star is yellow, the fainter, blue.

Kappa Geminorum (Pollux's right hand) is a challenging double star. The yellow primary is magnitude 3.7 while its fainter blue companion, magnitude 8.2, sits 7" to its southwest. With such a brightness difference, you'll need a large aperture with very high power – and very steady skies – to see them split.

find more at: www.cambridge.org/features/turnleft/seasonal_skies_january-march.htm

In Monoceros: An open cluster, NGC 2244, and the *Rosette Nebula*

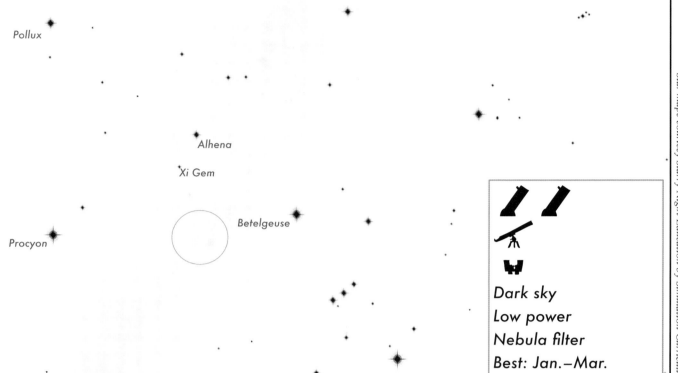

Star maps courtesy Starry Night Education by Simulation Curriculum

Dark sky
Low power
Nebula filter
Best: Jan.–Mar.

- Pleasant cluster even on a bright night
- Dark nights, the nebula envelops the cluster
- Rich star-formation region

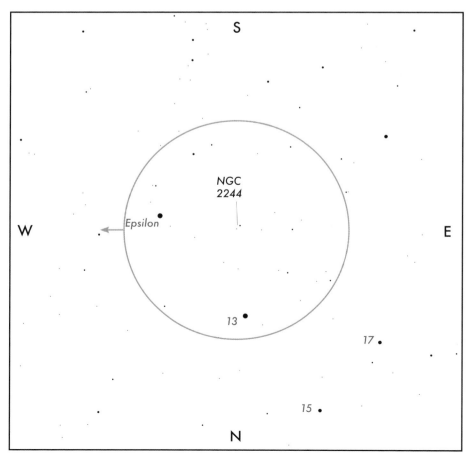

Where to look: Find Orion, high above the southern horizon, and find the very bright red star, Betelgeuse, which makes his shoulder (up and to the left of the three stars in his belt). Look eastward to the bright star Procyon, sitting to the right of Gemini, the Twins. Then find Alhena, the left foot of Pollux and the brightest of the stars in Gemini's feet.

You'll be looking at a point along the line between Procyon and Betelgeuse that sits due south of Alhena.

In the finderscope: The easiest way to find the nebula is to start at Alhena and move south. First you'll come to a pair of stars, Xi and 30 Geminorum, that make up Pollux's right foot. Center on them, and keep moving south. Next you'll come to the triangle of 15, 17, and 13 Monocerotis. Center on 13. Keep moving south from 13 until you are at a point even with Epsilon Monocerotis – due south of 13, due east of Epsilon. Look for the brightest members of the open cluster in your finderscope.

NGC 2244 in a star diagonal at low power

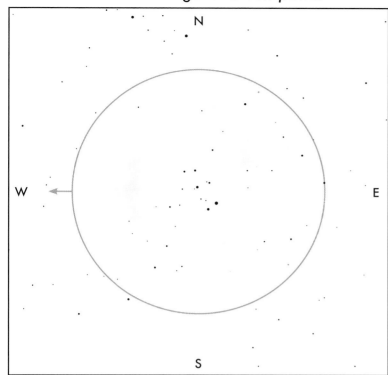

NGC 2244 and Rosette in a Dob at low power

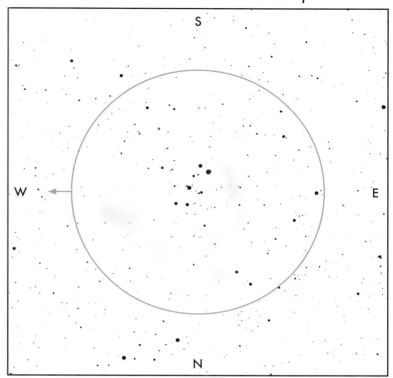

In a small telescope: The open cluster will appear as a small gathering of about half a dozen stars. If the night is extremely dark, you may be able to see a ghostly ring of light around the stars. If you can, try a nebula filter here.

In a Dobsonian telescope: On an ordinary night, only the open cluster will be easily visible, half a dozen stars with a sprinkling of fainter ones behind them. But looking more closely you may see a wide pale ring surrounding these stars. If the night is not particularly bright, the ring seems more like an absence of stars rather than a glow; but on a good dark night it can appear as a remarkable glow filling the field of view. A nebula filter can make all the difference here.

The **Rosette Nebula** is a region of star formation, much like the Orion Nebula (see page 55 for more on diffuse nebulae) and the open cluster **NGC 2244** consists of some of the stars formed in this region.

This nebula is so complex that the various bits of it have their own NGC catalog numbers, including NGC 2237, 2238, 2239, and NGC 2246. These were seen (and numbered) individually by different observers in the nineteenth century, before long-exposure photographs revealed that they were all different parts of the same structure.

The nebula itself is a large cloud of excited hydrogen gas, extending about 130 light years across, located some 5,200 light years from us. The gas is excited (and glows) from the radiation of the hot young stars being formed in its center. Indeed, the Chandra X-ray space telescope has discovered strong X-ray emissions from this region. Apparently the young blue O-type stars in the central cluster have heated the gas to a temperature in excess of 6,000,000 degrees. (That's degrees Celsius, but at that temperature it doesn't make much difference!)

We talk more about diffuse nebulae on page 55; for more on open clusters, see page 71.

Also in the neighborhood: Epsilon Monocerotis, the star in the finder field just west of the nebula and cluster (also called 8 Mon), is a pleasant double. The magnitude 4.4 primary has a 6.6 companion located 12" to its north-northeast. They're an easy split in a Dob, and within the range (on a good night) of a smaller telescope. The colors are a subtle tan and gray.

find more at: www.cambridge.org/features/turnleft/seasonal_skies_january-march.htm

In Monoceros: A multiple star, Beta Monocerotis, and an open cluster, NGC 2232

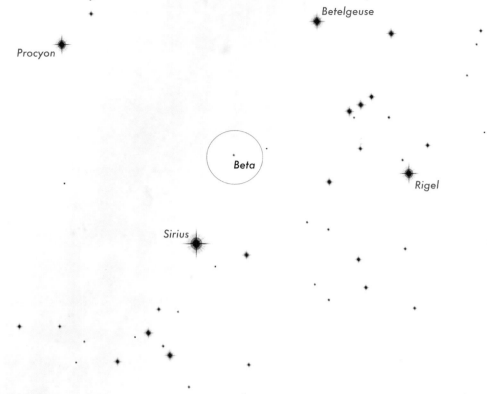

Star maps courtesy Starry Night Education by Simulation Curriculum

for NGC 2232

Any sky
Low/high power
Best: Jan.–Mar.

- Bright, easy to find
- Triple star: easy to split (if skies allow)
- Open cluster: nice in binoculars, low power

Where to look: Find Orion, high above the southern horizon, and find the very bright red star, Betelgeuse, which makes his shoulder (up and to the left of the three stars in his belt). Then, from Orion, turn left and follow the stars in Orion's belt which point to the southeast towards a dazzling blue star, Sirius. (At magnitude –1.4, Sirius is the brightest star in the sky, although some planets are brighter.)

A little less than halfway between Sirius and Betelgeuse you'll find two faint stars, lying in an east–west line. Aim for the one to the east, the one away from Orion. That's Beta Monocerotis.

In the finderscope, Beta Monocerotis: There are only two reasonably bright stars in this part of the sky; both should fit in the finder field. Beta is the one farther from Orion, to the east – the one *without* a pair of sixth-magnitude stars to its east.

NGC 2232: Look due north of Beta for a fifth-magnitude star, 10 Monocerotis. It's the brightest member of the open cluster.

NGC 2232 *in a star diagonal at low power*

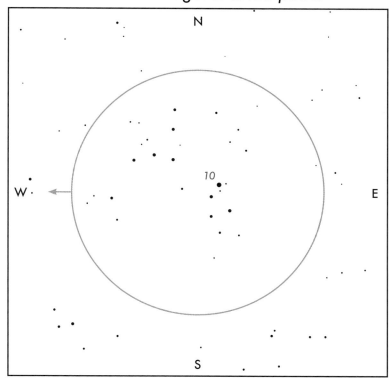

In a small telescope: The cluster is a bright, loose collection of stars, easy to see even on moonlit nights.

NGC 2232 *in a Dobsonian at low power*

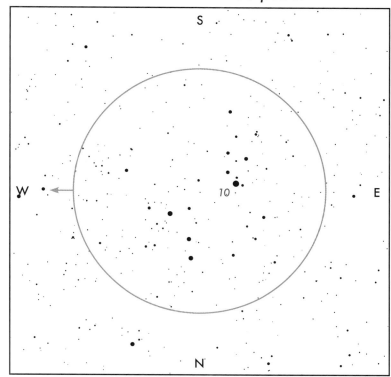

In a Dobsonian telescope: In a Dob you should be able to make out more than a dozen bright stars in a loose grouping. The rich Milky Way background adds to the appeal of the field, but can make it hard to figure out which stars are part of the cluster, and which are just background stars.

Since all three stars are bright, **Beta Monocerotis** can stand the highest magnification that the skies will allow. Unfortunately for northern hemisphere observers, winter skies can be very unsteady. If you notice that the stars tend to be twinkling a lot, relatively close doubles like the two blue ones in this group will be hard to split that night. Such *bad seeing* will limit just how high a power you can use effectively. Using lower power, you may only be able to make out two stars, as the blue ones can be hard to separate from each other. Even under such conditions, however, it still makes a pretty double star.

Beta Monocerotis is actually a quadruple star. Three of the stars are reasonably bright and close to each other, while the fourth is dimmer (too dim to be seen by a small telescope) and farther away from the others. They lie about 700 light years from us. The white star, A, is about 1,500 AU from B. Stars B and C are a bit over 600 AU from each other. Given their great distances apart, they move very slowly about each other, taking thousands of years to complete one orbit.

5' circle

NGC 2232: This open cluster is estimated to be about 25 million years old (see page 71). Recent X-ray observations suggest that most of the stars in this cluster do not have dust disks, unlike those seen in younger clusters; this puts interesting limits on how long such disks survive after the stars are formed, and thus how much time is available for these stars to form Earth-like planets from such disks of dust.

Though the cluster is now 300 light years from the Orion Nebula, there is some indication that it originated from the same large cloud of gas and dust.

Beta Monocerotis			
Star	**Magnitudes**	**Color**	**Location**
A	4.6	White	Primary star
B	5.0	Blue	7.2" SE from A
C	5.3	Blue	2.9" ESE of B

find more at: **www.cambridge.org/features/turnleft/seasonal_skies_january-march.htm**

In Monoceros: An open cluster, M50

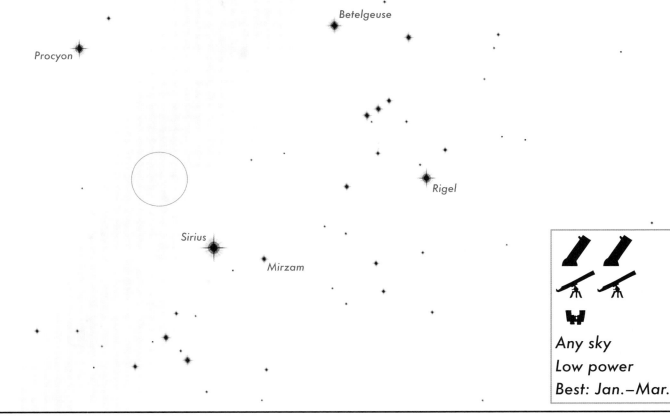

Star maps courtesy Starry Night Education by Simulation Curriculum

Any sky
Low power
Best: Jan.–Mar.

- Notable color contrast
- Wide range of star brightnesses
- Very young stars

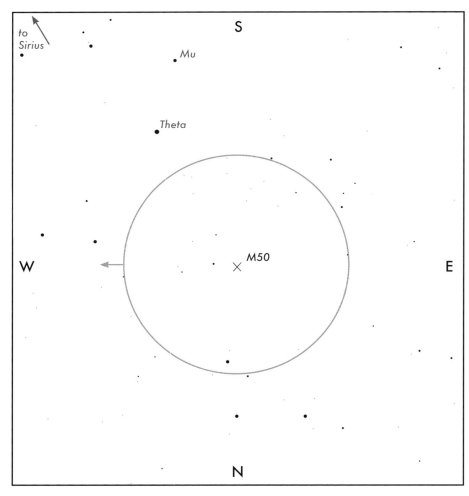

Where to look: Start at Sirius, the dazzlingly bright blue star to the left of Orion's belt. (Sirius is in fact the brightest star in the sky, although some planets get brighter.) To the west of Sirius is another bright star, called Mirzam. Call Sirius a clock hub with 12 o'clock at due north, and Mirzam is just shy of 4 o'clock; you will be looking in the region around 11 o'clock, out from Sirius in the direction to Procyon.

In the finderscope: Starting at Sirius, move a step northeast; you'll pass a triangle of fifth-magnitude stars. Continue moving northeast, the direction indicated by the short leg of the triangle, and you'll come to Theta Canis Majoris (with a slightly fainter star, Mu Canis Majoris, to its south). About one step farther in that direction beyond Theta brings you to the neighborhood of the open cluster. Look in the finder for three faint stars that point, like the hands of a clock, at 8 o'clock (or 2:30, depending which way is up) – to the south and just north of west). The cluster is just east of the clock hub. On a good night, it will be visible as a lumpy patch of light.

M50 in a star diagonal at low power

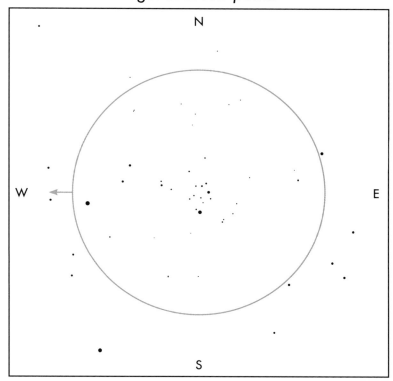

In a small telescope: The cluster will be easily visible in the low-power field. Look for a dim reddish star, at the southern end of the cluster. The rest of the cluster will look like a loose collection of about a dozen stars.

M50 in a Dobsonian at low power

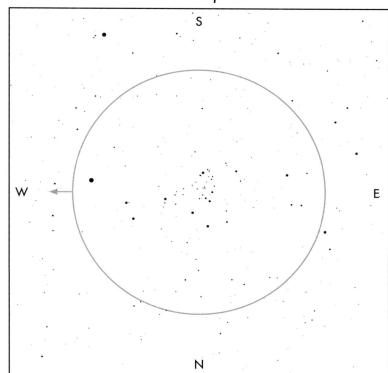

In a Dobsonian telescope: The cluster will be a loose collection of stars with a lovely color contrast. Notice the red star at the southern end of the cluster; compare it with the brighter blue stars at the northern edge.

The color contrast among the stars in **M50** is quite nice. The red star at the southern end, with a fainter star just to the northwest, is very distinctive. The three brightest of the stars at the northern end of the cluster, with a slightly dimmer one off to the side, form a flat Y shape; they all have a bluish color.

The middle of the cluster is populated by a sprinkling of dimmer stars, down to as faint as your telescope can see. No matter how good (or poor) the night is, you'll always have some stars just on the edge of being visible. Most nights, about a dozen stars are clearly visible in a small telescope, more on a very good night.

This cluster of roughly 100 stars lies about 3,000 light years from us. The main part of the cluster, visible here, is about 10 light years across.

The bright red star visible in the south indicates that this cluster has begun to evolve its largest stars into red giants; however, some of the bright blue stars are B-type, indicating that this cluster is even younger than the Pleiades, well less than 50 million years old.

For more on open clusters, see page 71.

*Also in the neighborhood: **Mu Canis Majoris** is a double star, a magnitude 5.3 primary with a magnitude 7.1 secondary star just 3" to the north-northwest. You'll probably need a Dob with very high power to split it. But it has lovely colors, the primary being a strong yellow and its companion very blue.*

*Speaking of double stars, it's worth pointing out that **Sirius** and **Procyon** are themselves doubles, but very tricky to split even with a Dob because the primaries are so bright that they tend to overwhelm their companions. Still, if you have very high power and a very steady night, it's worth a try. Sirius (magnitude −1.4) is orbited by a magnitude 8.5 star located 8.4" to its east. Procyon, magnitude 0.4, has a magnitude 10.8 star only 4.7" to its northeast.*

find more at: www.cambridge.org/features/turnleft/seasonal_skies_january-march.htm

In Canis Major: An open cluster, M41

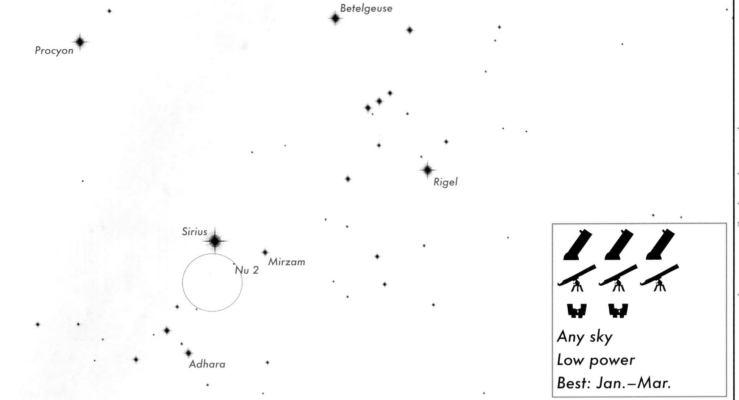

Star maps courtesy Starry Night Education by Simulation Curriculum

Any sky
Low power
Best: Jan.–Mar.

- Easy to find
- Wide range of star brightnesses
- Nice color contrasts

Where to look: Find Orion, high above the southern horizon, and turn left. The three stars in Orion's belt point to the southeast towards a dazzlingly bright blue star, Sirius. (Sirius is in fact the brightest star in the sky, although some planets get brighter.) A short distance to the west of Sirius is a bright star called Mirzam.

In the finderscope: Start at Sirius, and move towards Mirzam. On the way, you'll encounter a clump of stars, Nu 1, Nu 2, and Nu 3 Canis Majoris; Nu 2 and Nu 3 are significantly brighter than Nu 1. Step to the southeast from Mirzam to Nu 2, the southernmost and brightest of the clump. One more step and M41 should appear, looking like a lumpy patch of light.

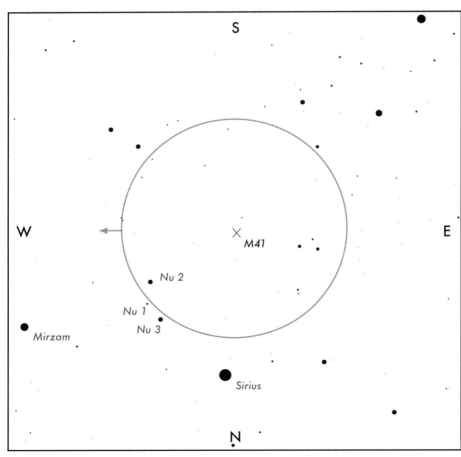

M41 in a star diagonal at low power

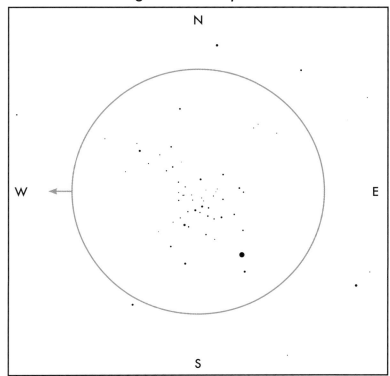

M41 in a Dobsonian at low power

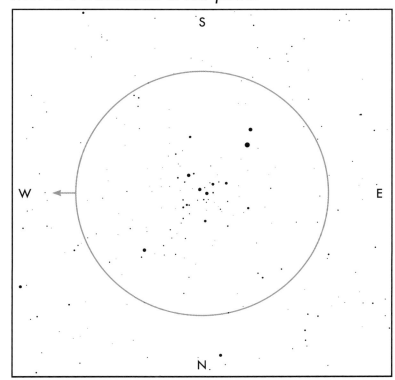

In a small telescope: The cluster is a rather loose collection of stars. A few dozen individual stars should be visible with a gradual range from brighter to dimmer stars.

In a Dobsonian telescope: The wide range of brightnesses within stars of this cluster make it look particularly rich in the larger aperture of a Dob.

M41 is a quite pleasing cluster to observe. Because of the range of brightnesses, this group is a good test of how dark the sky is; there'll always be some stars just on the edge of visibility.

Binoculars show a faint but large patch of hazy light; with a three-inch telescope you can pick out about three dozen stars, even on a night when the sky is not particularly dark. There are some particularly nice patterns of faint 10th- and 11th-magnitude white stars interspersed around a few brighter 7th- to 9th-magnitude stars near the center of the cluster. Most of the stars between magnitude 8 and 9 are blue; the brighter stars (magnitude 7 to 8) are reddish–orange giants. An even brighter 6th-magnitude star, 12 Canis Majoris, sits on the southeast edge of the cluster.

This cluster of roughly 100 stars forms a group 20 light years in diameter located 2,500 light years away from us. The bright orange–red star in the cluster is a K-type red giant, while most of the rest are blue stars, spectral types B and A. It's a fairly young cluster, on the order of a hundred million years in age. (For more on open clusters, see page 71.)

To find this cluster, we used the bright stars Sirius and Mirzam. Sirius, at magnitude –1.4, is the brightest star in the sky today. But recent measurements of stellar motions and distances, especially by the European Space Agency Hipparcos satellite, have allowed astronomers to calculate how the ap-

pearance of the sky has changed over the last five million years. It turns out that, four and a half million years ago, Mirzam was significantly closer to us; instead of 500 light years away, it was a mere 40 light years distant. At that time it would have had a magnitude of –3.6, comparable to Venus, seven times brighter than the brightest star in the present-day sky.

Adhara (in the hindquarters of the Dog, as shown in the naked eye view) is another star that in the astronomically recent past also came close to our Sun and planets. It would have been even brighter, magnitude –4, some 4.7 million years ago. As we note on page 87, Adhara is in fact a double star; back then when it was so close, we'd have seen its companion with a brightness of second magnitude and a separation of more than an arc-minute… a wonderful binocular object for time-travelers.

find more at: www.cambridge.org/features/turnleft/seasonal_skies_january-march.htm

In Puppis: Two open clusters, M46 and M47

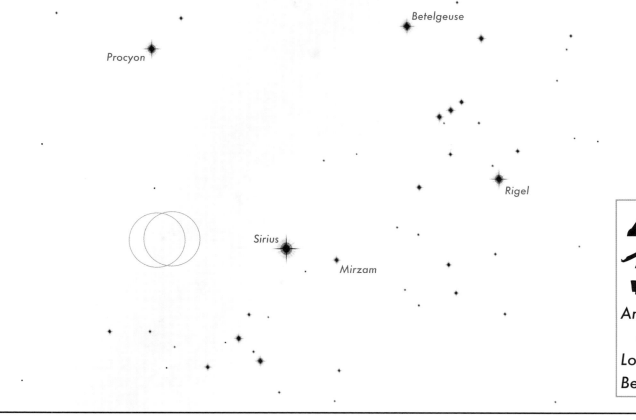

Star maps courtesy Starry Night Education by Simulation Curriculum

Betelgeuse

Procyon

Rigel

Sirius

Mirzam

Any sky (M47);
    dark sky (M46)
Low power
Best: Feb.–Mar.

- Two-for-one open clusters
- Contrast in richness and brightness
- Nearby, two more opens and a planetary

Where to look: First find Sirius, the brightest star in the sky, to the south and east of Orion (Orion's belt points straight towards it). The bright star to the right of Sirius is called Mirzam. Step left (eastward) from Mirzam, to Sirius; about two steps farther in this direction gets you to the neighborhood of M46 and M47.

In the finderscope: As you move east from Sirius, in the finderscope you'll see two fourth-magnitude stars with a northeast–southwest orientation. The star to the northeast, farther from Sirius, is Gamma Canis Majoris; it's about 40% of the distance to the clusters. Keep moving past Gamma and look for a flat triangle of stars. M46 and M47 lie on an east–west line, just inside the flat triangle. M47 is the easier one to see in the finderscope; it should be visible as a little cloud of light. M46 is just to the east of M47; it will be visible in the finderscope only if the sky is very dark.

S

← To Sirius

Gamma

W

E

M47   ✕ M46

N

M46 and M47 in a star diagonal at low power

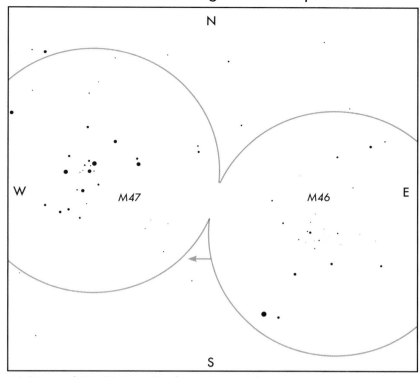

In a small telescope: M47 is a crisp and easily resolved group of stars, quite a few of which are fairly bright and blueish. The stars are too widely spaced and too well resolved to leave any haze, even though there are many faint stars in the group, too. There are about five fairly bright stars, eight somewhat dimmer ones, and 20 more stars faint but easily visible in a 3" telescope. Note a double star, a pair of magnitude 7 stars about 7" apart, near the center of the cluster.

M46 is not particularly conspicuous in a small telescope except under very dark skies. It looks like a hazy cloud of light with a few dim speckles in it. The haze starts to look like grains of dust if you use averted vision.

In a Dobsonian telescope: M47 should be easy, a dozen bright stars in a cloud of fainter ones. Look to the north and a bit east for another, fainter open cluster: that's NGC 2423. (And about 80 arc minutes southwest of M47, the width of a low power field of view, is the open cluster NGC 2414.) Moving farther east, M46 is less obvious; if you arrive at the two stars, 4 Puppis and 2 Puppis, in a northeast–southwest line, you've gone too far. (2 Puppis is a nice cat's-eye double.) On a dark night in a Dob you should see a haze of faint stars behind a group of about a dozen speckles.

Look for a slightly out-of-focus star at the northeast corner of M46. That is the planetary nebula NGC 2438 – located by chance in the same line of sight, but only half the distance between us and M46.

M46 and M47 in a Dobsonian at low power

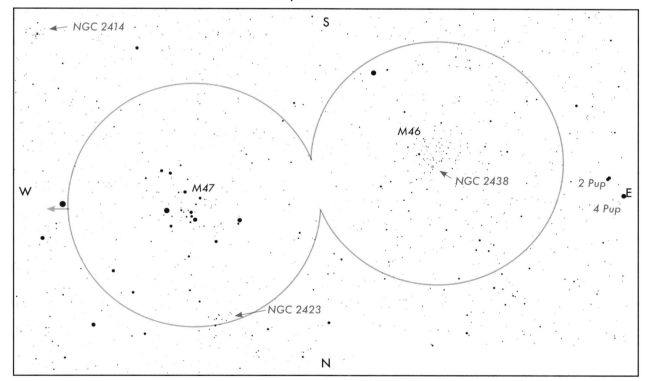

There are no bright guide stars to **M46** and **M47,** and they're in the neighborhood of many other faint Milky Way stars. This can make it tricky to find the clusters. M47 is easier to spot; it will clearly stand out as a cluster once you've found it. M46 is unimpressive by itself in a small telescope, but it's a nice challenge. In a Dobsonian, however, this cluster is actually more impressive than M47; it has many 11th- to 14th-magnitude stars which stand out nicely in the bigger scopes, but which are too faint to be seen in a three-incher.

M47 contains about 50 young stars, most of them blue in color, although there are one or two orange stars in the group. The whole cluster is about 15 light years in diameter, and it is located a little more than 1,500 light years from us.

M46 consists of several hundred young, bright, blue giant stars, all of roughly similar brightness, in a loose cloud about 40 light years across. It appears dim compared to M47, however, because it's three times as far from us, some 5,000 light years away.

Is the cluster we're calling M47 (known in the NGC catalog as NGC 2422) really the cluster Messier observed? Messier originally described an open cluster much like M47, but following his finding instructions literally brings you to a part of the sky where no such cluster exists. It's now generally believed that he made a mistake when he wrote down its position, and that in fact this is the cluster he observed.

For more on open clusters, see page 71.

find more at: www.cambridge.org/features/turnleft/seasonal_skies_january-march.htm

In Canis Major: An open cluster, NGC 2362, and the *Winter Albireo*, Herschel 3945

Procyon · Betelgeuse · Rigel · Sirius · Wezen · Aludra · Adhara

Star maps courtesy Starry Night Education by Simulation Curriculum

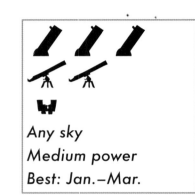

Any sky
Medium power
Best: Jan.–Mar.

- NGC 2362: easy to locate, challenging to see
- Very young cluster
- Winter Albireo: easy, colorful double

Where to look: Find Sirius, a brilliant blue star (the brightest in the sky), by drawing a line to the south and east from Orion's belt. To the south and a bit east of Sirius is a triangle of stars, the hindquarters of Canis Major, the Big Dog. The top (northernmost) star in the triangle is called Wezen; the one to the bottom left (southeast) is Aludra, and the one at the bottom right (southwest) is Adhara. Start at Wezen.

In the finderscope: You'll recognize Wezen because it has two fourth-magnitude stars just to its east. Look to the northeast of these stars for Tau Canis Majoris (Tau CMa for short).
You'll know that you have found Tau CMa, because there is a another, slightly less bright star (29 CMa) just to the north of it. Aim for Tau CMa; it lies right in the middle of the open cluster.
Note the location of M93, just to the east. After you've seen NGC 2362, you can find the M93 finder field (see page 88) easily by moving your finderscope east and a bit north.

S · Adhara · Aludra · Wezen · Tau · 29 · W · E · Winter Albireo · M 93 · N

NGC 2362 in a star diagonal at medium power

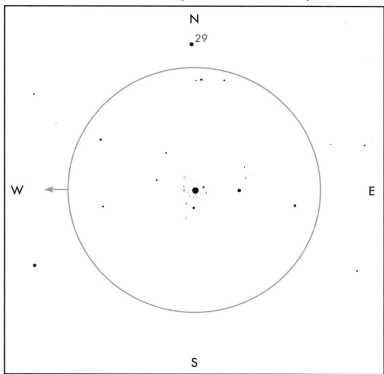

In a small telescope: The open cluster is a subtle sprinkling of stars, surrounding (and dominated by) Tau CMa. All the members are very faint.

NGC 2362 in a Dobsonian at medium power

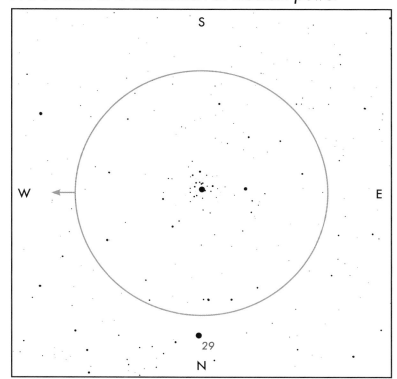

In a Dobsonian telescope: The cluster is especially rich in a Dob, dozens of stars surrounding the dominant star Tau CMa. Given the extra light-gathering power of a Dob, try using high power to pull out the details of the stars in this cluster.

NGC 2362 is possibly one of the youngest clusters in the sky. The stars include some of spectral class O, the kind of stars which have already evolved into red giants in other open clusters (see page 71). As a result, it is estimated that this cluster is very young, probably only about five million years old.

The cluster fills a region of space less than 10 light years across, located 5,000 light years away from us. About 40 stars can be easily picked out in photographs made by large telescopes, but there may be many more members in the cluster.

It's easy to find Tau CMa, but seeing the cluster surrounding it can be a good challenge to your stargazing skills. On a good night, however, you may be able to see a dozen or more stars with a small telescope.

The Winter Albireo: Move north off Tau CMa, past 29 CMa, and keep a lookout to the western edge of the low-powered telescope view for a fifth-magnitude star with a seventh-magnitude companion 26 arc seconds to the northeast. Even at low power this pair is easy to separate. The primary is a distinct red, while its companion may appear white or yellow.

It has the technical name of Herschel 3945, but it is more popularly known as the "Winter Albireo." Indeed, the color contrast and separation are reminiscent of Albireo, the much brighter and better known double star visible in the (northern hemisphere) summertime (see page 132).

It makes a fun double, especially since it's so easy to find from the Tau CMa open cluster. However, be aware that the colors may appear to be washed out if you're looking at it low in the sky, a perennial problem for northern hemisphere observers looking at objects to the south.

*Also in the neighborhood: Note that **Adhara** itself is a double star. The primary is magnitude 1.5 while its companion is much dimmer, magnitude 7.5, just under 7" to the south-southeast. It's a nice challenge in a Dob. And, as we mention on page 83, it would have been much closer (and spectacular) a mere 4.7 million years ago.*

The Winter Albireo			
Star	**Magnitude**	**Color**	**Location**
A	5.0	Red	Primary star
B	5.8	Yellow	26" NE from A

find more at: www.cambridge.org/features/turnleft/seasonal_skies_january-march.htm

In Puppis: An open cluster, M93

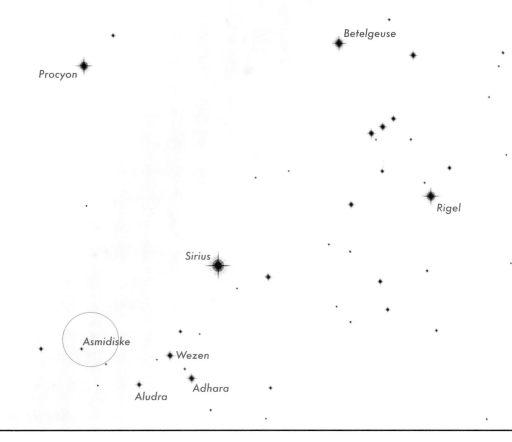

Star maps courtesy Starry Night Education by Simulation Curriculum

Any sky
Low/medium power
Best: Feb.–Mar.

- Easy to find
- Rich, pretty open cluster
- Changes appearance as sky conditions change

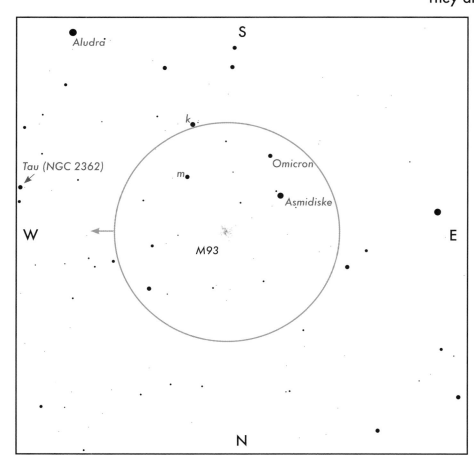

Where to look: Find Sirius, a brilliant blue star (the brightest in the sky), by drawing a line to the south and east from Orion's belt. To the south and a bit east of Sirius are a triangle of stars: Wezen, Aludra, and Adhara. They are the hindquarters of Canis Major, the Big Dog. Call the length of one side of this triangle equal to one step. Starting from the topmost star of this triangle, Wezen, step away from Sirius (southeast) to Aludra. Then, make a right turn (towards the northeast) and follow in this direction for one and a half steps. Now you are in the neighborhood of Asmidiske.

In the finderscope: In a rather rich field of stars, Asmidiske is distinguished by having a companion star (two magnitudes dimmer than itself) just to its southwest, which makes it look like a double star in the finderscope. It's the northeast corner of a box of stars including Omicron Puppis, m Puppis, and k Puppis. Look for M93 just outside this box. Think of the line from Asmidiske to m Puppis as a mirror; M93 is at the spot to the north that mirrors Omicron's position south of that line. On a good night it may be visible in the finderscope.

M93 is, in fact, rather close to NGC 2362 (see page 86). From here, you can move your finderscope due west until you see Tau Canis Majoris, which is located amid the stars of that cluster.

M93 in a star diagonal at low power

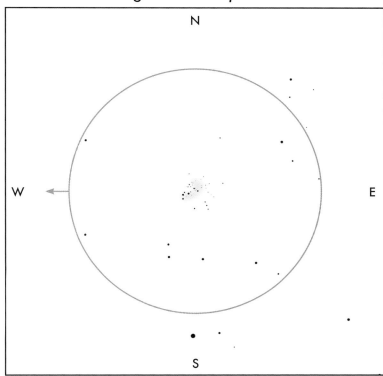

M93 in a Dobsonian at medium power

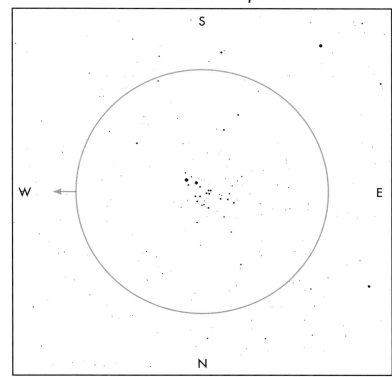

In a small telescope: The cluster is a rich, concentrated band of about 20 stars, many of them quite dim, with an impression of a graininess suggesting that there are many more stars just beyond the edge of visibility. If the night is good, try it at higher power to see how much of the graininess you can resolve into individual stars.

In a Dobsonian telescope: Use a medium-power eyepiece to pull out the individual stars. The graininess suggested in the small telescope view becomes revealed as a rich collection of stars in a Dob.

M93 is quite a pretty open cluster, but it's in a part of the sky that often gets overlooked, especially in comparison to Orion. The cluster will be most easily observed later in the season (or later in the evening), when Orion has moved towards the western horizon. Although we use (northern) winter stars to find it, in some ways this cluster can be thought of as a harbinger of spring (or, in the southern hemisphere, the end of summer).

Averted vision helps bring out more stars; the more you look, the more you see. On a good night you may be able to pick out two dozen stars, or more, with a small telescope.

This cluster consists of 60 or so stars easily distinguished in photographs, with perhaps a couple of hundred fainter stars as members as well. The cloud of stars is 20 light years across, and lies about 3,500 light years away from us. From the spectra of the stars present, one can estimate its age at roughly 100 million years.

For more on open clusters, see page 71.

*Also in the neighborhood: Three degrees southwest of Asmidiske is a fourth-magnitude star, **k Puppis**. High power reveals it to be a close but evenly matched pair of magnitude 4.5 stars, oriented northwest–southeast, making an elegant "cat's eyes" set. The separation, just under 10 arc seconds, is ideal for a small telescope; even on marginal nights you'll eventually have a moment when the twinkling stops, just for an instant, and you can see dark sky between the stars.*

find more at: www.cambridge.org/features/turnleft/seasonal_skies_january-march.htm

Seasonal skies: *April–June*

The Milky Way lies along the horizon during this time of year, and so the sorts of objects that are associated with our Galaxy, such as open and globular clusters, are few and far between. On the other hand, this means that it's an ideal time to look at objects outside our Milky Way. April through June is the best season for looking at other galaxies.

The worst problem that northern hemisphere observers have during this time of year is probably mud. If the place where you observe is prone to get swampy during the spring, be sure to stake out a dry spot before it gets dark, and bring something to kneel on.

A more subtle difficulty for northern hemisphere observers is that as the days get longer, twilight comes a minute or more sooner and lasts a few minutes longer every night. Since every star sets four minutes earlier each night, every passing day gives you about seven minutes less time to catch objects in the west.

Southern hemisphere observers, of course, get an extra few minutes of darkness every night during these months.

Looking west

Finding your way:
April–June sky guideposts

Find the *Big Dipper*. In northern latitudes it will be high in the sky, almost directly overhead, but even as far south as Rio de Janeiro it will still be visible to the north. The two stars at the end of the bowl are called the *Pointer Stars*. Follow the line they make towards the northern horizon, and you'll run into **Polaris**, the pole star. Face this star, and you face due north.

Turn yourself around so that you're facing south, so that you have to bend over backwards to see the Dipper; then straighten up, looking south, about one dipper's length below the dipper, and you'll see a collection of stars that looks like a backwards question mark, something like " "). Some people call this the Sickle. Since this collection of stars is part of the constellation of *Leo*, the Lion, a more imaginative way is to say these stars represent the lion's mane.

The bright star at the foot of this mirrored question mark, or mane, is called **Regulus**. It is a bright, first-magnitude star.

From the Lion, turn and face the west. Standing up on the western horizon will be the two stick men who make up *Gemini*, the twins. The bright stars that make up their heads are **Castor** (the twin on the right, to the north) and **Pollux** (on the left, to the south). South of them is an even brighter star, Procyon. In April, the bright stars of Orion and Canis Major,

However, the approach of winter often brings extra dew on your lenses. (Finderscopes get dewy, too!) If everything in your telescope looks like a nebula, warm up your lenses; bring them back indoors for a few minutes, if necessary. We talk more about fighting dew on page 15.

In the northern hemisphere, don't miss the Ursa Major galaxies (pages 200, 204–207) now at their highest.

In the southern hemisphere, some of your best telescope objects arrive this season. Look for Eta Carinae and its nearby clusters (pages 228–231), the clusters in Musca and Crux (pages 234–237), the double stars of Crux and Centauri (pages 232–233), and Omega Centauri with a wealth of wonderful "neighborhood" objects (pages 238–239).

To the west: Many of the best January–March objects are still easily above the western horizon, especially the clusters in Puppis and Canis Majoris. Be sure to catch these objects before they disappear:

Object	Constellation	Type	Page
Orion Nebula	Orion	Nebula	52
M35	Gemini	Open cluster	72
Beta Mon.	Monoceros	Triple star	78
M41	Canis Majoris	Open cluster	82
M46, M47	Puppis	Open clusters	84
NGC 2362	Canis Majoris	Open cluster	86
M93	Puppis	Open cluster	88

Looking east

including Sirius, the brightest star of them all, are low in the west at sunset. As the season proceeds, these stars are lost to the twilight, one by one.

Return to the Big Dipper, and now "arc to Arcturus and spike to Spica." In other words, follow the arc of the Big Dipper's handle as it curves to the south and east. The first brilliant star you'll find along this arc is **Arcturus**. A magnitude zero star, it is one of the brightest stars in the sky, and it has a distinct orange color. Continuing on this arc farther south, you'll come across a very bright blue star, called **Spica**. It's in the constellation *Virgo*. These two stars are our main guideposts to the east. (From the southern hemisphere, if you can't see the Dipper use the Southern Cross: spike northward

from Acrux to Gacrux, across the zenith overheard, to Spica. Orange Arcturus will sit below it, to the right, just above the northern horizon.)

The Twins, Leo (with Regulus), and Virgo (with Spica) are all zodiac constellations. That means that planets will often be visible among these stars. Any bright "star" among these constellations that doesn't appear on the charts above is probably a planet, and well worth looking at.

Finally, rising on the eastern horizon will be the **Summer Triangle** of **Deneb, Vega**, and **Altair**. And just peeping over the horizon, in the zodiac constellation of *Scorpius*, is the reddish rival of Mars, **Antares**. These will be our guideposts for later in the year – or later in the night.

In Cancer: The *Beehive*, an open cluster, M44; and *Tegmine*, Zeta Cancri, a double star

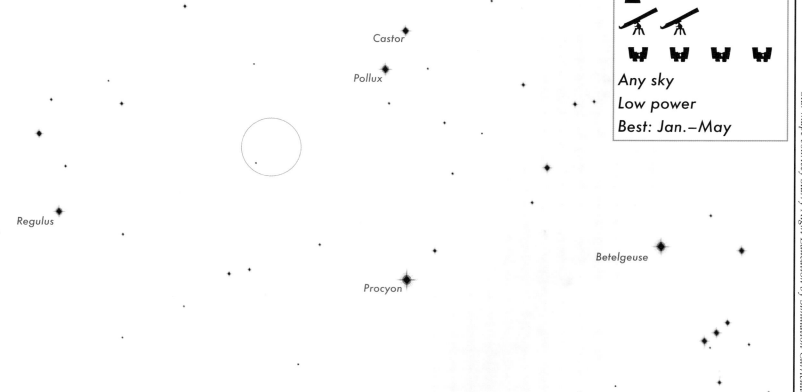

Star maps courtesy Starry Night Education by Simulation Curriculum

- Cluster: easy and glorious in finder, binoculars
- Subtle colors in stars – blue and yellow
- Double star: fun challenge for small scope

Where to look: Find the Twins, the very bright stars Castor and Pollux, to the west. The twin on the right (north) is the blue star Castor, the yellow one to the left (south) is Pollux. Call the distance from Castor to Pollux one step; continue in the direction from Castor, to Pollux, to three steps farther on. From this spot, turn right and go up a step. This should bring you to a spot roughly halfway between Pollux and Regulus. You should be able to see two stars, lined up north–south, Gamma and Delta Cancri. The Beehive is a tiny bit west of the midpoint between these two faint stars. On a good night it's visible to the naked eye as a small fuzzy patch of light.

In the finderscope or binoculars: Ideal in binoculars, the Beehive should be easily visible as a lumpy patch of light between and slightly to the west of two stars oriented north and south. You will probably be able to resolve some members of the cluster in your finderscope or binoculars.

The Beehive in a star diagonal at low power

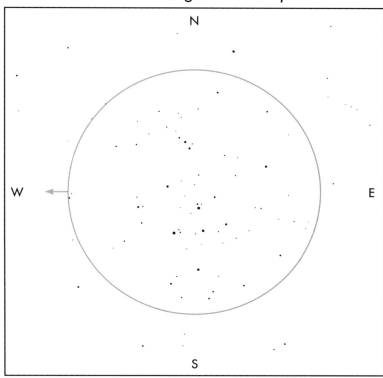

The Beehive in a Dobsonian at low power

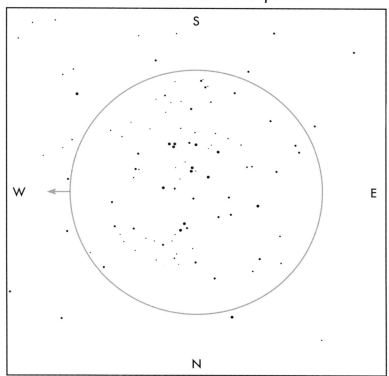

In a small telescope: About 50 stars will be visible, including many doublets and triplets. Many of these stars are quite bright, seventh and eighth magnitudes, and some of the brightest are distinctly yellow or orange–yellow class-G giants. The cluster will probably extend beyond the field of view of your telescope, unless you have a very low-power eyepiece.

In a Dobsonian telescope: There are many bright stars visible, some with a notable yellowish color; but this cluster is just too spread out to be seen to good advantage in a Dobsonian. Any eyepiece wide enough to contain most of the cluster in a Dob will probably also show coma, a distortion of the pinpoints of starlight into smudges at the edge of a wide-angle field of view (see pages 21, 244).

M44: A big, bright cluster, the Beehive (M44) has few of the faint, barely resolved stars that add a sense of richness to other open clusters. It actually looks better in the finderscope. A couple of the brighter stars are orange; the rest are blue.

In total, this open cluster of stars consists of about 400 stars in a loose, irregular swarm. Most of the stars are in a region some 15 light years across. The cluster is located only about 500 light years away, just a bit farther from us than the Pleiades. The bright orange stars are ones that have had time to evolve into red giants. From evidence like this (see page 71 for more on open clusters and how they can be dated), one can conclude that this cluster is relatively advanced in age, about 400 million years old.

Like the Pleiades, this cluster is visible to the naked eye. It figured in Greek mythology as a manger (in Greek, "praesepe") flanked by asses; it is still often called the Praesepe.

Tegmine: Note the four stars in the finderscope view that box in the Beehive. Step across the two southern stars, east to west, Delta to Theta, and continue another slightly larger step westward. That should carry you about five degrees west (and a bit south) of M44. There you'll find a fifth magnitude star, reasonably bright in the finderscope, Tegmine (Zeta Cancri).

It's a multiple-star system, with at least three Sun-like stars. In a small telescope, it will look like a magnitude 5 star with a 5.8 companion 6 arc seconds away. But the primary is itself a double star, a 5.3 and 6.3 pair separated by only 1.1 arc seconds. With steady skies you can split it in a Dobsonian at very high power. The more distant companion, Zeta Cancri C is also a close double, but too close for amateur instruments. A, B, and C are all yellow dwarf stars much like our own Sun. The system is 83 light years distant from us; A and B are only 30 AU apart, while C is 180 AU from this pair.

Tegmine (Zeta Cancri)			
Star	**Magnitude**	**Color**	**Location**
A	5.3	Yellow	Primary star
B	6.2	Yellow	1.1" NE from A
C	5.8	Yellow	6.3" ENE from A

Also in the neighborhood: Look in your finderscope about five degrees northwest of the Beehive for a string of stars running east–west. The one star out of line, north of the string, is the magnitude 6.9 star **24 Cancri**. It's a nice challenge in a Dob, a seventh-magnitude primary with a magnitude 7.5 companion only 5.6" to the northeast. The pair are located 250 light years from us. The secondary is actually a pair of magnitude 8.5 stars, too close together to split in a small telescope.

A reddish star east southeast of Delta, shown in the finderscope view, is the variable star **X Cnc**. It changes from magnitude 5.7 to 6.9 over a six month period.

find more at: www.cambridge.org/features/turnleft/seasonal_skies_april-june.htm

In Cancer: An open cluster, M67, and a variable star, VZ Cancri

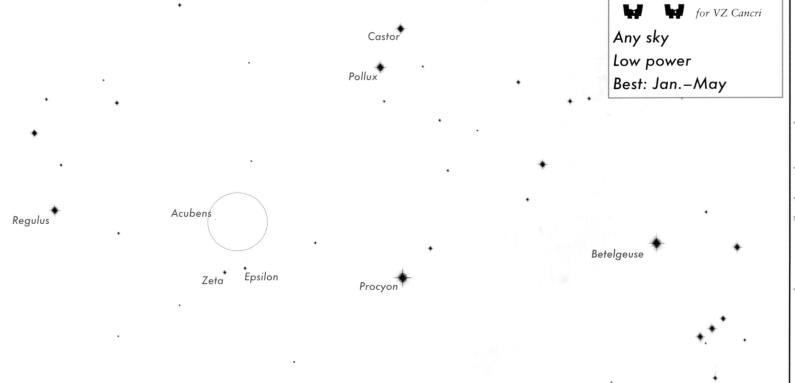

Star maps courtesy Starry Night Education by Simulation Curriculum

for VZ Cancri

Any sky
Low power
Best: Jan.–May

- M67: pleasant grainy look in small telescope
- One of the oldest known open clusters
- VZ Cancri: rapidly varying brightness

Where to look: About halfway between Regulus (the foot of the Lion's Mane) and Procyon (the bright star down and to the left of the Twins), look for a pair of third-magnitude stars, Zeta and Epsilon Hydrae. They're south and just a bit east of the stars around the Beehive, the brightest stars in that region of the sky.

In the finderscope, M67: In the finderscope you should see four stars in the group around Zeta and Epsilon. Head due north from Zeta until you find a star of similar brightness, called Acubens; it has a dimmer neighbor, 60 Cancri, to its southwest. Now move your finderscope about half a finder field west until you see a pair of stars, 50 Cancri (the southeast of the pair) and 45 Cancri (to the northwest). Halfway between 60 Cancri and 50 Cancri you should see a small cloud of light. That's the open cluster.

VZ Cancri: South of 45 and 50 Cancri in the finderscope is another star of similar brightness called 49 Cancri. VZ Cancri is a faint star to the west of 49 Cancri. It lies halfway between 49 Cancri and a dim pair of stars called 36 and 37 Cancri.

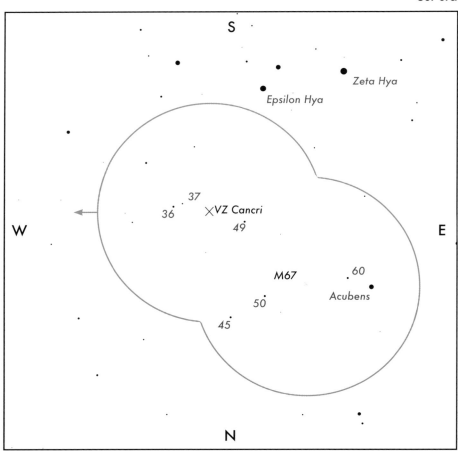

M67 in a star diagonal at low power

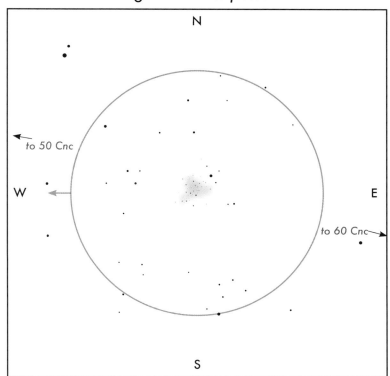

M67 in a Dobsonian at low power

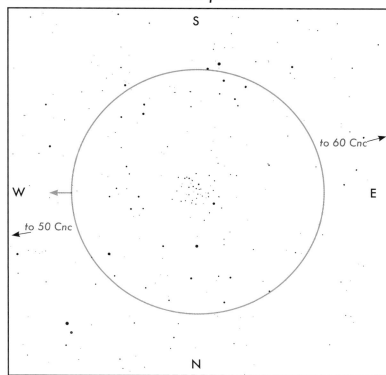

In a small telescope: M67 looks like a small blob of light sprinkled with individual stars. Note the relatively bright eighth-magnitude star on the edge of the cluster. There are a few ninth-magnitude stars in M67, another dozen or so of about tenth magnitude, plus a sprinkling of even fainter stars that give the cluster an overall look of a dim, grainy haze of light.

In a Dobsonian telescope: The grainy look remains even with a Dobsonian, which resolves several dozen cluster members. The cluster shows a lovely, even sprinkling of stars. One sees relatively few of the bright blue stars that are the sign of a young cluster.

M67: Although not overpowering, this cluster is quite a nice object. The stars are mostly quite faint, too dim to make out any colors with certainty in a small telescope (3" or smaller).

Most open clusters are formed in the plane of the Galaxy; during the course of their orbits about the galactic center, the gravity of other stars eventually disperses these clusters. The orbit of M67, however, is unusually tilted out of the galactic plane (it is currently some 1,500 light years above the plane). It spends most of its time far from the disruptive gravitational pull of other stars, and so it has stayed together as a cluster for an unusually long period of time. Judging from the number of its stars which have evolved into their red-giant stage, it is estimated that this cluster may be five to ten billion years old, older than the age of our Solar System; it is one of the oldest open clusters known.

M67 has about 500 member stars, forming a rough ball a bit more than 10 light years in diameter, about 2,500 light years from us. (See page 71 for more information about open clusters.)

VZ Cancri: Look for a faint star sitting by itself about one degree (two full-Moon radii) to the east, and a little north, of the pair 36 and 37 Cancri. This is VZ Cancri, arguably the most dynamic variable star visible in a small telescope. It is a fairly dim star (magnitude 7.9 at its dimmest), but it doubles in brightness in the space of two hours, and then dims back down to its minimum. When you first go out observing in the evening, compare its brightness against the pair 36 and 37 Cancri; then look again about an hour later. The variable will always be considerably dimmer than the magnitude 6.5 star 36 Cancri. However, 37 Cancri, which is magnitude 7.4, makes an excellent comparison as VZ varies between magnitudes 7.2 and 7.9 during its four-hour cycle.

VZ Cancri is an example of an *RR Lyrae-type* variable star. Apparently, the interiors of such stars are somewhat unstable, going through a cycle of heating, expanding, cooling, and contracting, all in a bit over four hours. At its brightest, VZ Cancri is almost 40 times as luminous as our Sun, but it appears so dim because it is about 600 light years away from us.

Also in the neighborhood: Epsilon Hydrae *(see the finderscope view) is a nice double star for a Dobsonian: the magnitude 3.5 primary has a magnitude 6.7 companion just 2.8" away, orbiting with a 990-year period. The primary itself is actually a close double, split in very large telescopes, with a 15-year period.*

find more at: www.cambridge.org/features/turnleft/seasonal_skies_april-june.htm

The Cancer and Leo double stars

Where to look: Find Regulus, a bright star high in the south, at the bottom of six stars making a large backwards question mark shape called the Sickle or the Lion's Mane. The third of those stars, counting up starting with Regulus, is Algieba. After Regulus, it's the next brightest star in the question mark.

Then look west for the Twins, Castor and Pollux. Between Regulus and the Twins is the Beehive, visible on a dark night as a fuzzy patch flanked by two dim stars (see page 92). The Beehive is sometimes called the Manger, and the two faint stars to the north and south of

the manger represented two donkeys, the Southern Ass (Gamma Cancri) and the Northern Ass (Delta Cancri). Call the distance from the Southern Ass to the Northern Ass one step. Go from the Southern, to the Northern, to two steps farther north. Look slightly west of this spot for a moderately dim star, somewhat brighter than its immediate neighbors. This star is Iota Cancri.

Algieba and Iota will be our anchors for finding the other nearby double stars. And don't forget, Castor itself is a nice double (see page 74).

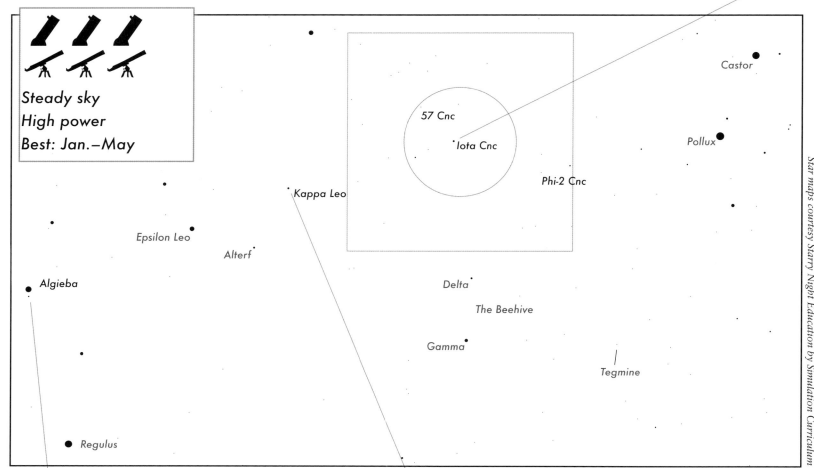

Steady sky
High power
Best: Jan.–May

Star maps courtesy Starry Night Education by Simulation Curriculum

Algieba is a bright star, but it's in a region with lots of bright stars; if you have the right one, you'll see another, dimmer star in the finderscope just to the south of it. The primary star is golden yellow in color, while it is more difficult to assign a color to its companion – some people see it as orange. Because it is bright, it is a nice object to look at during twilight while you're waiting for the dimmer objects to become visible. Seeing it against a twilight sky also helps the colors stand out.

Algieba is located 125 light years from us, and its companion orbits in an elliptical path more than 300 AU in diameter. It takes roughly 500 years to complete an orbit, and so over the past 100 years astrono-

mers have been able to see this star slowly move relative to the primary. It is currently close to its maximum separation, as seen from Earth. Recent evidence also indicates the presence of a planet roughly ten times Jupiter's mass with a 1.2 AU-radius orbit and a 429-day period.

Kappa Leonis is a fourth-magnitude star between the Sickle and Iota Cancri. Aim your finder at Epsilon Leo, the star at the end of the Sickle, and move west; Kappa is the northwest corner of a triangle with Epsilon and the fourth-magnitude star Alterf. It's a challenging double in a Dob since its companion is so faint, 10th magnitude. The pair lie 210 light years from us, separated by 150 AU.

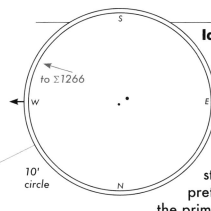

to Σ1266

10'
circle

Iota Cancri is the brightest, and most southwesterly, of a scattering of fainter stars visible in the finderscope. While only fourth magnitude, it is brighter than anything else near it. Colorful and easily split, it's an easy double star (once you find it) and very pretty. Its colors are quite striking; the primary star is yellow, like our Sun, but it's a giant star, about ten times our Sun's radius and 60 times as bright. Its companion is a bit bigger than our Sun, and hotter as well, hence its blue color.

Iota Cancri is located 300 light years away from us, and its stars are separated by some 2,800 AU, orbiting roughly once in 50,000 years. Because the stars are so far apart from each other, anyone living around star A would see its companion as a very bright star, about half as bright as the full Moon. From a planet around B, star A would look four times brighter than a full Moon.

Struve 1266 sits in the same low-powered field of view as Iota Cancri, 40" (a bit more than a full Moon diameter) to the west-southwest. It should be easy to find from Iota.

Five degrees to the west-southwest of Iota is **Phi-2 Cancri,** an equally matched cat's eye pair of sixth-magnitude stars, 5.2" apart, aligned northeast–southwest. It lies 275 light years from us.

Two degrees to the northeast of Iota Cancri is **57 Cancri**. It's a magnitude and a half fainter, and much more difficult to split, a pair of sixth-magnitude stars, well suited for a Dob at high power. A third star, ninth magnitude, lies nearly an arc minute from this pair. The system is 300 light years from us. The closer companion is 140 AU from star A; star C is 5,000 AU away from them.

Note 53 Cancri, a wide double that's visible, and split, in the finder, a degree and a half east-southeast of Iota Cancri.

West of the Beehive is the double star Tegmine; see page 92 for more details.

Also note that Regulus itself is a double star, but its companion is faint and far from the primary.

Finally, beware of planets moving through these constellations!

Algieba (Gamma Leonis)

Star	Magnitude	Color	Location
A	2.4	Yellow	Primary star
B	3.6	Orange	4.6" SE from A

Kappa Leonis

Star	Magnitude	Color	Location
A	4.6	White	Primary star
B	9.7	White	2.4" SSW from A

Iota Cancri

Star	Magnitude	Color	Location
A	4.1	Yellow	Primary star
B	6.0	Blue	30" NW from A

Struve 1266

Star	Magnitude	Color	Location
A	8.8	White	Primary star
B	10.0	White	24" ENE from A

Phi-2 Cancri

Star	Magnitude	Color	Location
A	6.2	White	Primary star
B	6.2	White	5.2" SW from A

57 Cancri

Star	Magnitude	Color	Location
A	6.1	White	Primary star
B	6.4	White	1.5" NW from A
C	9.2	White	54" SSW from A

5'
circle

5'
circle

Finderscope view around Iota Cancri

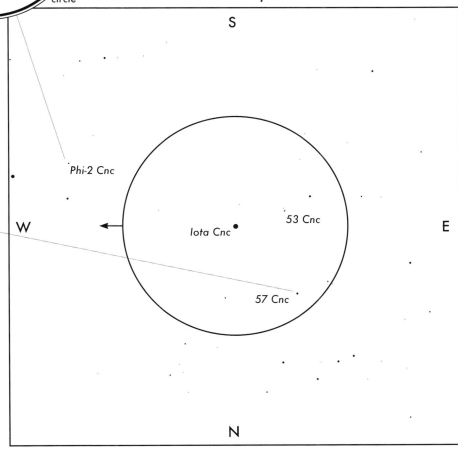

Dark sky
High power
Nebula filter
Best: April–May

In Hydra: The *Ghost of Jupiter*, a planetary nebula, NGC 3242

Regulus

Procyon

Alphard

Nu   Mu

Corvus

Sirius

Star maps courtesy Starry Night Education by Simulation Curriculum

- Fun challenge to find
- Slightly flattened disk looks like a (small) Jupiter
- Superb with larger aperture and nebula filter

Where to look: Half the fun is finding this nebula in a dim, remote part of the sky. The general area you'll be looking in is due south of Regulus (follow the backwards question mark southwards), almost as far south as Sirius. There's only one second-magnitude star in this part of the sky between Sirius and Regulus; it's called Alphard (Arabic for "the solitary one").

Now find Procyon, the bright star down and to the left of the Twins. Take one large step from Procyon to Alphard; an even larger step gets you to a distinctive box of third-magnitude stars, the constellation Corvus, the Crow. Halfway between Alphard and Corvus, look for a third-magnitude star, Nu Hydrae. Start there.

In the finderscope: Aim the finderscope at Nu. In the western edge of your finder field you should see a fifth-magnitude star, Phi Hydrae. Step westward from Nu to Phi and slowly continue one step farther. When you have passed Phi you will see the brighter (fourth-magnitude) Mu to its west. From Mu, move due south half a finder field. The nebula sits in the center of a box made by Mu and three sixth-magnitude stars.

S

W   ✕ NGC 3242   E

Mu   Phi

Nu

N

Ghost of Jupiter in a star diagonal, low/high power

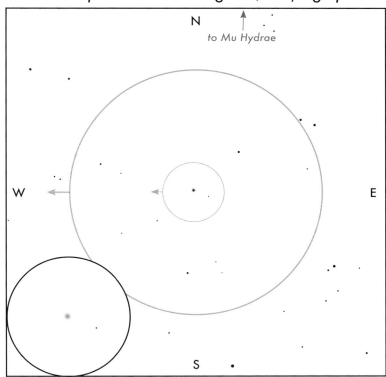

Ghost of Jupiter in a Dobsonian at low/high power

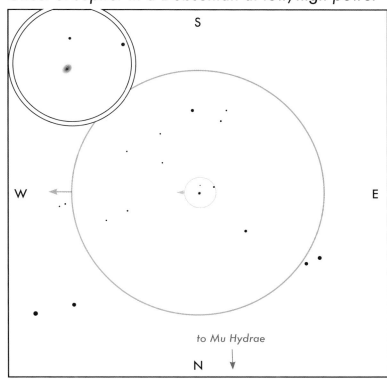

In a small telescope: A challenge to find and to see, it looks like a small, faint blue–green disk. Locate it with low power, then confirm you've got it by looking at the fuzzy spot at high power. (The insert circle shows a 20' high-power field of view; note this is less magnified than the 10' Dob insert, to the right.)

In a Dobsonian telescope: Aperture makes a huge difference here. Hunt for the nebula with low power, then once you have centered on it, switch eyepieces; it is bright enough to stand your highest power. Notice the flattening that gives it its name. With averted vision, look for a "two-step" effect, a disk within the disk. Then, with your highest power, look directly into the disk for the star in the center of the nebula.

The Ghost of Jupiter (NGC 3242) lies some 1,400 light years from us. From the size of its inner disk and the rate at which it is seen to be expanding, we can infer that this nebula was formed from the collapse of a star some 1,600 years ago.

The nebula gets its name from the shape of its flattened disk, reminding early observers of how Jupiter appears in a telescope. But unlike in Jupiter's case, where the flattening is the result of centrifugal force pulling its equator out from a spherical ball of gas, the flattening seen in this planetary is probably the result of viewing a disk of gas from something less than directly head-on.

About planetary nebulae

Late in the life of a star, when most of the lighter elements in its core have become fused into heavier elements (it's this fusion that causes the star to shine), the temperature and pressure in the central region of the star start to oscillate. As the star runs short of nuclear fuel, it cools; but this cooling causes the star to contract, which makes the interior warm up again, and expand.

The swings between expanding and contracting interiors take thousands of years. The extremes grow larger and larger, until eventually a big collapse into the center releases enough energy that the outer layers of the star are blown out into space. When this happens, the gases form an expanding and cooling cloud around the core, while the core turns into a small, hot, white-dwarf star. These central stars are generally too faint to be seen in small telescopes (but see the Blinking Nebula, page 134, for an exception to this rule).

If the star had previously ejected a relatively dense ring of gas outward from its equator, the new, expanding gas cloud would be partially blocked by the ring. In that case, it would balloon into a pair of lobes above and below the ring. (These might be what we see in the Dumbbell Nebula, page 136.) If the disk is viewed face-on, we see the ring and not the bulges of expanding gas (see the Ring Nebula, page 130).

The gas in the nebula glows because it is irradiated by the hot white dwarf at the center of the cloud. The color of the glow is a mixture of red and green. Unlike our eyes, CCD chips and color photographic film tend to pick up the red more than the green, so images of these nebulae tend to show them as red clouds; but to our eyes, in small telescopes they look like little bright green disks – hence, well seen in a green nebula filter.

These nebulae were first seen in the late eighteenth century, about the same time that William Herschel discovered the small, greenish disk of the planet Uranus. Since the two look similar in a telescope, these clouds were given the name "planetary" nebulae. Other than their appearance, however, they have nothing to do with planets.

find more at: www.cambridge.org/features/turnleft/seasonal_skies_april-june.htm

In Leo: The *Leo Trio*, three galaxies, M65, M66, and NGC 3628

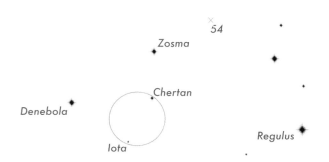

Castor

Pollux

Zosma

Chertan

Denebola

Regulus

Iota

54

- Wonderful Dobsonian objects
- Subtle challenge for a small telescope
- They look, to us, like we'd look to them...

Dark sky
Low power
Best: Feb.–June

Star maps courtesy Starry Night Education by Simulation Curriculum

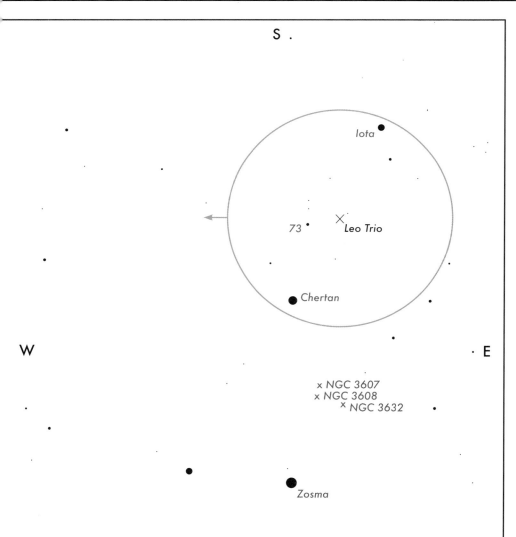

S

Iota

73 · × Leo Trio

Chertan

W

E

x NGC 3607
x NGC 3608
× NGC 3632

Zosma

N

54

Where to look: Find Regulus and the Sickle, the mirror-flipped question mark shape that makes up the Mane of Leo the Lion. Off to the left (eastwards) are three stars which make up the lion's hindquarters. They make a right triangle; the star at the corner with the right angle, the lower right (southwest) corner, is called Chertan. You'll see a dimmer star, down and a bit to the left (south and a bit east) from Chertan. It's called Iota Leonis. The galaxies are halfway between these two stars.

In the finderscope: You may just be able to fit both Chertan and Iota Leonis in the finderscope. Aim for the spot halfway between the two stars. A fifth-magnitude star, 73 Leonis, is just to the west of this point; it is the northernmost of a line of three stars, the others about a magnitude and a half fainter than 73 Leo. Aim the finder-scope at 73, then find it in your low-power telescope eyepiece. From 73 move slowly eastwards.

A sure-fire way to these galaxies is to aim at 73 Leonis and wait. Within two minutes, the galaxies will be drifting into your telescope's low-power field of view; by four minutes they'll be in the center of your eyepiece.

The Leo Trio in a star diagonal at low power

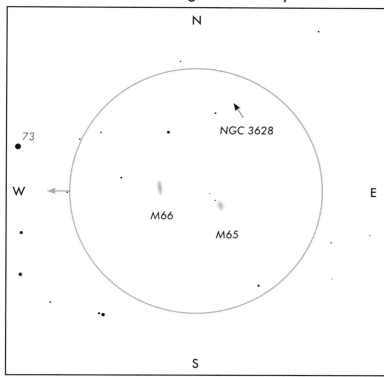

The Leo Trio in a Dobsonian at low power

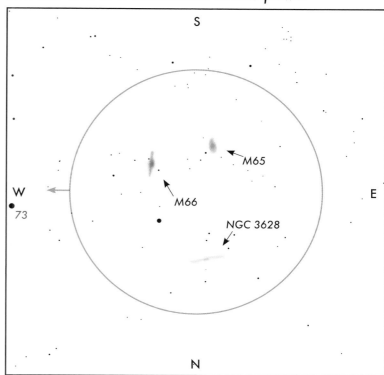

In a small telescope: The galaxies will appear as faint smudges of light, of about the same size. M65, the one to the southeast, is a slightly elongated blob, dimmer than M66, with a bar of light running through it. M66 is of more uniform brightness, more elongated than the disk of M65 but not as extreme as M65's inner bar.

In a small telescope, NGC 3628 is a challenge to find. Imagine M65 and M66 to be the base of a triangular dunce cap twice as tall as it is wide, pointing roughly northward. NGC 3628 is where the point of this triangle would lie.

In a Dobsonian telescope: A Dobsonian on a good dark night should be able to show significant detail in all three galaxies. M65 is quite bright, the roundest of the three (though still elongated, about 2:1, oriented north–south). Look for a hook (a spiral arm) on the south side. M66 is also quite bright, a 4:1 glow with a nucleus that looks a bit off-axis to the east. NGC 3628 may be the most interesting of the three: a long, thin, and somewhat off-kilter slash of light running east–west, with a little lump in the middle.

You need a good dark night to appreciate **M65**, **M66**, and **NGC 3628**, and your eyes should be well adapted to the dark, but as a close trio of galaxies they are well worth the effort.

These are good objects to try practicing the trick of averted vision: stare at a nearby dim star (there are plenty in this telescope field) and the edge of your vision, which is more sensitive to faint light, will pick up more detail of the galaxies. Try glancing in different directions, at different dim stars in this field, to see which corner of your eye works best for you.

All three galaxies are spiral galaxies, located about 20 million light years away from us. Each galaxy is roughly 50,000 light years in diameter; in a small telescope, we only see the central cores, which are about 20,000 light years across. These galaxies are somewhat smaller than our own Milky Way.

If we were in one of these galaxies, looking back towards Earth, we'd see our Milky Way and the Andromeda Galaxy (M31 – see page 176) as two splotches of light, about two and a half degrees apart, comparable in brightness to how these galaxies appear to us. From that vantage point, the Milky Way would be a pretty face-on spiral, while M31 would be seen edge-on, looking like a lumpy bar of light.

For more about galaxies, see page 109.

Also in the neighborhood: On a good dark night, hunt for other members of the Leo galaxy cluster! Another trio of them is nearby, lying halfway between Chertan and Zosma: NGC 3607, 3608, and 3632, as indicated in our expanded finderscope view. They're small roundish lumps of light, best seen in a Dob. More of the Leo group can be found on the next page…

The double star **54 Leonis** can be found easily from the hindquarters triangle of Chertan, Denebola, and Zosma. Step from Denebola to Zosma, and half a step farther brings you to 54 Leonis, a fourth-magnitude star. At high power, this divides into a magnitude 4.5 primary and its magnitude 6.3 secondary, lying 6.4" east.

find more at: www.cambridge.org/features/turnleft/seasonal_skies_april-june.htm

In Leo: M105 and the Leo I galaxies

Dark sky
Low power
Best: Feb.–June

Castor

Pollux

Chertan

Regulus

Rho

Star maps courtesy Starry Night Education by Simulation Curriculum

- Six galaxies in one small area
- Range of brightnesses
- Part of the Virgo Supercluster

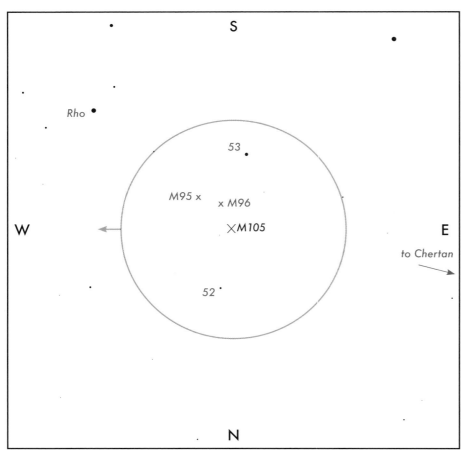

S

Rho

53

M95 x   x M96

×M105

W   E

to Chertan

52

N

Where to look: Find Regulus and the Sickle, the backwards-oriented question mark that makes up the Mane of Leo the Lion. Off to the left (eastwards) are three stars which make up the lion's hindquarters. They make a right triangle; the star at the corner with the right angle, the lower right (southwest) corner, is called Chertan. Start at Chertan and move towards Regulus, west and a bit south.

In the finderscope: Between Chertan and Regulus is a fourth-magnitude star, 52 Leonis – the first star of that brightness you'll come across from Chertan – and another fifth-magnitude star to the south of 52, 53 Leonis. Aim halfway between these two stars. If your finder (and the night) is good, you may see a seventh-magnitude star nearby.

Leo I galaxies in a star diagonal at low power

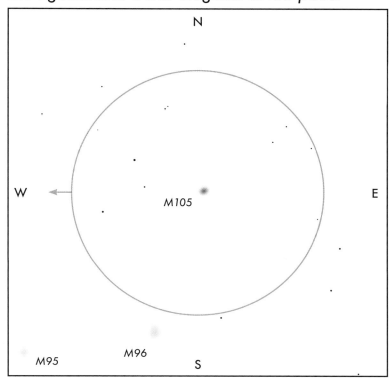

Leo I galaxies in a Dobsonian at low power

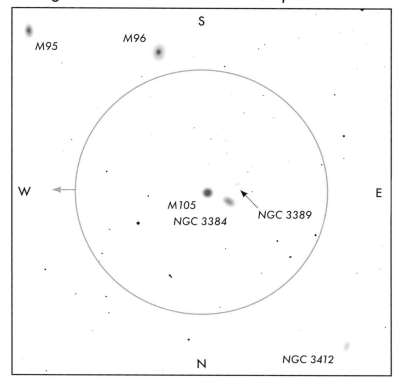

In a small telescope: Wait for your darkest night; if you have trouble seeing the Leo Trio, then these will be too faint to see.

M105 is a small, featureless circle of faint light. To the south and west, look for the smudges of light that are M95 and M96. The other galaxies are extremely challenging for a small telescope.

In a Dobsonian telescope: M105 is nice and fairly bright; it looks circular and almost featureless except for a gradual central brightening. NGC 3384 is easily seen as an oval, twice as long as it is wide, running southwest–northeast. NGC 3389 is a challenge, small and hard to see unless the conditions are good.

Move south from M105, and M96 should come into your field of view. It's a large oval, about 50% longer than it is wide, oriented almost north–south. You'll first see its bright, fuzzy nucleus; watch for a while and the rest of the galaxy will come into view, gradually fading away from the center.

From here, look for M95 on the western edge of your low-power field of view. Its oval is shaped and oriented much like M96's, but a bit smaller and fainter – almost ghostly – with a more concentrated nucleus.

From here, head back northeast to M105 and keep moving the telescope. As M105 moves out of your field of view, look for NGC 3412. It's a small, dim oval of light oriented northwest–southeast.

These galaxies and the Leo Trio (see the previous pages), are part of the **Leo I Group**, located about 20 million light years from us. (At least 11 other galaxies, too faint to include here, are also included in the group.) As you might expect from the name, there are other groups of galaxies visible in the constellation Leo as well. They are farther away than this group, ranging from 60 to 95 million light years distant. All these galaxy groups along with the large Virgo Cluster (see pages 104–109) make up what is known as the Virgo Supercluster.

The fact that galaxies come in clusters and superclusters is one of the characteristic traits about the structure of the Universe that puts limits on the possible ways it could have formed. It's a fundamental trait that any successful version of a model for the Big Bang must be able to account for.

find more at: www.cambridge.org/features/turnleft/seasonal_skies_april-june.htm

In Virgo: The Virgo galaxy cluster

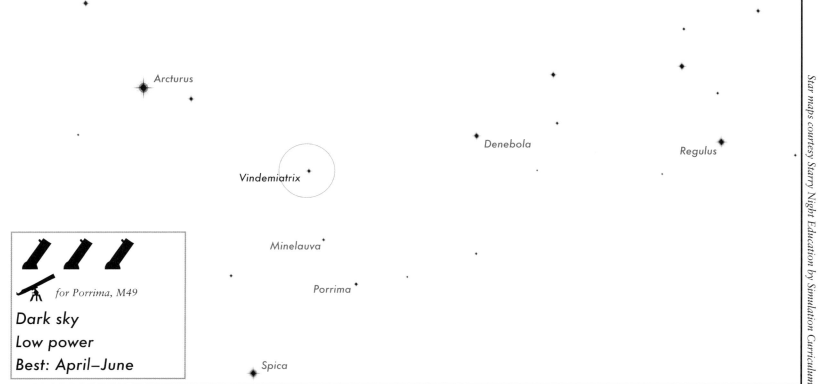

Arcturus

Denebola

Regulus

Vindemiatrix

Minelauva

Porrima

Spica

for Porrima, M49

Dark sky

Low power

Best: April–June

Star maps courtesy Starry Night Education by Simulation Curriculum

- Classic example of a galaxy cluster
- More than 20 galaxies (in a Dob)
- See many different galaxy types

Where to look: First, find Arcturus and Spica. (Following the arc of the Big Dipper's handle, you'll "arc to Arcturus and spike to Spica.") Next, look west to Leo; the mane of the Lion is the backwards question mark Sickle while three bright stars to its east make up the Lion's hindquarters. The brightest star, at the Lion's tail, is Denebola.

Arcturus, Spica, and Denebola make a nice equilateral triangle. In the middle of the triangle (start looking halfway between Arcturus and Denebola, then move south towards Spica) you'll see a third-magnitude star, Vindemiatrix.

Vindemiatrix is a Latin translation of a much older name, denoting a woman who gathers grapes. One theory is that, wherever and whenever this star was first named, it rose with the Sun when grapes were ready for harvest. Today it is most notable as the gateway star to the Virgo cluster.

Also in the neighborhood: Step south from Vindemiatrix to the third-magnitude star Minelauva; one step farther to its southwest is the double star **Porrima**.

Porrima and its companion do a very eccentric dance about each other. It only takes about 170 years for them to complete one orbit, during which time they'll move from barely 3 AU apart to more than twenty times that distance. In 1836, the greatest telescopes of that day could not split them; by 1920, they were at their maximum separation and an easy double to see even in the smallest telescope. Because orbiting stars move their slowest when they are farthest apart, for the next 50 years they moved very little. But since the mid 1970s they have been moving quite rapidly again.

The first edition of this book, written in the 1980s, said that star B was 4" west-northwest

of A. By the second edition we were referring to it as the "shrinking double," almost too close to split, and the third edition didn't bother listing it at all. But it reached its closest approach, about 0.4", in 2005. Since then it has become a "growing double." They're still slowly separating; over the next twenty years or so, star B can be found be about 3" to 3.5" north of A.

These two stars are located 35 light years from us, putting them among our closest neighbors. They are both F-type stars, a bit bigger and brighter than our Sun. Any planet orbiting either star would certainly see extreme variations of climate; but because of the stars' eccentric orbit, it's unclear whether any planet could find a stable orbit around either of them.

10"(!) circle

Porrima (Gamma Virginis)			
Star	**Magnitude**	**Color**	**Location**
A	3.6	Yellow	Primary star
B (2018)	3.6	Yellow	2.7" N from A
B (2024)			3.4" N from A
B (2030)			4.0" N from A

In the finderscope: Like most galaxies (except for Andromeda and the Magellanic Clouds) these are too faint to be seen in a finderscope, so in fact we will be galaxy-hopping with the low-power telescope field itself. Fortunately, this field is so rich in galaxies that, once you've found the first one, you can step onwards to the next. Thus, we'll use the finderscope only to find the good starting points for this string of galaxies.

Start by centering the finderscope at Vindemiatrix. Look for the sixth-magnitude stars 34 and 41 Virginis, in the northwest corner of the finder field. Move to a point halfway between Vindemiatrix and 34 Virginis, just south of 41 Virginis. Rho Virginis may be just at the western edge of the finder field. If you have a good finder, you may see nearby (just a bit south of the line connecting 41 Virginis with Vindemiatrix) a faint seventh-magnitude star, Struve 1689. (It's a double: a magnitude 7.1 primary with a magnitude 9.1 companion, 30" to the southwest.)

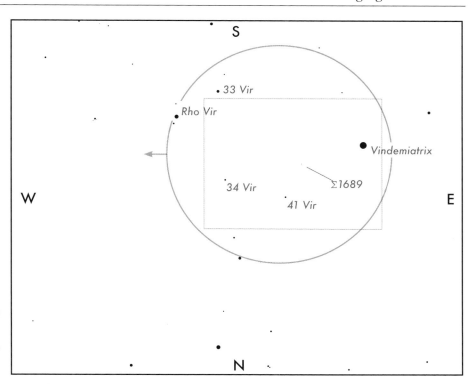

Chart 1: The eastern Virgo cluster galaxies in a Dobsonian

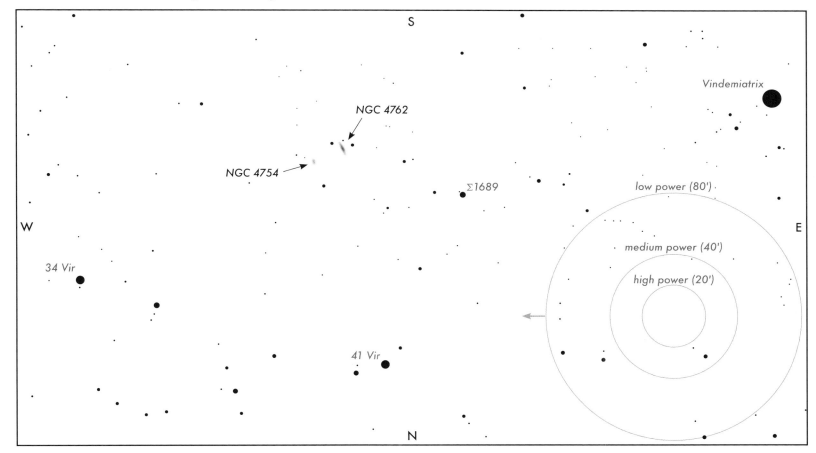

Chart 1: In a Dobsonian, NGC 4762 and NGC 4754: With Struve 1689 in your low-power field, move southwest until the galaxy NGC 4762 appears; it is a slash of haze running in the northeast–southwest direction, with a distinct central brightening. From it, look just to the west for a fainter smudge of light, NGC 4754. It is smaller and fainter than NGC 4762, and clearly more circular; we are seeing the disk of the galaxy more face-on whereas NGC 4762 is edge-on, from our point of view.

Once you spot the galaxies, try observing them with higher powers. High power can dissipate the sky brightness and so increase the relative contrast between the galaxies and the background sky.

In our telescope view above, we have indicated (in an uninteresting corner of the sky) typical fields of view for low-, medium-, and high-power eyepieces. You should be able to fit both galaxies into a medium-power eyepiece; whether both are visible under high power depends on the precise field of view afforded by your particular telescope and eyepiece combination, but it's worth a try.

find more at: www.cambridge.org/features/turnleft/seasonal_skies_april-june.htm

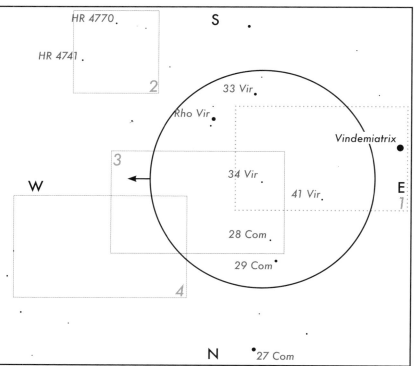

In the finderscope: Stepping from Vindemiatrix to 41 Virginis, keep moving west to a sixth-magnitude star, 34 Virginis. Aim there, then follow the directions given in the captions for each of the three Dobsonian views on these pages.

Chart 2: M49 and companions in a Dobsonian

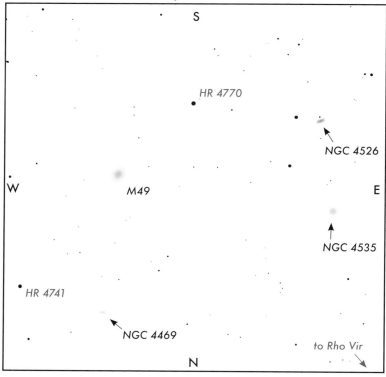

Chart 2: In a Dobsonian, M49 and neighbors (above): From 34 Virginis, step southwest to the fourth-magnitude star Rho Virginis; one step farther brings you to a pair of sixth-magnitude stars, HR 4770 and HR 4741. Halfway between them is the large, relatively bright galaxy M49. Due east, about one low-power field away, are the galaxies NGC 4535 and NGC 4526. The same distance north of M49 is NGC 4469.

Chart 3: In a Dobsonian, M60 to M90 (opposite page): Just east of 34 Virginis, find a seventh-magnitude star, HIP 62536. Step from 62536 to 34 Virginis; two steps westward brings you to M60. It is a ghostly oval, oriented east–west. With averted vision, it looks rounder. With its almost starlike nucleus it is easy to see even on less-than-perfect nights.

NGC 4638 is just beyond M60, in the same direction from 34 Virginis; it is dim, small, and ghostly.

West of M60, to the northwest of NGC 4638, is M59. It is dimmer than M60, but brighter than NGC 4638. Observe how it grows brighter from edge to center.

Keep moving west; as M59 leaves the field of view, M58 should enter it, looking like a small but fuzzy spot, relatively bright, next to an eighth-magnitude star.

Move north from M58. As it leaves your low power field of view, find the round galaxy M89: featureless except for slight brightening at the center. Then move farther north and a bit east to the northeast–southwest oval of M90. It's perhaps the most interesting of this group, with a somewhat lumpy looking disk of light.

Chart 4: In a Dobsonian, M90 to Markarian's Chain (opposite page): From 34 Virginis, follow the directions in Chart 3 to M89. Now move westwards until you see the round galaxy M87; it sits just south of an eighth-magnitude star.

If that star is the hub of a clock and the galaxy is at 6 o'clock, move the telescope in the 2 o'clock direction. This moves you northwest. As M87 departs from the southeast corner of your finder field, you should just see a string of faint galaxies moving into your field of view.

As you can see in the telescope view to the right, there are seven galaxies in a curved line here. This part of the Virgo cluster was first described by the Armenian astronomer Benjamin Markarian and is named for him. The largest and brightest of these galaxies are at the western end of the Chain, M86 and M84. Once you have found the chain, move west until these two round galaxies are in your field; then, slowly move back east and, eventually, north, picking off the galaxies one by one. After M84 and its slightly larger companion M86, you will arrive at a pair of smaller galaxies, NGC 4438 (running northeast–southwest) and its small round companion NGC 4435. Next, beginning the northward curve of the chain, is the small oval NGC 4461. After that, see the round galaxy NGC 4473 and finally NGC 4477. You can catch several at a time in low power but, as we noted before, switching to medium power may help make the background sky look darker, and allow you to recognize the fuzzy spots of galaxy amid the sharper stars.

From the end of the chain, in low power, step from NGC 4473 to NGC 4477, and take another three steps; you will arrive at the relatively large round galaxy M88. Moving due east from here as M88 leaves your field of view, the round galaxy M91 should come into view.

Chart 3: The Virgo cluster galaxies, M60 to M90, in a Dobsonian

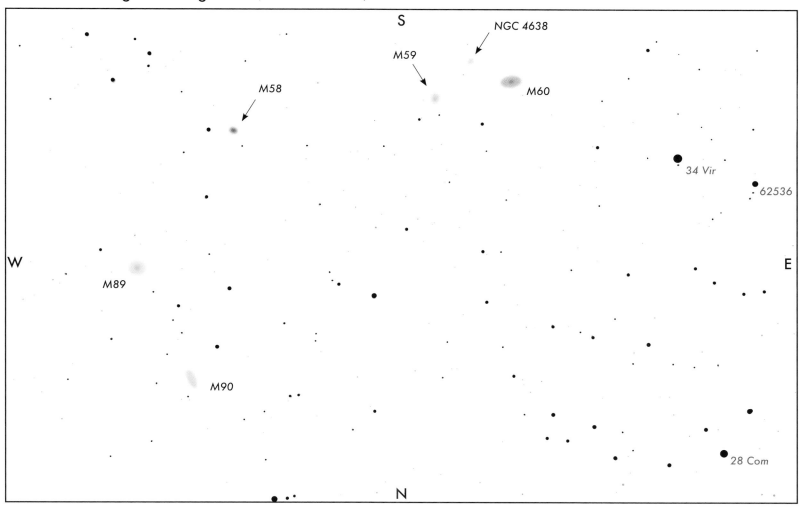

Chart 4: The Virgo cluster galaxies, M90 to Markarian's Chain, in a Dobsonian

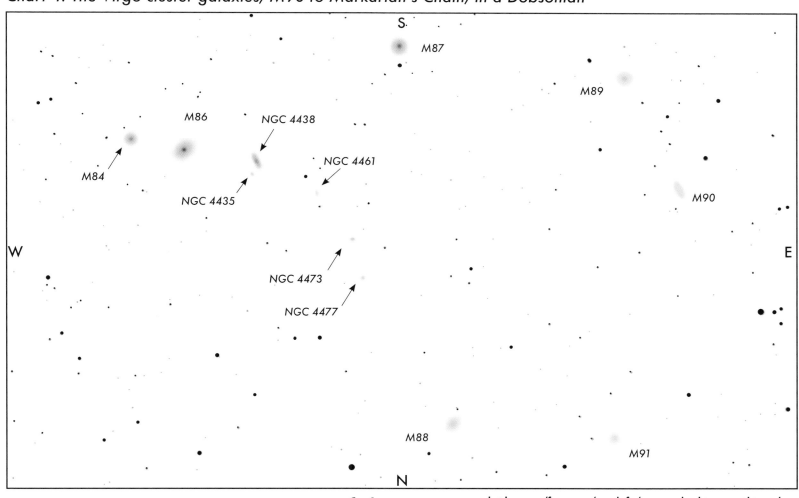

find more at: www.cambridge.org/features/turnleft/seasonal_skies_april-june.htm

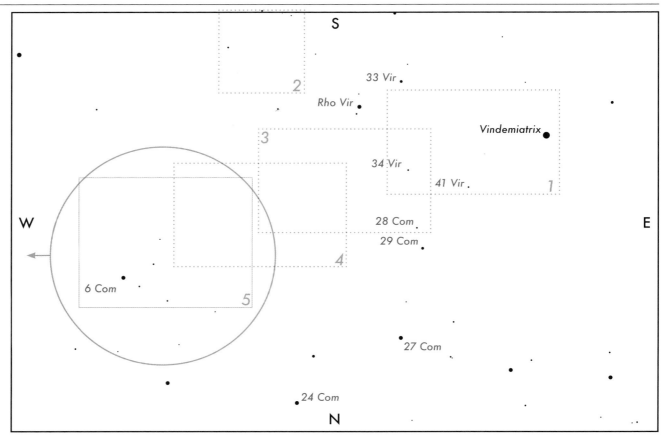

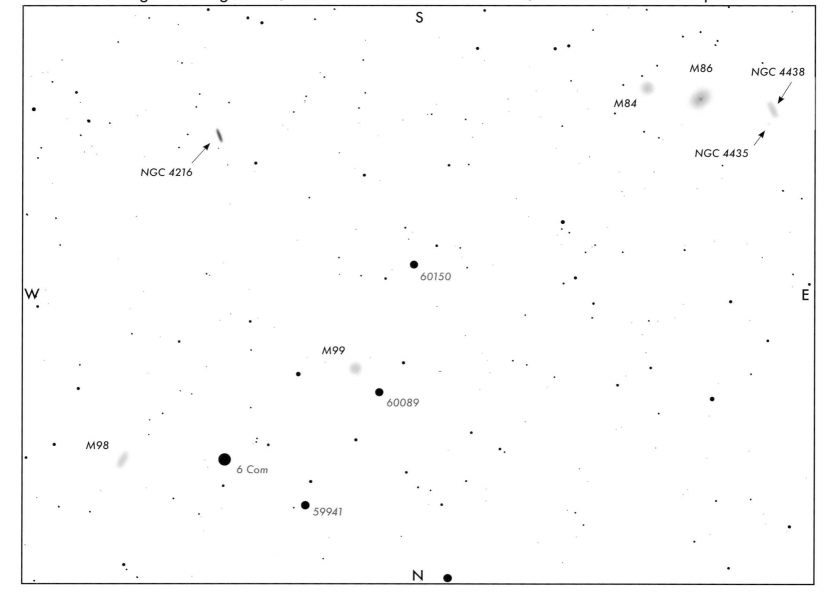

Chart 5: The Virgo Cluster galaxies, Markarian's Chain to NGC 4216, in a Dobsonian at low power

In the finderscope (opposite page): You should be able to find the fifth-magnitude star 6 Comae Berenices (Com), with a set of stars to its east looking like the letter L (or the number 7, depending on your orientation). One galaxy, M98, is to the west of 6 Com; another, M99, to the west of the middle star in the long stoke of the 7/L.

Chart 5: In a Dobsonian, Markarian's Chain to NGC 4216 (opposite page): If you've been observing the galaxies in the order described here, then begin at M84, at the end of Markarian's Chain, as described on the previous pages. From here, move your telescope northwest, towards the fifth-magnitude star 6 Comae Berenices (Com). En route, you will pass three stars in a row to the east of 6 Com, running roughly south to north: HIP 60150 at seventh magnitude, HIP 60089 at magnitude 6.5, and HIP 59941 at sixth magnitude. The last star is just northeast of 6 Com.

Look just to the west of the middle star, 60089; between it and an eighth-magnitude star you should see the disk of M99. Likewise, to the west of 6 Com you will find the disk of M98. These are two of the brighter galaxies in the Virgo cluster.

From M98, move due south. As first M98, then M99, move out of your field of view, keep an eye out for a thin sliver of light: NGC 4216.

The Virgo Cluster galaxies

Galaxy	Class	Chart	Appearance
NGC 4762	SB	1	Thin oval NE–SW
NGC 4754	SB	1	Round
M49	E4	2	Slight oval NW–SE
NGC 4535	SB	2	Round
NGC 4526	S0	2	3:1 oval NW–SE
NGC 4469	SB	2	3:1 E–W
M60	E1	3	Round
NGC 4638	S0	3	3:2 oval NW–SE
M59	E3	3	Slight oval N–S
M58	SAB	3	Oval NE–SW
M89	E0	3,4	Round
M90	SAB	3,4	5:2 oval NE–SW
M87	E1	4	Round
NGC 4461	SB	4	2:1 oval N–S
NGC 4473	E5	4	Slight oval E–W
NGC 4477	SB	4	Round
M88	SA	4	2:1 oval NW–SE
M91	SB	4	Round
NGC 4438	SB	4,5	5:2 oval NE–SW
NGC 4435	SB	4,5	Slight oval N–S
M86	S0	4,5	3:2 oval NW–SE
M84	E1	4,5	Oval N–S
M99	SA	5	Round
M98	SAB	5	2:1 oval NW–SE
NGC 4216	SAB	5	4:1 oval NE–SW

About galaxies

For reasons that are still poorly understood, after the initial expansion of the Universe in the Big Bang, it seems to have fragmented into discrete lumps of matter, each with enough mass to make billions of stars. A typical lump would then condense into a galaxy, with a cloud of globular clusters swarming erratically around the galactic center, and a disk of stars orbiting the center in much the same way as the planets orbit the Sun. Galaxies are the basic units of the Universe.

Galaxies come in three general forms. Elliptical galaxies look like large, somewhat flattened globular clusters. Irregular galaxies, as their name implies, are irregular collections of many billions of stars. Most beautiful are the spiral galaxies, whose stars are organized into two or more arms that twist around their galactic center. The Whirlpool (page 204), the Pinwheel (page 206), and our own Milky Way are examples of spiral galaxies. The Whirlpool's companion may be an elliptical galaxy. M82, in Ursa Major, is an irregular galaxy, while its companion M81 is a spiral (see page 200).

Galaxies are observed to be clumped together into **groups** *(up to about 50 galaxies) and* **clusters** *(hundreds of galaxies), each tied by the gravity of the others into a cloud moving together through space. The Virgo Galaxies listed here are an example of a cluster. Andromeda and its companions (page 176), the Triangulum Galaxy (page 178), and the Milky Way and its companions (the Magellanic Clouds, pages 210 and 216) are all part of our group, the* Local Group. *The galaxies in Ursa Major (pages 200, 204, and 206) and in Leo (pages 100 and 102) are examples of members of other groups. The clusters all appear to be moving away from each other, implying that they are the fragments of the Big Bang that took place about 13.7 billion years ago.*

Clusters themselves are associated into clusters of clusters called **superclusters**. *Are these superclusters independent entities, like raisins in a pudding, or connected together like the stuff in a sponge around "bubbles" of empty space? We still don't know. But the answer holds an important key to understanding just what went on during the Big Bang, when the Universe was created.*

Galaxy classes are defined:
SA: Spiral galaxy, no central bar
SB: Spiral galaxy with central bar
SAB: Spiral galaxy, intermediate between SA and SB
S0: Lens-shaped, but no visible spiral arms
E0: Elliptical, nearly spherical
E5: Elliptical, elongated

Galaxy classes are judgment calls, and various sources may disagree. What is listed to the left is our best guess at the consensus classification for each of these galaxies.

Elliptical galaxies have a designation of the form En, where the number n after the E represents the oblateness of the ellipse. If its long axis is a and its short axis is b, then n = 10(a − b)/a.

find more at: www.cambridge.org/features/turnleft/seasonal_skies_april-june.htm

In Coma Berenices: A globular cluster, M53, and the *Black Eye Galaxy*, M64

Star maps courtesy Starry Night Education by Simulation Curriculum

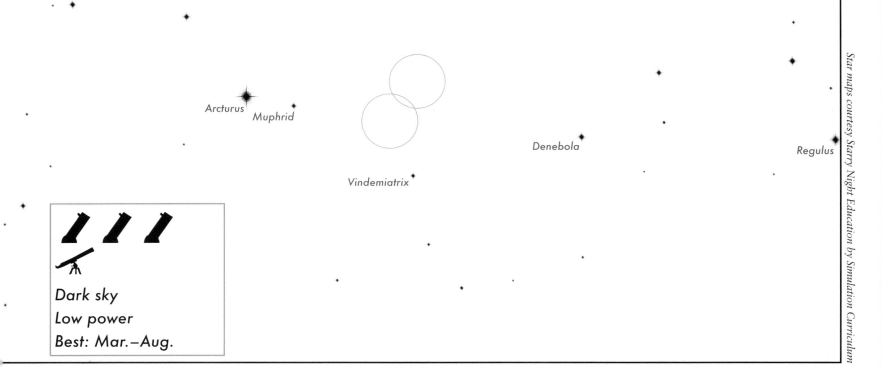

Arcturus
Muphrid
Denebola
Regulus
Vindemiatrix

Dark sky
Low power
Best: Mar.–Aug.

- Two-for-one: globular and galaxy
- Fun challenge for a small telescope
- Glorious in a Dobsonian

Where to look: Step from Arcturus to Muphrid, the star just to its west; two more steps farther and a bit north, you'll find a fourth-magnitude star, Alpha Comae Berenices. (It's north and a shade east of Vindemiatrix.) Point your finderscope at Alpha.

In the finderscope, M53: From Alpha, M53 is about one degree (two Moon diameters) to the north and east, a bit more than halfway between Alpha and a clump of fainter stars. **M64:** Put Alpha on the southeast corner of your finderscope view, and look in the northwest corner of the finderscope for a star almost as bright as Alpha, called 35 Comae Berenices. Move to 35, and look for a pair of stars, HIP 62653 (magnitude 6.5) and HIP 63008 (magnitude 7). Aim just southeast of 63008. (*Also in the neighborhood, note the nice, if faint, cat's-eyes double Struve 1685 in the southwest edge of the finder field.*)

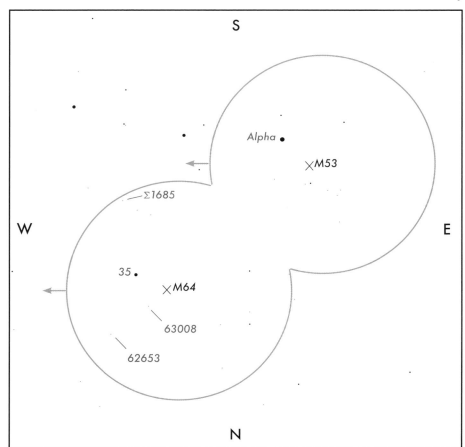

M53 is a cloud of ancient stars, maybe over 100,000 of them, in a cluster whose outer reaches extend almost 300 light years across, located some 65,000 light years from us. The center (which is all you can see in a small telescope) is itself about 60 light years across.

M64 is noted for the black band across the galaxy's center. This is due to two clouds of dust that obscure parts of the nucleus and the spiral arm above the nucleus. Seen with a large telescope, these dark

M53 in a star diagonal at low power

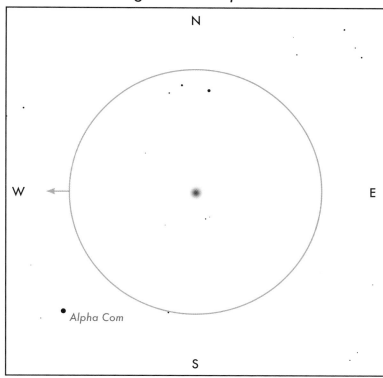

M53 in a Dobsonian at low power

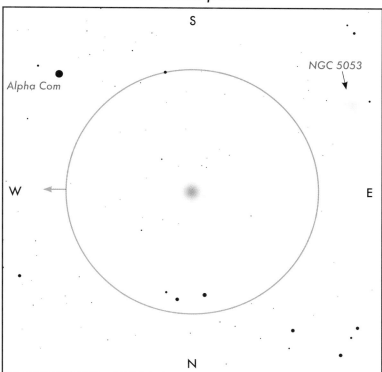

In a small telescope: M53 is a small, fuzzy, almost perfectly round disk; its center is slightly brighter than its edges. **M64** is not very conspicuous, a faint patch of light looking like a dim globular cluster, but slightly elongated: about half again as wide as it is long, in the west-northwest–east-southeast direction.

In a Dobsonian telescope: You should see a graininess in the center of **M53** and the fainter surrounding disk of light that marks the outer regions of this cluster. A degree and a half southeast is a fainter globular, NGC 5053. In **M64**, look for a dark band across center of an elongated patch of light, which gives this galaxy its name.

M64 in a star diagonal at low power

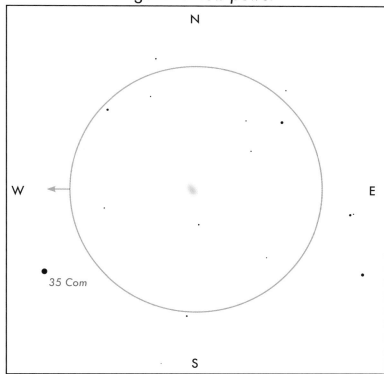

M64 in a Dobsonian at low power

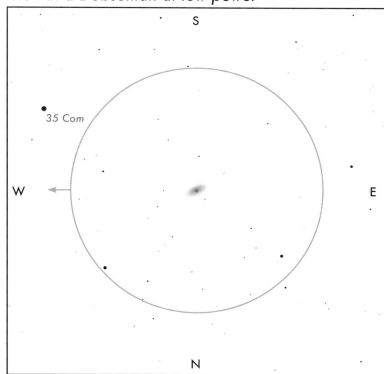

patches on the egg-shaped galaxy do give it something of the appearance of a black eye.

The gas in the outer galaxy rotates opposite to the gas and stars in the inner galaxy. One explanation is that a satellite galaxy collided with this one about a billion years ago.

Exactly how large and luminous we calculate M64 to be depends on how far away it is; estimates range from 13 to 55 million light years away! The smaller values are favored at present, but if the larger estimates turn out to be correct, it may be intrinsically as large and bright as the Andromeda Galaxy.

find more at: www.cambridge.org/features/turnleft/seasonal_skies_april-june.htm

In Canes Venatici:
A galaxy, M94,
and *Cor Caroli,*
a double star

The Big Dipper

for Cor Caroli

for La Superba

Dark sky
Low power
Best: March–August

Cor Caroli

Arcturus

Denebola

Regulus

Star maps courtesy Starry Night Education by Simulation Curriculum

- Bright but small galaxy, nice in a Dob
- Cor Caroli: Easy, colorful double star
- Nearby, *La Superba* – a very red star

Where to look: Find the Big Dipper, and look for the three stars that make up the handle of the dipper. Inside the bow of the dipper's handle (away from the North Star) are two stars clearly visible to the naked eye. They represent Canes Venatici (CVn), the Hunting Dogs. The brighter of these, directly under the handle, is Alpha CVn: *Cor Caroli,* "The Heart of Charles" (the tragic King Charles I of England).

Alternately, if you are approaching from the south, look for the star that makes an equilateral triangle with Arcturus and Denebola.

In the finderscope: To the northwest from Cor Caroli is another fairly bright star, Beta CVn. Imagine a line connecting these two stars, and point the telescope to the spot on this line halfway between them. Move at right angles away from this line towards the north and east, a distance equal to about a third of the distance between the two stars.

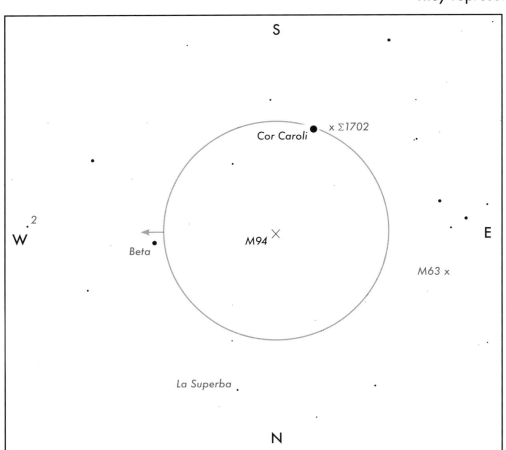

M94 in a star diagonal at low power

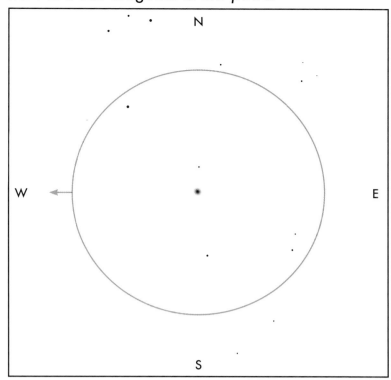

M94 in a Dobsonian at low power

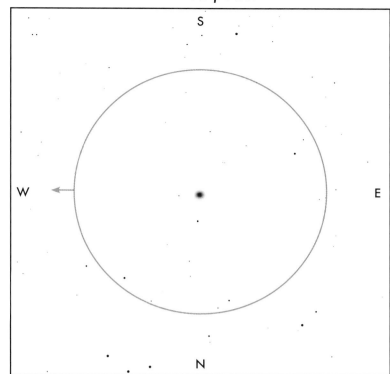

In a small telescope: As you move your telescope away from the line from Cor Caroli to Beta CVn, keep an eye out in the low-power field for **M94**, looking like a tiny round patch of uniform brightness. Once you've found it, try medium power.

In a Dobsonian telescope: The bright round core of **M94** is surrounded by a fainter haze of light, slightly elongated east–west. Once you have it, try observing with higher powers.

M94 is a relatively small, peculiar barred spiral galaxy, about 15 million light years away from us. In a small telescope, all that's visible of M94 is the central core of the galaxy. It's quite small; at low power it might be mistaken for an out-of-focus star. Find it at low power, then try it with medium power.

The galaxy is roughly 30,000 light years in diameter, and shines with the brightness of nearly 10 billion Suns. For more on galaxies, see page 109.

Cor Caroli is a rather elegant double star, easy to split and with a subtle shading in color. The primary is bluish white, while the color of its companion is usually described as yellow or orange. It is 110 light years from us, and the stars orbit each other at a distance of more than 650 AU with a period of over 10,000 years. The bright star is about three times as massive as our Sun and perhaps 70 times as bright; the dimmer star is about a third bigger than our Sun, has two thirds more mass, and is six times brighter.

5' circle

Cor Caroli			
Star	**Magnitude**	**Color**	**Location**
A	2.8	Blue	Primary star
B	5.5	Yellow	19" SW from A

Also in the neighborhood: Just to the east (about half a degree, or one Moon radius) from Cor Caroli is **Struve 1702.** *It's an eighth- and ninth-magnitude pair oriented roughly east–west with almost twice the separation as Cor Caroli, 36 arc seconds. It's easy to find: just center the field on Cor Caroli and wait. Two and a half minutes later, the rotation of the Earth will bring Struve 1702 to near the center of your field of view.*

Step from Cor Caroli to Beta, turn left (north-northeast), take another step, and this brings you near a wonderfully deep red star, Y CVn. It's a variable, going from magnitude 5 to 6.5 with a 160 day period. Because of its remarkable color, it was called **La Superba** *by Fr. Angelo Secchi, the nineteenth-century Jesuit priest–astronomer who discovered that stars like these are rich in carbon. It's well worth a look with binoculars or a small telescope.*

Step about a finder field east from Beta and look for a blue sixth-magnitude star, **2 CVn.** *It has a yellow ninth-magnitude companion 11" to the west, a sharp color contrast.*

East and a bit north of M94 is another galaxy, M63, the Sunflower Galaxy. Move the finderscope eastward from M94 until you see a triangle of stars; then move a degree north. Most nights it will be too faint for a small telescope, but a nice challenge for a Dob. On a really good night, however, it may even be visible in the finder. In a Dob you'll see an oval roughly twice as wide as it is tall, slightly brighter in the center. It's the brightest of a cluster of galaxies in this region of the sky, the others are mostly too faint for a small telescope, but they make nice challenges for a Dob on a dark night.

In Canes Venatici:
A globular cluster, M3

Star maps courtesy Starry Night Education by Simulation Curriculum

The Big Dipper

Cor Caroli

Arcturus

Muphrid

Regulus

Dark sky
Medium power
Best: Mar.–Aug.

- Classic bright globular cluster
- Grainy center even in a small scope
- Resolve individual stars in a Dob

Where to look: Find Cor Caroli, the brighter of two stars easily visible inside the curve of the Big Dipper's handle; and arc to Arcturus, the brilliant orange star off to the southeast from the handle. To the west of Arcturus is a second-magnitude star, Muphrid. Start here.

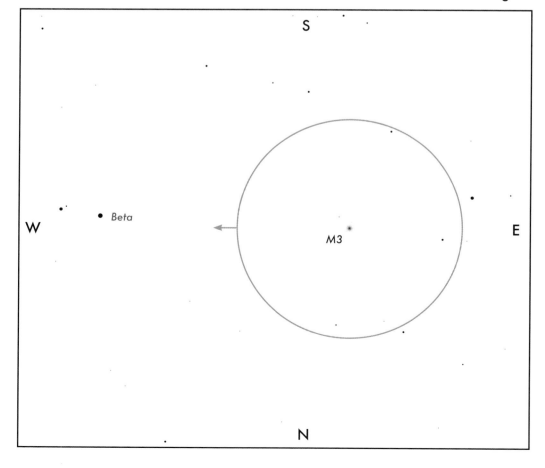

In the finderscope: From Muphrid, move slowly northwest towards Cor Caroli. About halfway to Cor, you should come across a fourth-magnitude star, Beta Comae Berenices. (If the night is dark it should be visible to the naked eye.) In a fairly poor field of stars, Beta will be the only prominent one visible. In your finderscope, you should see a second star just to the west of it; that's how you'll know you're on the right track.

Once you have found Beta, move to the east until it is just at the edge of the finderscope field of view. As it moves out of your field of view, M3 should be just appearing at the opposite side of the field, looking like a faint fuzzy star in the finderscope.

M3 *in a star diagonal at medium power*

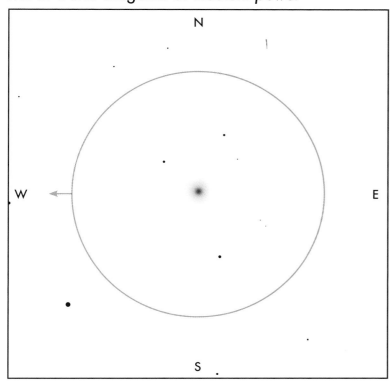

In a small telescope: The globular cluster will appear like a compact, bright, and perhaps somewhat grainy ball of light.

M3 *in a Dobsonian at medium power*

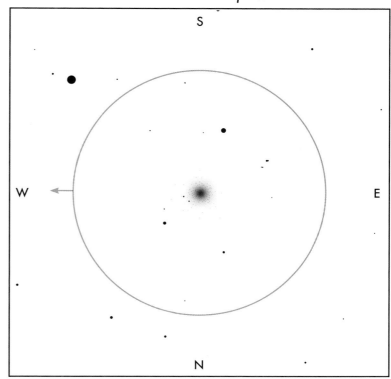

In a Dobsonian telescope: Medium power shows a grainy cluster with outlier stars, making this quite an impressive sight.

M3 is one of the brightest globular clusters visible in the northern sky. The nucleus is distinctly brighter than the surrounding cloud of light. With low power on a good dark night, the center seems grainy; you might even be able to resolve individual stars at the edge of the cluster, if your telescope aperture is four inches or larger, but even in a smaller telescope the cluster still looks quite pretty. Higher power helps resolve the stars, but you lose a lot of brightness – you may have a hard time seeing it at all if you increase the magnification too much.

The cluster is a collection of stars in a ball more than 200 light years in diameter, held together by the mutual gravitational attraction of the individual stars ever since they were formed, some 10 billion years ago. Large telescopes have resolved nearly 50,000 individual members; given its size and brightness, it's estimated that there may be up to half a million stars, total, in this cluster.

M3 is located 40,000 light years from us, and so even though it has the intrinsic brightness of 300,000 Suns, it appears to be about as faint as a sixth-magnitude star.

About globular clusters

A globular cluster like M3 contains some of the oldest stars in the Universe. One can find the age of these clusters much as is done with open clusters (see page 71) by noting which stars have had time to evolve into red giants. Such calculations indicate that globular clusters are at least 10 billion years old, as old as the Galaxy itself.

However, there's another peculiarity about these stars. Stars in globular clusters are generally made only of the gases hydrogen and helium, with little of the iron or silicon or carbon that we find in our Sun (and which is needed to make planets like Earth). If these elements exist in globular cluster stars, they must be buried deep inside the stars. In fact, current theories suggest that heavy elements, like iron, must actually be created in the center of very large stars by the nuclear fusion of hydrogen and helium atoms.

If the heavy elements are made deep inside stars, how do these elements get out to where they can make planets? When a star has used up all its hydrogen and helium fusion fuel, sometimes it explodes, making a supernova *(see page 67). In this way, all the heavy elements get thrown out into space, where they mix in with the hydrogen and helium gases that form new stars.*

But globular cluster stars don't have any of these heavy elements, except for what they've made themselves, hidden deep inside the stars. Wherever they came from, they must have been made from gas that had never been contaminated by supernova debris. In other words, they must have been formed before any supernovae had gone off. That would make them the oldest stars in the Galaxy.

find more at: www.cambridge.org/features/turnleft/seasonal_skies_april-june.htm

In (and near) Boötes: The Boötes double stars

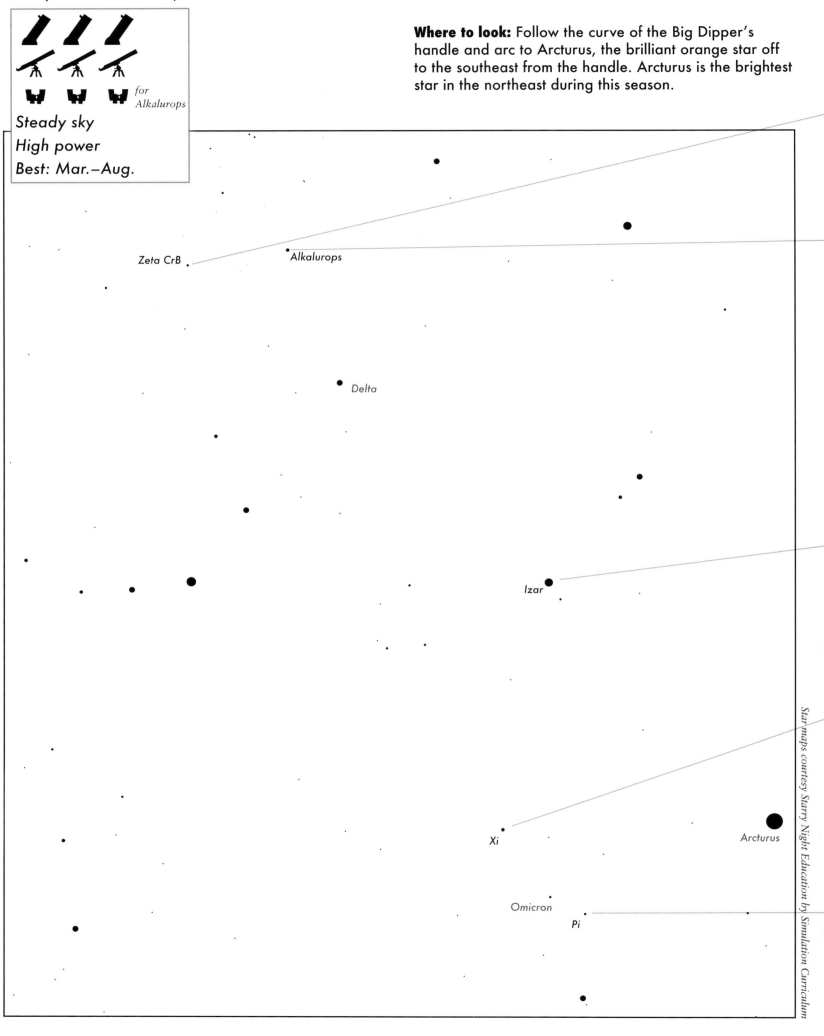

for *Alkalurops*

Steady sky
High power
Best: Mar.–Aug.

Where to look: Follow the curve of the Big Dipper's handle and arc to Arcturus, the brilliant orange star off to the southeast from the handle. Arcturus is the brightest star in the northeast during this season.

Zeta CrB

Alkalurops

Delta

Izar

Xi

Omicron

Pi

Arcturus

Star maps courtesy Starry Night Education by Simulation Curriculum

Zeta Coronae Borealis: Find Arcturus, a brilliant (zeroth-magnitude) orange star high in the west. Izar is the second-magnitude star about ten degrees north-northeast of Arcturus. Continuing on the same line from Izar is a third-magnitude star, Delta. Aim your finder at Delta and then move northeast until a pair of fourth-magnitude stars come into view: Alkalurops (itself a double) and Zeta Coronae Borealis. It's a busy field in the finderscope; these two are distinctive for having a pair of sixth-magnitude stars halfway between them. Zeta Coronae Borealis is the star to the west.

The Zeta Coronae Borealis pair are fairly closely matched in magnitude, 5 and 6, and rather tightly separated, with subtle but lovely colors: a yellow–blue primary and a greenish–blue secondary. It's easy in a Dob but a challenge in a small telescope, just possible if the night is still.

This pair of stars lie about 470 light years from us.

Xi Boötis: From Arcturus, head south and east to a crooked line of four third- and fourth-magnitude stars running roughly north–south. The northernmost star, almost due east of Arcturus, is Xi Boötis, a lovely yellow–orange pair that's a fun challenge in a small scope but easy in a Dobsonian; the larger aperture brings out the colors.

This star is a near neighbor of ours, only 22 light years away. Infrared telescopes have found evidence of a belt of comets around the primary, similar to our Sun's Kuiper Belt out beyond Neptune, while subtle motions of the secondary star indicate that it may have a Jupiter-like planet. These are, in fact, among the closest Sun-like stars to our own Solar System, making them an obvious target for interstellar probes sometime in the next millennium…

The orbiting star is moving rapidly. By 2025, it will be two thirds of its 2010 distance from its primary (under 5"), and at a more westerly position.

Pi Boötis is the third from the top (north) of the four stars in a crooked line east of Arcturus. In the finder field it sits southwest of a star of similar brightness, Omicron Boötis.

It's an easy split in a Dob ("easy as Pi"), though a tight pair in a smaller telescope.

The pair are located about 320 light years away from us.

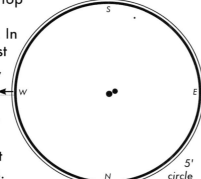

Alkalurops is the fifth-magnitude star about 3° to the west of Zeta Coronae Borealis; aim at Zeta, and Alkalurops should be just at the western edge of your finder field of view, with a pair of sixth-magnitude stars lying between them. It is a wonderful object for binoculars, which can split the wide pair easily; and also for a Dobsonian with a high-powered eyepiece, which can just split the secondary. If the night is very steady, even a small telescope can split the secondary – despite the distraction of the brighter primary so close by.

The name Alkalurops refers to the shepherd's crook of Boötes, the Herdsman. They lie about 120 light years from us and the B/C pair orbits more than 4,000 AU from the primary star. The motions of B and C themselves are well known; they go around each other in an eccentric dance every 260 years. The closest approach was in 1865.

Izar is the second-magnitude star about ten degrees north-north-east of Arcturus. It is an especially pretty double; a dazzling yellow primary orbited by a fifth-magnitude star that's pale blue with just a touch of green. However, the separation between the two is only 3"; and Izar is so bright it tends to overwhelm its partner, making this a difficult split. High power helps, though that can wash out the colors. Izar is about 210 light years from us. The pair, several hundred AU apart, have moved only slightly since their discovery in 1829.

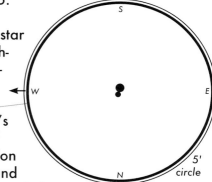

Star	Magnitude	Color	Location
Alkalurops (Mu Boötis)			
A	4.3	White	Primary star
B	7.1	Yellow	109" N from A
C	7.6	White	2" NNE from B
Xi Boötis			
A	4.8	Yellow	Primary star
B	7.0	Orange	6.4" NW from A
Pi Boötis			
A	4.9	Blue	Primary star
B	5.8	Yellow	5.6" ESE from A
Zeta Coronae Borealis			
A	5.0	Yellow	Primary star
B	5.9	Blue–green	6.3" NW from A
Izar (Epsilon Boötis)			
A	2.6	Yellow	Primary star
B	4.8	Blue	3" NNW from A

Seasonal skies: *July–September*

The center of the Milky Way is in Sagittarius, at its best visible during these months. That means that there are more good things to see in a small telescope now than in any other season. Many wonderful objects are to be found in the southern part of the sky, in the Milky Way; when setting up, try to find a place with a good dark southern sky.

Northern hemisphere observers will be experiencing their summer, the most comfortable time to observe; but even so, be prepared for a chill in the air, dew on the grass – and on your lenses! – and insects hovering over your head. (Don't underestimate the value of a good insect repellent.) You also have fewer hours to observe; not only are the nights shorter, but also with daylight saving time it may not be dark enough to see faint stars until 10 p.m. or later, depending on where you live. But one good thing about the summer is that it's the time for vacations, when you can stay up as late as you want. The stars we describe under October–December will be visible as early as July if you can hold out to about 3 or 4 a.m.

Looking west

Finding your way:
July–September sky guideposts

In the west, find **Arcturus**, a bright orange star; if you are at a latitude where the *Big Dipper* is visible, to the northwest, you can spot it using the adage, "arc to Arcturus" – follow the arc of the Big Dipper's handle on an imaginary curve bending southwards until you see a brilliant orange star, Arcturus. The two stars at the edge of the Big Dipper's bowl, of course, will also point you towards Polaris, the North Star. Arcturus is our signpost in the western sky.

Move eastwards, and high in the sky you'll find the *Summer Triangle* of very bright stars. The one to the south is a blue

first-magnitude star called **Altair**. It's the center and brightest of three stars in a short row. The other two Summer Triangle stars lie almost directly overhead (in northern latitudes). To the east is the bright star **Deneb**. It appears at the top of a big bright cross of stars, sometimes called the *Northern Cross*. (Unlike its smaller but brighter southern namesake, the Northern Cross is not by itself an official constellation; rather it is the neck and wings of *Cygnus*, the Swan.) And the brightest of the triangle stars, highest in the sky, is **Vega**. It's a brilliant blue–white star. Compare its color to the equally bright orange of Arcturus and the deep red of Antares to its south.

In the south you'll see **Antares**, a very bright red star. (Observers in the southern hemisphere will see it directly overhead during this time of year.) It is in the constellation

Be sure to take your telescope with you if you go vacationing out in the countryside, far away from city lights. Clusters and nebulae that just look like fuzzy patches of light in the city will stand out in surprising detail once you get to clear, dark skies. Observing stars over an ocean or a lake may be disappointing, however, since the air there can be humid and misty.

Southern hemisphere observers have the chills of winter to deal with. Be sure to dress with an extra layer of clothes, and invest in a pair of lightweight gloves; remember, the clearest nights are also the coldest. But unlike your northern cousins, you have the advantage of extra long nights to observe all the wonderful objects available during this time of year… and the rich Milky Way region around Sagittarius is high overhead!

To the west: *Above the western horizon, many nice April–June objects are still easily seen. Try to catch these just after sunset:*

Object	Constellation	Type	Page
Leo Trio	Leo	Galaxies	100
Virgo cluster	Virgo	Galaxies	104
M53	Coma Berenices	Globular cluster	110
Black Eye	Coma Berenices	Galaxy	110
Cor Caroli	Canes Venatici	Double star	112
M94	Canes Venatici	Galaxy	112
M3	Canes Venatici	Globular cluster	114
Izar	Boötes	Double star	116
Alkalurops	Boötes	Double star	116

Looking east

Scorpius, which is in the zodiac, so look out for planets here. Because it lies in the path of the planets and is reddish in color, Antares can be confused with the planet Mars: "Antares" is Greek for "rival of Mars." However, Antares twinkles more than any planet does. (We explain why planets don't twinkle on page 40.)

To the west of Antares are three bright stars in a vertical row; they're easier seen in southern climes. These are sometimes called the claw of the Scorpion.

Rising to the east of Antares is a set of stars that looks like the outline of a house; look more closely for other slightly fainter stars nearby, and you can turn it into the shape of a teapot. Indeed, when the sky is really dark, you can see a bright knot of the Milky Way above it that looks like a cloud

of steam coming out of its spout! This is *Sagittarius*, another zodiac constellation and so another place to look for planets. It's also the constellation containing the center of our Milky Way galaxy, and so the richest region of the sky to look for clusters and nebulae.

Rising in the east are the four stars of the Great Square, looking more like a diamond than a square during this season, and to their north the large W shape of Cassiopeia. These stars will be used later in the year (or later in the night) as guideposts; see the October–December chapter starting on page 162.

The *Milky Way* makes a path of light from the northeast to the south. On a good dark night it's nice just to scan your telescope through it. Look especially around Sagittarius, and near Deneb and the Northern Cross.

find more at: www.cambridge.org/features/turnleft/seasonal_skies_july-september.htm

In Hercules: The *Great Globular Cluster*, M13

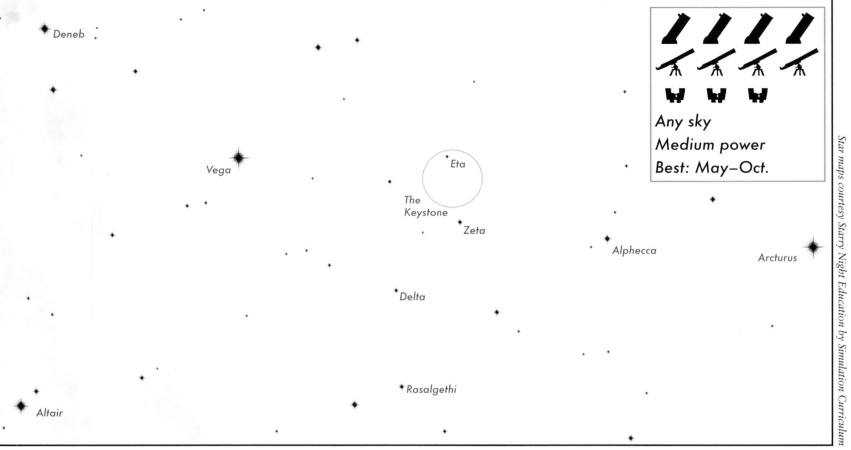

Star maps courtesy Starry Night Education by Simulation Curriculum

- Easy to find and see even on bright nights
- Best globular cluster visible from northern sites
- Extra Dobsonian bonus: the galaxy NGC 6207

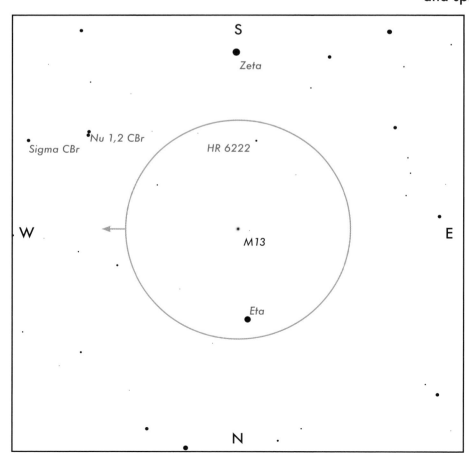

Where to look: Find Vega, a brilliant blue star, the westernmost of the stars in the Summer Triangle; and Arcturus, the brilliant orange star that the handle of the Big Dipper points to. Draw a line between these two stars, and split it into thirds. A third of the way from Arcturus to Vega is a half circle of faint stars, the constellation Corona Borealis (the Northern Crown), in which the bright star Alphecca is set.

Two thirds down the line from Arcturus to Vega (or half the distance from Alphecca to Vega), high overhead on a summer evening, are four stars that make a somewhat lopsided rectangle called the *Keystone*. Find the two stars of the western side of this box, the side towards Alphecca and Arcturus. Go to the point halfway between them, and aim a bit north.

A classic test for a clear, dark night is to try to see M13 with the naked eye.

In the finderscope: Find the star that makes the northwest corner of the Keystone, Eta Herculis. Move south; about a third of the way from Eta to the star in the southwest corner, Zeta Herculis – or halfway between Eta and the sixth-magnitude star HR 6222 – will be a faint, fuzzy patch of light. That's M13.

M13 in a star diagonal at low power

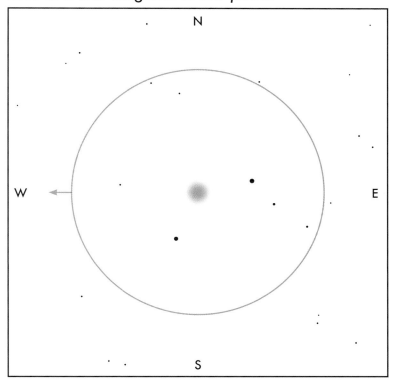

In a small telescope: The globular cluster looks like a ball of light, with the center brighter than the outer edges. It is flanked by two seventh-magnitude stars.

M13 in a Dobsonian at medium power

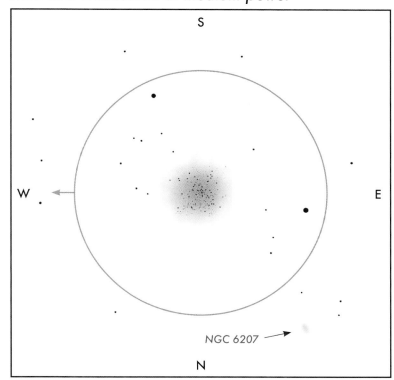

In a Dobsonian telescope: You will see an indescribable mass of uncountable stars, superimposed on a ball of light that looks as if it were dusted with diamonds.

Look to the northeast for the faint galaxy NGC 6207, as described below.

M13 is the best globular cluster to observe from northern latitudes, one of the five best in the sky. (The others are M22, page 158, best seen from latitudes south of 35° N – Florida or farther south; Omega Centauri, page 238, which only rises above the horizon at such latitudes; and 47 Tucanae, page 212, and NGC 6752, page 240, both objects seen well only from the southern hemisphere.)

The small bright core will be visible even on a poor night, and the darker the night, the more of the cluster you'll be able to see. In fact, the bright central part is only about a fifth of the total radius of the cluster. A small telescope will show a slightly grainy texture to the cluster; a six-inch telescope can start to resolve an incredible number of individual stars along the edges.

This is a globular cluster of an estimated 300,000 stars. (For more information on globular clusters see page 115.) The central part of this cluster is over 100 light years in diameter and it's located 25,000 light years away from us.

Large telescopes have been able to resolve about 30,000 stars on the edges of this cluster, but in the center they're too close together for anyone to pick out individual stars. "Close together" is a relative term, of course; even in the densely populated center of the cluster these stars are still roughly a tenth of a light year apart, so collisions are probably very rare.

In fact, as far as we can tell, these stars may have been clustered together like this for some 10 billion years, more than twice as long as our Solar System has existed, and possibly dating back to the origin of the Galaxy itself.

Also in the neighborhood: In the northeast corner of your low-power telescope field of view, about half a degree away from M13, is the galaxy **NGC 6207**. *Even with a Dob, you may need averted vision to see it as an elongated, misty cloud of light; it's only 11th magnitude, concentrated into a 3.5'×1.5' disk. However, on a really dark night it's a sweet sight.*

A fun game if you're observing with a buddy is to center your telescope on NGC 6207 with a high enough power eyepiece to exclude M13 from the field of view, then call your friend over to share the view. After they see the galaxy, have them nudge the scope southwest, and watch their reaction as M13 comes into view!

From M13, move your finder southeast and look for a pair of fifth-magnitude stars in a north–south line, Nu-1 and Nu-2 Coronae Borealis. Just to their west is another fifth-magnitude star, **Sigma Coronae Borealis**. *It's a lovely double, a pale yellow primary (magnitude 5.6) with a yellow–blue magnitude 6.5 partner 7.1" to its west.*

Step from the northwest star of the Keystone (Zeta) to the southeast star; one step farther brings you to **Delta Herculis**, *a nice challenging double star for a Dob. The magnitude 3 primary has an 8th magnitude companion 11" to the west-southwest.*

Continuing another step in this direction, you will see a pair of relatively bright stars, Rasalhague (second magnitude) and Rasalgethi (third magnitude, and distinctly reddish in color). Rasalgethi is a colorful double star; see page 124 for details.

find more at: **www.cambridge.org/features/turnleft/seasonal_skies_july-september.htm**

In Hercules: A globular cluster, M92

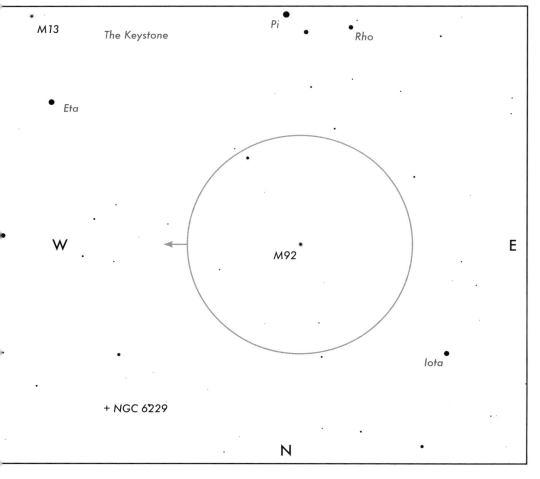

Dark sky
Medium power
Best: May–Oct.

- Slightly tricky to find
- Pretty cluster to observe
- The longer you look, the more you see

Where to look: Find the brilliant blue star Vega, the westernmost of the Summer Triangle stars; and Arcturus, the brilliant orange star off the handle of the Big Dipper. Draw a line between these two stars, and split it into thirds. A third of the way from Arcturus to Vega is a bright star called Alphecca. Two thirds of the way along the Arcturus-to-Vega line (or half the distance from Alphecca to Vega), high overhead on a summer evening, are four stars that make a somewhat lopsided rectangle called the *Keystone*. The northern two stars of the Keystone are Eta and Pi Hercules. Northeast of them is a third-magnitude star, Iota Hercules. (Don't confuse it with either of the two brighter stars to its north.) Start at Pi, and move due north until you hit the line connecting Eta and Iota.

If you're starting from M13 (page 120), move your finder northeast past Eta towards Iota.

In the finderscope: Relatively bright (sixth-magnitude), M92 should be easily visible in the finderscope. However, this cluster can still be tricky to find because there are few other stars nearby to guide you.

M92 *in a star diagonal at low power*

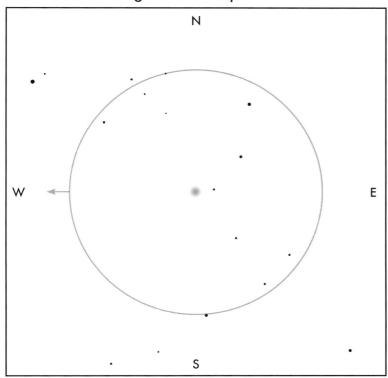

M92 *in a Dobsonian at medium power*

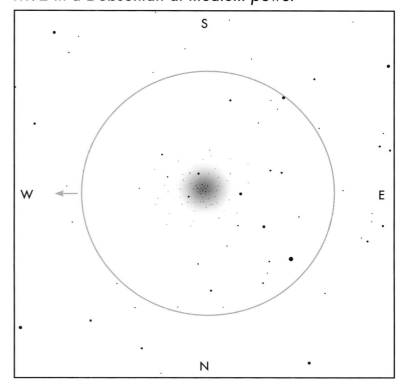

In a small telescope: We show the low-power field here, to help you find it; but once found, switch to medium power. The cluster is a small but unmistakable ball of light. The light is somewhat uneven, almost lumpy looking, without being grainy. There is a distinct nucleus, but it is not particularly prominent. Averted vision (turning your eye away from the nebula, since the corner of your eye is more sensitive to faint light) helps bring out the full extent of the cluster; as your eye gets adapted to seeing it, it tends to look bigger and brighter.

In a Dobsonian telescope: Almost anywhere else in the sky this would be considered a marvelous globular cluster; but, unfortunately for M92's reputation, its nearby neighbor M13 has already appropriated the title of the Great Hercules Cluster. Look for a bright, easily resolved cluster with a strong central concentration of stars. At higher power, you can see an asymmetry among the distribution of stars surrounding the center.

M92 is a collection of at least several hundred thousand stars, in a ball roughly 100 light years in diameter. For all its size, it appears dimmer than M13 because it is so much farther away, lying about 35,000 light years from us.

More information about globular clusters can be found on page 115. As we discuss there, the fact that stars in globular clusters do not have a high abundance of heavier elements implies that they are primitive stars, and indeed detailed calculations of stellar evolution suggest that stars in these clusters may be 13 to 15 billion years old.

These calculations, based on some well-established bits of physics, at one time led to an interesting conundrum. Until recently, a different set of observations and calculations on the Hubble constant describing the expansion of the Universe suggested that the Big Bang itself only occurred 12 billion years ago – and it seemed unlikely, to say the least, that there would be stars within the Universe older than the Universe itself!

Fortunately for common sense, observations starting in 2001 by the WMAP satellite measuring the anisotropy of cosmic background microwave radiation (essentially, how the energy from the Big Bang echoes to us from one direction of the Universe compared to another direction) and other improvements in our understanding of the evolution of the Universe now put its age at 13.7 billion years. That's just long

enough to allow these clusters to exist. Still, it is interesting to note that these old-timers date back to the first generation of stars in our Universe.

Also in the neighborhood: Just east of Pi Herculis, the northeast star of the Keystone, are a pair of fourth-magnitude stars; the one farther from Pi to the east is **Rho Herculis**. *It's a pleasant tight double, a magnitude 4.5 primary with a bluish magnitude 5.4 companion just 4" to the northwest. You'll want your highest power to split it.*

About seven degrees northwest of M92, as seen in the finderscope view, is a small but sweet globular cluster, NGC 6229. Even by globular cluster standards it is quite far away from us, some 100,000 light years; that's three times as far as M92, four times as far as M13. At that distance, it's no surprise that it is totally unresolved. Its small size and decent brightness give it a pretty good surface brightness, while the lack of resolution (not even graininess) makes it look comet-like. It makes a great contrast to its closer neighbors M92 and M13.

find more at: www.cambridge.org/features/turnleft/seasonal_skies_july-september.htm

In Serpens: A globular cluster, M5
In Hercules: A double star, *Rasalgethi*, Alpha Herculis

Arcturus

Rasalgethi
Rasalhague

Zeta

Unukalhai

Epsilon

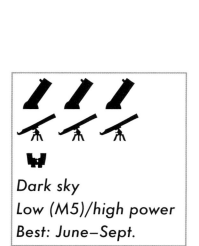

Dark sky
Low (M5)/high power
Best: June–Sept.

Star maps courtesy Starry Night Education by Simulation Curriculum

- M5: Tricky-to-find globular; but worth it!
- Unusually old cluster
- Rasalgethi: Unusual colors in a double star

Where to look: Find Arcturus, a brilliant (zeroth-magnitude) orange star high in the west. To the southeast from Arcturus is a third-magnitude star, Zeta Boötis. Step from Arcturus to Zeta Boötis; two more steps in this direction will get you at a spot of dark sky to the southwest of a star called Unukalhai, a reasonably bright star with two dimmer stars, Lambda and Epsilon Serpentis, nearby. Start at Unukalhai. The globular cluster, M5, will be to the southwest; the double star, Rasalgethi, to the northeast.

In the finderscope: Step from Lambda to Unukalhai and keep moving in that direction until you see a triangle of reasonably bright stars. Two of the stars are in an east–west line (and the third is a bit to the south, nearer to the western star). Aim at the western star of that line, 5 Serpentis. You may see M5 as a fuzzy dot just to its northwest.

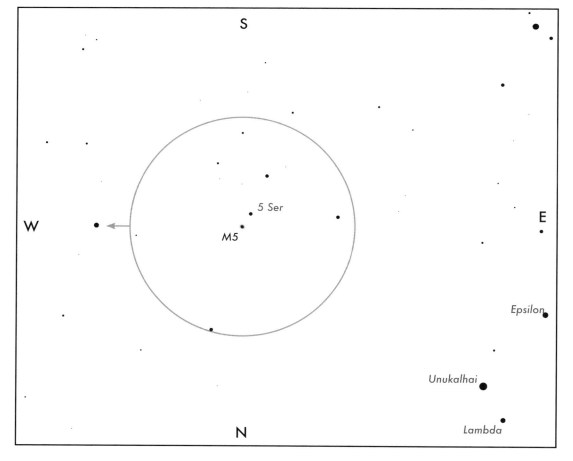

M5 in a star diagonal at low power

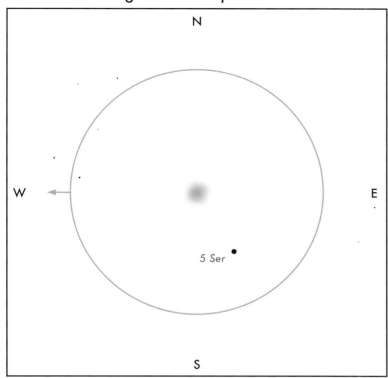

M5 in a small telescope: The globular cluster will look like a ball of light, somewhat brighter towards the center. To the south-southeast of the cluster, a fifth-magnitude star, 5 Serpentis is visible.

Don't expect to see any marked structure to the cluster, just a bright round ball of light that gradually fades into the background sky. In a 2"–4" telescope, the light at the outer edge of the ball may appear to be clumpy; what you're seeing there are little groups of stars in the outer regions of the cluster.

M5 in a Dobsonian at low power

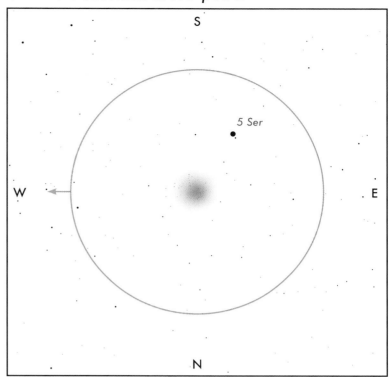

M5 in a Dobsonian telescope: Partially resolved at low power, under dark skies at high power in a Dob the cluster is very impressive: hundreds of stars resolved, with one particularly bright star within a clump on the western edge of the cluster.

M5: Since there are no easy naked-eye stars near M5, it can be hard to find at first. But, being one of the brighter globular clusters, it should be easily visible in the finderscope. You might even see it with the naked eye on a particularly dark night.

This globular cluster may be one of the oldest known. Judging from the types of stars visible (even low-mass, long-lived stars have begun to expand into red giants, the final phase of their existence) some estimates make this cluster to be as much as 13 billion years old, nearly three times as old as our Solar System. It probably contains close to a million stars, gathered in a slightly elliptical ball some 100 light years in diameter, located 27,000 light years from us. In a large telescope, the center appears bright and densely packed with stars. It is this central cluster of stars that we can see in our smaller scopes.

More information about globular clusters can be found on page 115.

Rasalgethi: To the northeast of Epsilon and Unukalhai, halfway to the Summer Triangle, you will see a pair of relatively bright stars, Rasalhague (second-magnitude, to the east) and Rasalgethi (third-magnitude, and distinctly reddish in color, to the west). Rasalgethi is a colorful double star.

The primary is a red-giant star, and a variable, ranging (irregularly, over a three-month period) from third to fourth magnitude. Its fifth-magnitude companion, 4.9" to its east,

appears green (perhaps by contrast with its red partner). On a good night even a small telescope should be able to split it; in a Dob it is quite impressive, the primary star showing a distinct orange–yellow color. Use high power.

Rasalgethi is a huge but distant red-giant star. Best estimates locate this star at about 360 light years from us, and it is roughly 400 times as wide as our Sun. In other words, if the Sun were placed at its center, all the planets out to Mars (including Earth) would be inside it. Its companion star is more than 500 AU from Rasalgethi and takes about 3,600 years to complete an orbit. The companion is itself also a double star.

Rasalgethi (Alpha Herculis)			
Star	**Magnitude**	**Color**	**Location**
A	3.0*	Red	Primary star
B	5.4	Green	4.9" ESE from A
	*variable (3.0–3.9)		

find more at: www.cambridge.org/features/turnleft/seasonal_skies_july-september.htm

In Ophiuchus:
Two globular clusters, M10 and M12

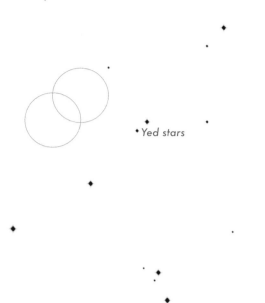

Arcturus

Rasalgethi

Zeta Boö

Yed stars

Antares

Dark sky
Medium power
Best: June–Sept.

Star maps courtesy Starry Night Education by Simulation Curriculum

- Tricky to find, easier to see
- See individual stars in larger scopes
- Nice contrast in structure between M10, M12

Where to look: Find Arcturus, the brilliant orange star in the western sky. There's a dimmer star east and a bit south of Arcturus, called Zeta Boötis. Step from Arcturus to Zeta Boötis, then three more steps, until you find two stars of near equal brightness in an 8 o'clock–2 o'clock orientation, Yed Prior and Yed Posterior. Start at the Yed stars.

In the finderscope, M10: Move east from the Yed stars until two stars come into view. Stop at a spot about a degree (two Moon diameters) west of 30 Ophiuchi and north (about three times that distance) of the fainter star, 23 Ophiuchi. On a good night M10 itself should be visible in the finderscope as a tiny smudge of light. **M12:** From M10, move northwest. Think of a clock face, where the location of M10 is the hub, 30 Ophiuchi is at 3 o'clock, and 23 Ophiuchi is at the 11 o'clock position; move in the direction of the 8 o'clock position, and twice as far as the distance from 23 to 30 Ophiuchi. The finder field at M12 itself is very faint, and in the finderscope M12 will be a bit harder to see than M10.

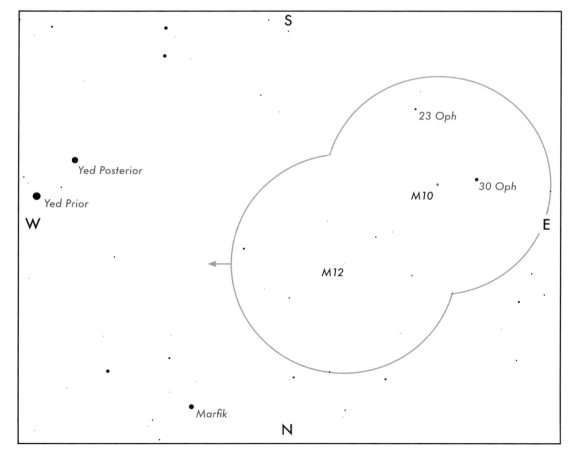

S

23 Oph

Yed Posterior

30 Oph

Yed Prior

M10

W

E

M12

Marfik

N

M10 in a star diagonal at low power

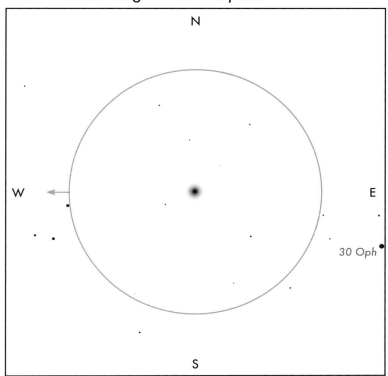

M10 in a Dobsonian at medium power

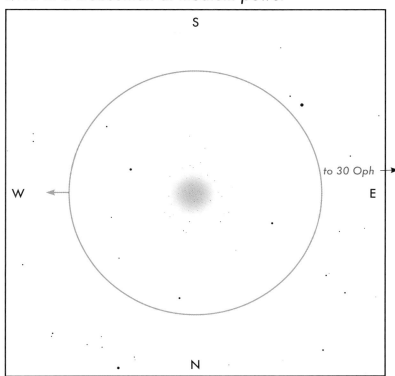

In a small telescope: Find in low power (shown here), observe in medium. **M10** looks like a fairly bright hazy ball with a bright, relatively large core displaced a bit towards the southwest. **M12**'s disk of light looks somewhat larger but dimmer than M10.

In a Dobsonian telescope: You need a 6" telescope, or larger, to resolve individual stars in **M10**. At high power, look for lines of starlight in the grainy light. **M12** is especially delightful in a Dob; it is possible to resolve individual stars if the night is dark enough.

M12 in a star diagonal at low power

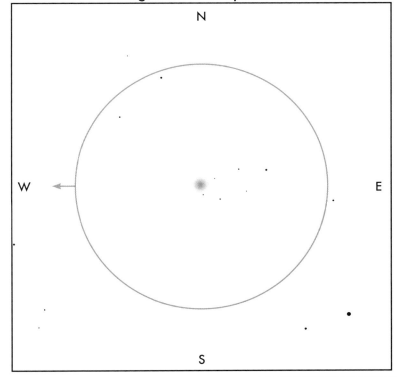

M12 in a Dobsonian at medium power

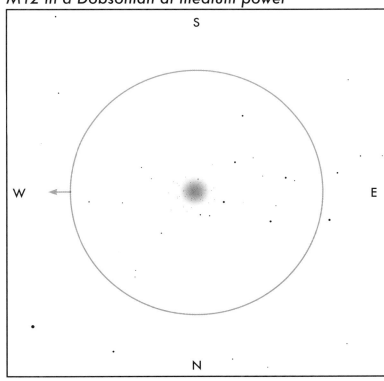

M10 and **M12** are relatively close neighbors to each other, as globular clusters go; they probably lie barely 1,000 light years from each other. The pair are about 20,000 light years from us. M10 is a ball of hundreds of thousands of stars, 80 light years in diameter; M12 is only slightly smaller, about 70 light years across.

It's easy to confuse M10 and M12. To tell the two apart, note that M12 has a clump of stars, magnitudes 9.5–10, trailing it (to the east). M10 is smaller but a bit brighter than M12 and more easily resolved into stars. M12 is somewhat looser in structure, and less concentrated in its core.

For more information on globular clusters, see page 115.

find more at: www.cambridge.org/features/turnleft/seasonal_skies_july-september.htm

Double stars near the Summer Triangle

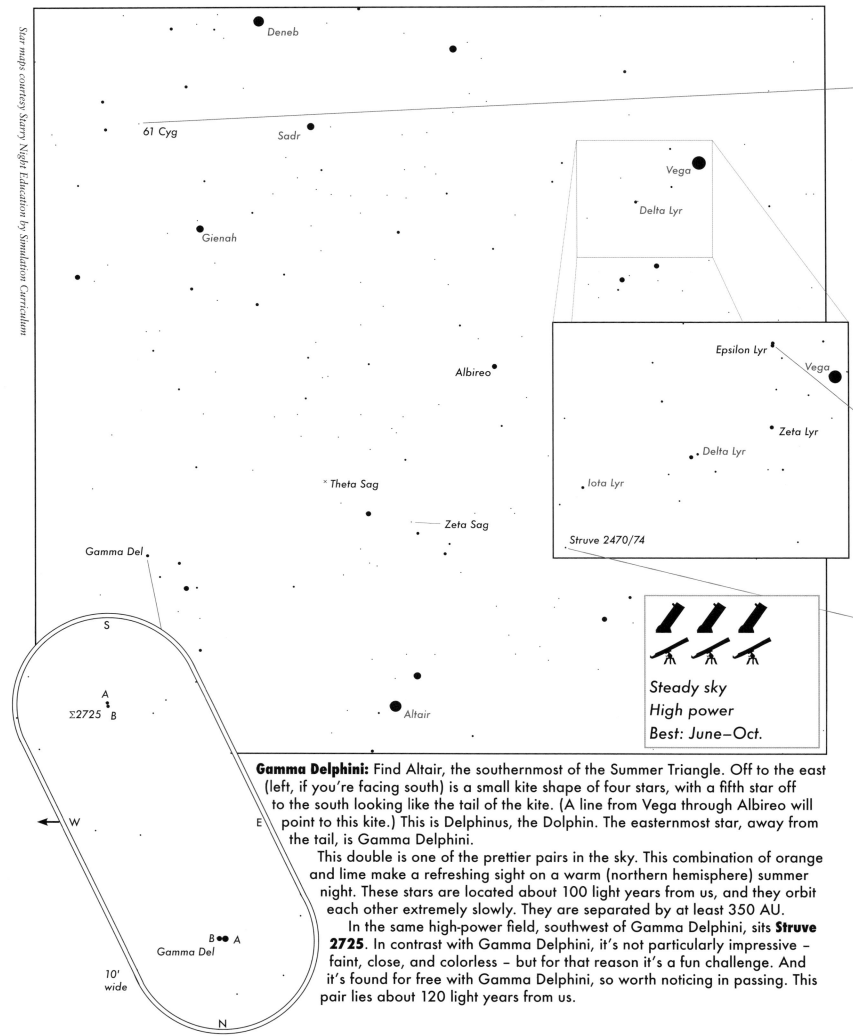

Star maps courtesy Starry Night Education by Simulation Curriculum

Deneb

61 Cyg

Sadr

Gienah

Albireo

× Theta Sag

Zeta Sag

Gamma Del

A
Σ2725 B

Altair

B ●● A
Gamma Del

S

W

E

N

10'
wide

Vega

Delta Lyr

Epsilon Lyr

Vega

Zeta Lyr

Delta Lyr

Iota Lyr

Struve 2470/74

Steady sky
High power
Best: June–Oct.

Gamma Delphini: Find Altair, the southernmost of the Summer Triangle. Off to the east (left, if you're facing south) is a small kite shape of four stars, with a fifth star off to the south looking like the tail of the kite. (A line from Vega through Albireo will point to this kite.) This is Delphinus, the Dolphin. The easternmost star, away from the tail, is Gamma Delphini.

This double is one of the prettier pairs in the sky. This combination of orange and lime make a refreshing sight on a warm (northern hemisphere) summer night. These stars are located about 100 light years from us, and they orbit each other extremely slowly. They are separated by at least 350 AU.

In the same high-power field, southwest of Gamma Delphini, sits **Struve 2725**. In contrast with Gamma Delphini, it's not particularly impressive – faint, close, and colorless – but for that reason it's a fun challenge. And it's found for free with Gamma Delphini, so worth noticing in passing. This pair lies about 120 light years from us.

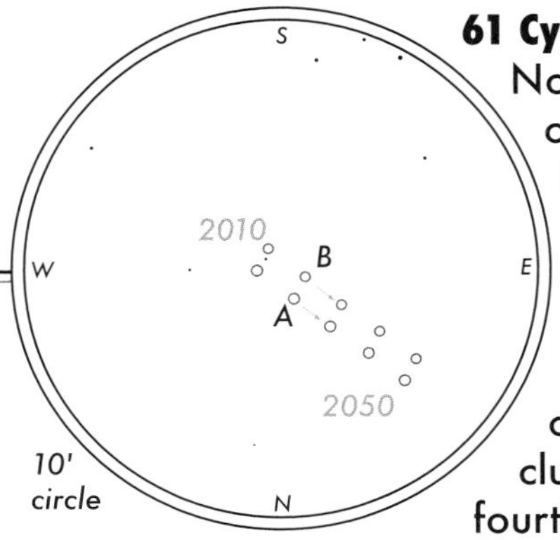

10' circle

61 Cygni: Find Deneb, top of the Northern Cross. Where the crosspiece meets the body is the star Sadr; the star of the left (southeast) arm is Gienah. Imagine a lopsided box with Gienah, Sadr, and Deneb making three of the corners. There is a cluster of faint stars where the fourth corner of the box ought to be. Of the three stars of roughly equal brightness you'll see there, aim for the one closest to the Cross. 61 Cygni is easily split; and note the contrast of their orange colors against the background blue-and-white Milky Way.

One of the closest stars in our sky, 61 Cygni can be seen to move northeast against the background stars at 5" per year. The open circles in the figure here show how these stars move from 2010 to 2050. Note the 11th-magnitude background star that was between them in 2010 position; draw a careful picture of where you see these stars now, compared to it, and check their positions again next year.

Over the course of a year 61 Cygni also appears to shift position slightly back and forth; actually, what we're seeing is the motion of the Earth going around the Sun. From this parallax one can compute how far away the star is, quite precisely: 11.4 light years. 61 Cygni was the first star to have its parallax measured.

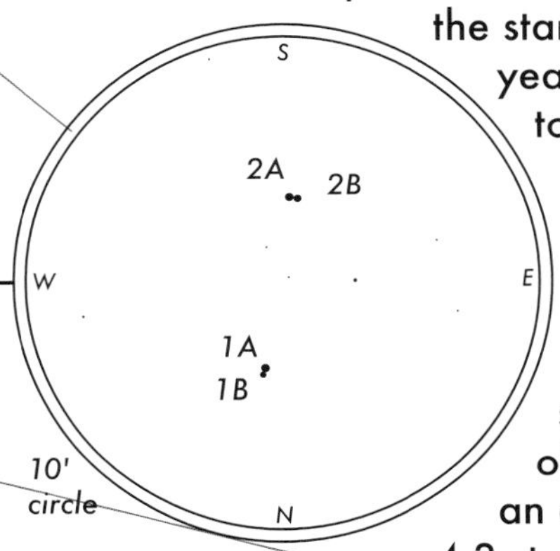

10' circle

The double-double (Epsilon Lyrae) is a test of just how sharp your telescope and your eyes are. Put Vega in the finderscope; you'll see a triangle of stars with Zeta Lyrae (itself an easy double – a magnitude 4.3 star with a magnitude 5.7 companion, 44" south-southeast) south of Vega, and Epsilon Lyrae, a close pair of stars, to its northeast. Epsilon is the double-double. The primary pair you can split in the finderscope, but each star itself is a tight double requiring your highest power.

The double-double is a complex multiple-star system, a collection of stars 160 light years away from us. The pair to the north, 1A and 1B, are separated by about 150 AU, and they take over 1,000 years to orbit each other. The pair to the south, 2A and 2B, are also about 150 AU apart, but they take about 600 years to orbit about their center of mass. This pair orbits more quickly than the other pair because the stars are more massive. The positions of both pairs have been seen to change significantly over the past century.

The two pairs are separated by nearly 0.2 light years, and so it must take half a million years for them to orbit about their common center of mass.

Star	Magnitude	Color	Location
Gamma Delphini			
A	4.4	Orange	Primary star
B	5.0	Lime	9.1" W from A
Struve 2725			
A	7.5	White	Primary star
B	8.4	White	6.0" N from A
61 Cygni			
A	5.2	Orange	Primary star
B	6.0	Orange	31" SE from A
The double-double (Epsilon Lyrae)			
1	4.7	White	208" N from 2
2	4.5	White	Primary star
Split:			
1A	5.2	White	Primary star
1B	6.1	White	2.3" N from 1A
2A	5.3	White	Primary star
2B	5.4	White	2.3" E from 2A
The double-double's double			
Struve 2470:			
A	7.0	White	Primary star
B	8.4	White	13.8" E of A
Struve 2474:			
A	6.8	Yellow	Primary star
B	8.1	Yellow	16.1" E of A

Can't split the double-double? Then try splitting the **double-double's double**! Go back to Vega, then move your finderscope southeast to Zeta Lyrae, the other corner of our triangle. One step farther gets you to Delta Lyrae, visible as a pair of stars in the finderscope. One step more brings you to Iota Lyrae. Find it in your telescope at low power, and move a degree and a half due south (three full Moon diameters) until you see a pair of seventh-magnitude stars, **Struve 2470** and, to its south, **Struve 2474**. Under high power, each star splits into a pair, but their separations are much easier than the double-double's.

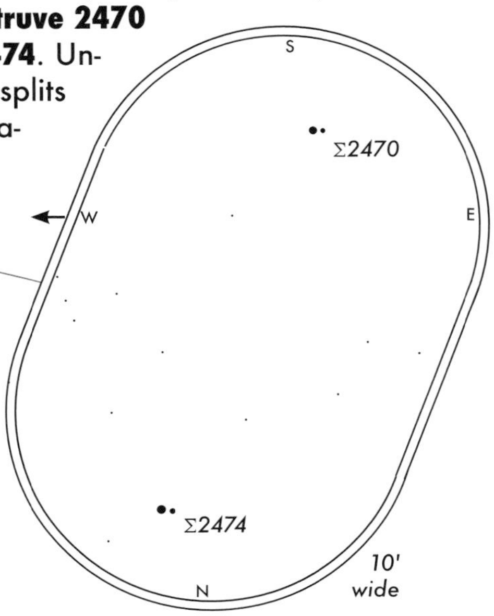

Unlike the double-double, the two pairs aren't actually associated with each other; Struve 2474 is only 155 light years away, while Struve 2470 is much farther and may itself be only an optical double, not a real pair.

10' wide

And don't forget these doubles:
Albireo: See page 132 for details.
Zeta and Theta Sagittae: See page 139 for details.

In Lyra: The *Ring Nebula,* a planetary nebula, M57

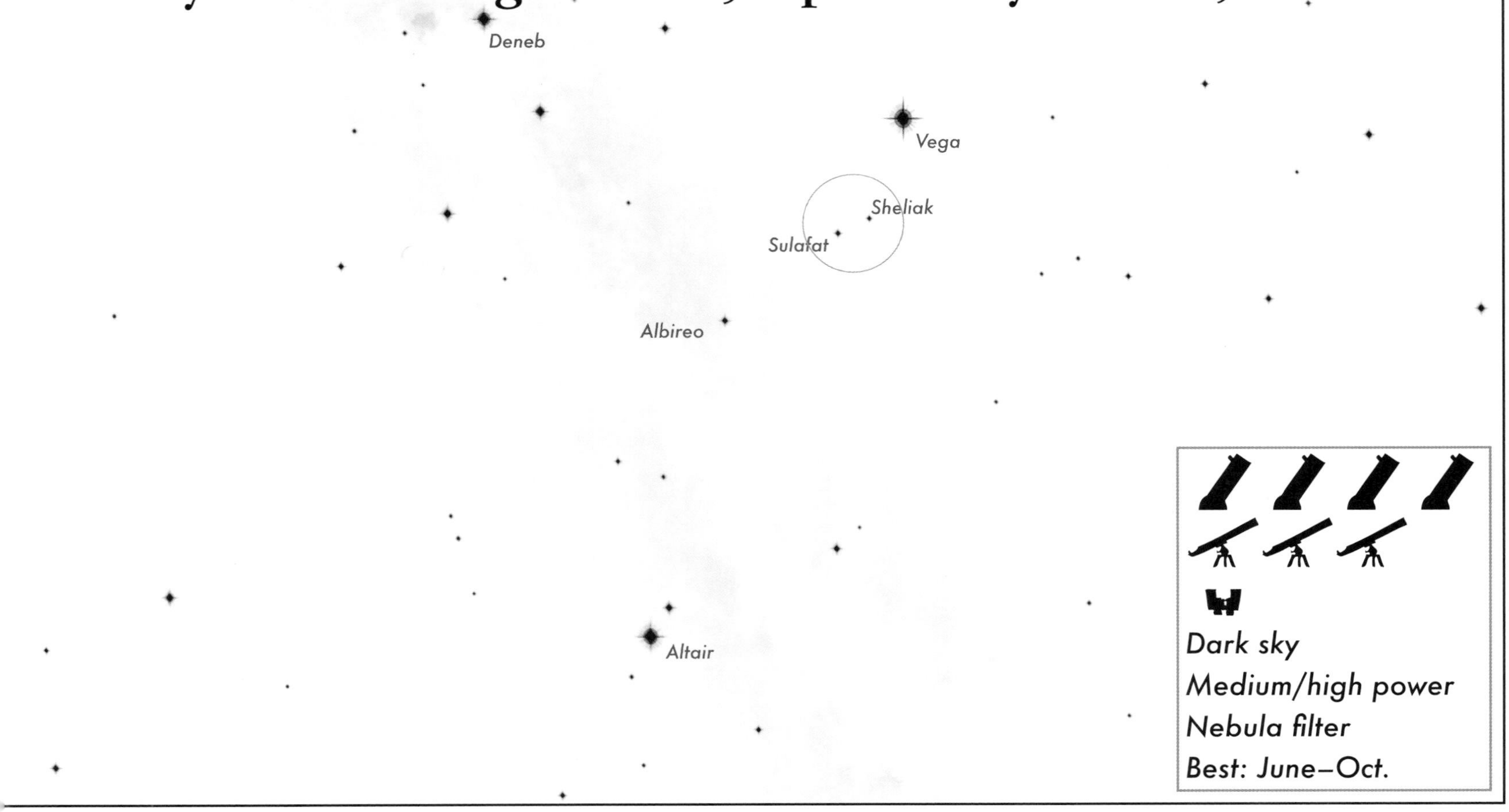

Deneb

Vega

Sheliak

Sulafat

Albireo

Altair

Dark sky
Medium/high power
Nebula filter
Best: June–Oct.

Star maps courtesy Starry Night Education by Simulation Curriculum

- Famous planetary nebula
- Easy to find
- See the ring with averted vision

Where to look: Find the Summer Triangle of Vega, Deneb, and Altair, high overhead. Point your telescope towards Vega, the brightest of the three, in the northwest corner of the triangle. To the south of Vega, about halfway between Vega and Albireo, you'll see two reasonably bright stars. They're named Sheliak and Sulafat. Point your telescope just between these stars.

In the finderscope: Sheliak and Sulafat are easily visible in the finderscope. Notice a third star, Burnham 648, lying on a line between these two, much closer to Sulafat. The nebula is halfway between Burnham 648 and the other star, Sheliak.

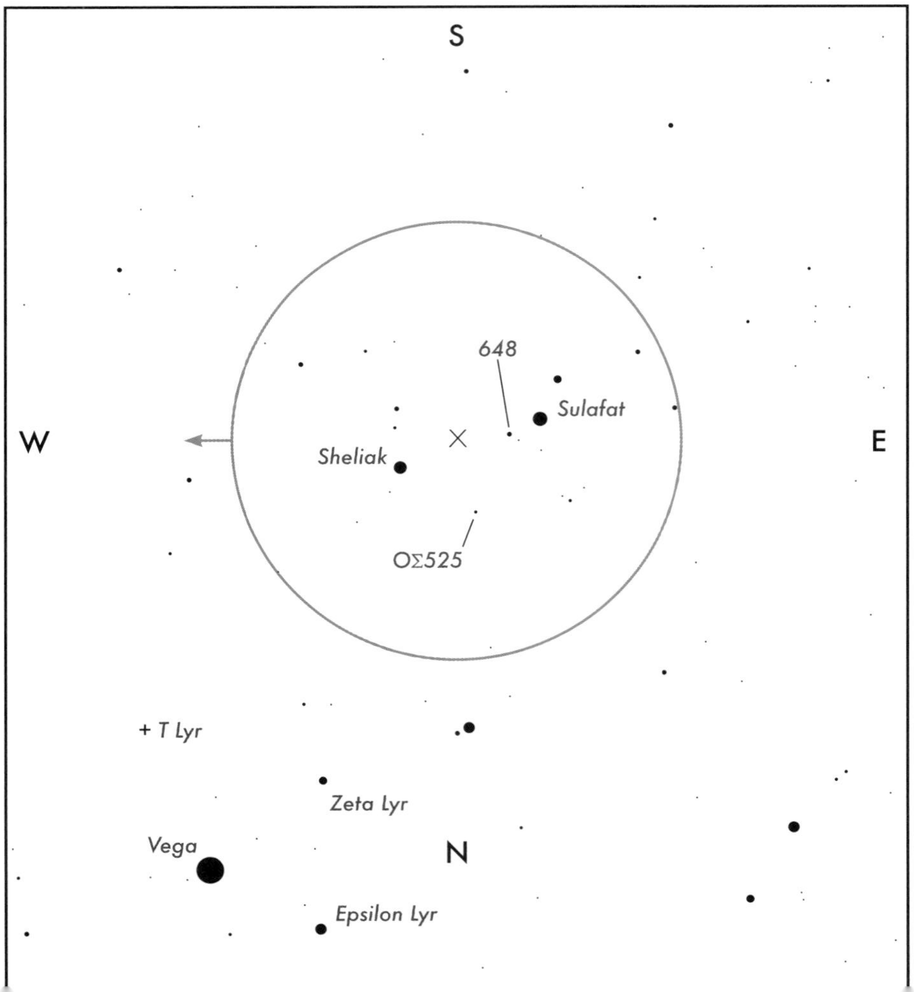

S

648

Sulafat

W

Sheliak

E

O∑525

X

+ T Lyr

Zeta Lyr

Vega

N

Epsilon Lyr

M57 and OΣ525 in a star diagonal at low power

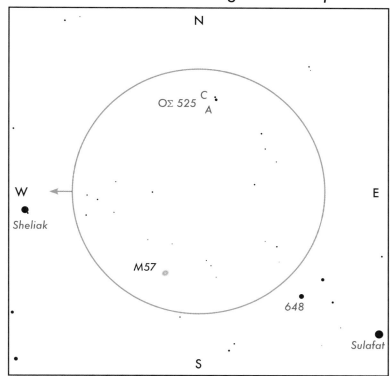

M57 in a Dobsonian at medium/high power

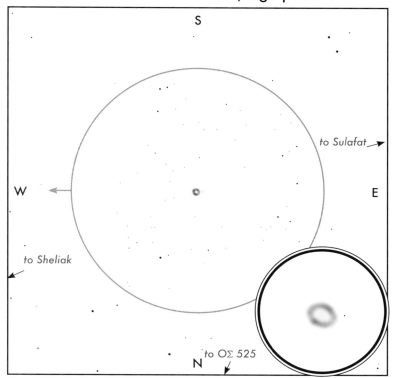

In a small telescope: Find in low power (shown here), observe in medium power. The ring looks like a very small disk, bright but hazy, in contrast to the bright pinpoint stars surrounding it. At higher magnification, it looks like a somewhat flattened disk of light, darker in the center. Note the easy and colorful double star Otto Struve 525, which fits with M57 in the same low-power field of view.

In a Dobsonian telescope: The ring shape can be seen beautifully in a Dobsonian. At high power, in a larger telescope, the ring looks slightly flattened; you may begin to make out irregular variations in the light from the ring itself. Look about one low-power field to the north for Otto Struve 525; a Dob can split it as a triple star.

The **Ring Nebula** (M57) is perhaps the most famous planetary nebula, though not as impressive in a small telescope as the Dumbbell (M27; see page 136). Like the other planetary nebulae, it is smaller but brighter than most of the diffuse nebulae or star clusters listed in this book. There should be no trouble seeing it in the telescope, but in your lowest power it may be hard to distinguish from a star at first. Because it is reasonably small but bright, it is worth looking at with higher magnification than you'd use for most other nebulae. Use averted vision, and eventually you may be able to see it as a tiny "smoke ring," even in a 2.4" or 3" telescope. In a Dob, the ring jumps out at you.

This nebula is a cloud of cold gas, mostly hydrogen and helium, and very tenuous, having less than a quadrillionth of the density of air on Earth. This gas is expanding away from a small hot central star, too faint to be seen in any telescope less than 12" in aperture, which provides the energy to make the gas cloud glow.

Estimates place its position anywhere from 1,000 to 5,000 light years away from us, and so the ring itself might be about a light year in diameter. Some observations suggest that this gas cloud is expanding at a rate of 12 miles per second; if it's been expanding at this rate since it was formed, then it would take roughly 20,000 years to grow to the size we see today.

For more on planetary nebulae, see page 99.

Also in the neighborhood: Sheliak is a famous variable star. It is a very close binary; indeed, its stars are so close that they are virtually touching – gas from one star can flow onto the other star. As they orbit each other, every 12.9 days, the stars pass in front of each other. Normally, with both stars visible, Sheliak is magnitude 3.4, almost as bright as **Sulafat** *(magnitude 3.3). When the dimmer star is partly covered, its brightness drops for a couple of days, reaching magnitude 3.7. When the brighter (but smaller) star is obscured it drops all the way to magnitude 4.3. Thus, if you notice that Sheliak is considerably dimmer than Sulafat, you know that you've caught it in eclipse. (Sheliak also has a more distant orbiter: a magnitude 7.8 star a full 47 arc seconds to the south-southeast.)*

South and a bit west of Vega is a faint (8th magnitude) but impressively red carbon star, **T Lyrae.** *T Lyrae lies about as far from Vega as Zeta Lyrae; the three make a right triangle, with Vega at the corner.*

Less than a degree north-northeast of M57 is the colorful triple star **Otto Struve 525.** *The AC pair is easy, a sixth-magnitude yellow primary with a magnitude 7.6 blue companion, 46" to the north. Star B is ninth magnitude and only 1.8" to the southeast, however; a real challenge in a Dob.*

find more at: www.cambridge.org/features/turnleft/seasonal_skies_july-september.htm

In Cygnus: *Albireo*, a double star, Beta Cygni; and in Lyra: M56, a globular cluster

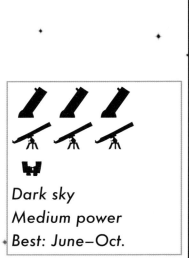

Star maps courtesy Starry Night Education by Simulation Curriculum

Dark sky
Medium power
Best: June–Oct.

- Albireo: colorful, easy double
- M56: Tricky to find, but in a beautiful setting
- Wander through the Milky Way around here!

Where to look: Find the Summer Triangle of Deneb, Vega, and Altair, stretching from high overhead off to the south. The easternmost of the stars is Deneb, which sits at the top of the Northern Cross (or, if you prefer, at the tail of Cygnus the Swan). The cross runs to the south and west, going right between the other two stars of the triangle, Vega and Altair. Albireo is the star at the foot of the cross. Start there.

In the finderscope: Albireo is easy to spot in the finder, and a lovely double star. **M56** is about half the distance from Albireo to Sulafat. Moving to the northwest of Albireo (in the direction of Sulafat and Sheliak, the two rather bright stars south of Vega near M57) you will first come across a star about two magnitudes dimmer than Albireo, called 2 Cygni. One step farther in that direction is an even dimmer star, HR 7302. Aim for that star, and look in your low telescope field to the southwest of that second star for M56.

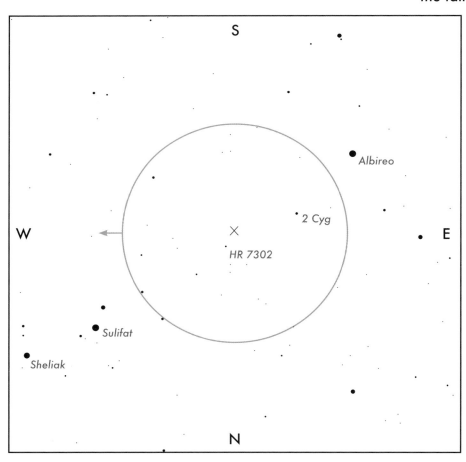

M56 in a star diagonal at low power

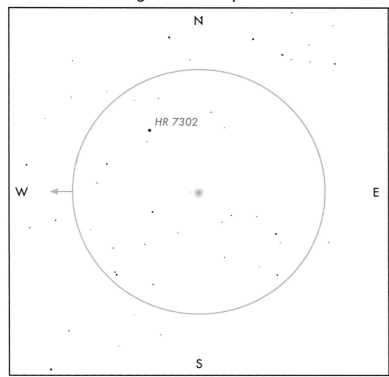

M56 in a Dobsonian at medium power

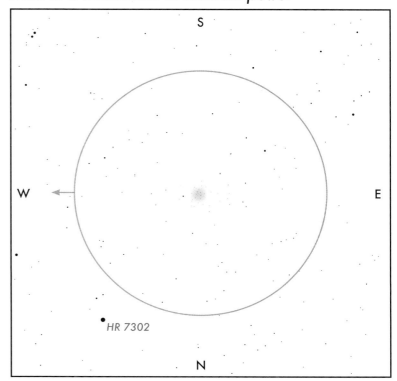

In a small telescope: Find with your low power (as shown), then observe with medium power. The cluster looks like a small, hazy disk of light in a rich field of Milky Way stars.

In a Dobsonian telescope: A Dobsonian resolves many individual stars in this small, grainy cluster, making it look rather loosely assembled compared to other globular clusters.

Albireo: Easy to find; easy to split; and if you have ever doubted that stars have colors, Albireo should remove any question. Albireo is made up of a giant orange (spectral-type-K) star, orbited by a hot blue (spectral-type-B) star. Star B lies at least 4,000 AU away from A, and so it must take roughly 100,000 years to complete an orbit about it – much too slow for us to actually observe any motion. (Each star also has a faint companion, much too close to split.) The two stars lie a bit less than 400 light years away.

M56: By itself, M56 is not a particularly striking globular cluster; but sitting in the rich field of Milky Way stars, it takes on a special charm and beauty even in a small telescope. This globular cluster consists of about 100,000 stars in a ball 10 light years in diameter. It has been estimated to be as much as 13 billion years old, three times the age of our Solar System, dating back to the beginnings of the Universe itself. It lies 40,000 light years away from us.

For more information on globular clusters, see page 115.

Also in the neighborhood: When you're finished observing Albireo and M56, you may wish to stroll through the Milky Way in this part of the sky. On a dark night, there is a breathtaking background of dim stars in this region. Wander up the cross towards Deneb with your lowest-power eyepiece.

*Work your way up towards Sadr, the star marking the crosspoint of the cross. Place Sadr near the northern edge of your finderscope field of view. In your low-power eyepiece, look here among the rich background of Milky Way stars for a small grouping of faint stars looking something like a tiny Pleiades. This is the open cluster **M29**. It contains about 20 stars, of which about half a dozen are visible in a small telescope. It lies perhaps 5,000 light years from us.*

*Sauntering farther up the Milky Way, past Deneb towards Cassiopeia, you may encounter the open cluster **M39**, a handful of stars visible against the background haze of the Milky Way.*

Albireo (Beta Cygni)			
Star	**Magnitude**	**Color**	**Location**
A	3.2	Orange	Primary star
B	5.4	Blue	34" NE from A

find more at: www.cambridge.org/features/turnleft/seasonal_skies_july-september.htm

In Cygnus: The *Blinking Planetary Nebula*, NGC 6826, and a double star, 16 Cygni

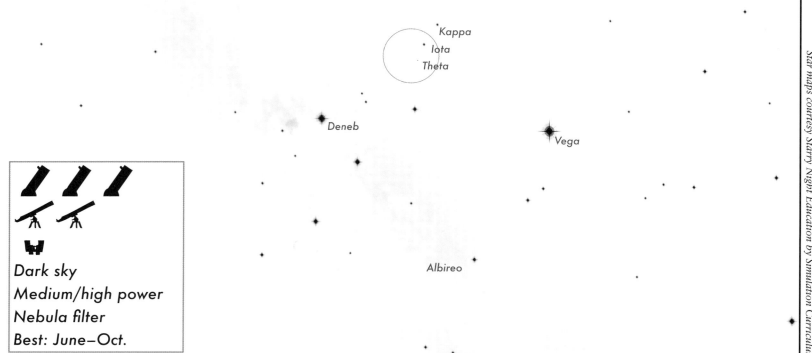

Star maps courtesy Starry Night Education by Simulation Curriculum

Dark sky
Medium/high power
Nebula filter
Best: June–Oct.

- Challenging to find
- Intriguing come-and-go nebula
- Double star: both stars are like our Sun

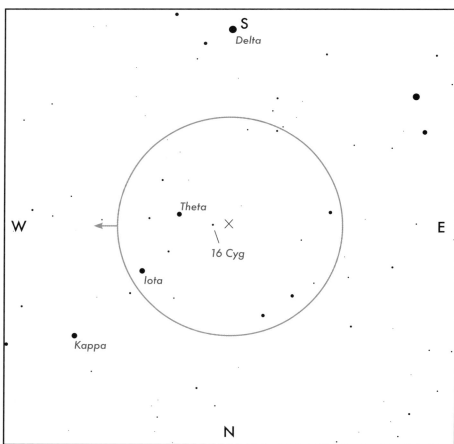

Where to look: Find the Summer Triangle of three brilliant stars stretching from high overhead off to the south. Deneb, the star to the east, sits at the top of the Northern Cross; it's also known as the tail of Cygnus the Swan.

Three bright stars in a line running southeast to northwest form the crosspiece of the Cross; consider them as the leading edge of Cygnus' wings. Look now for the trailing edge of the left wing, the one stretching out to the northwest. You can trace this edge: start at Deneb; then go past two stars northwest of Deneb; then west to a line of three stars, Theta, Iota, and Kappa Cygni, running to the northwest. These three make the tip of the swan's wing. Aim at Theta, the star to the southeast, closest to the body of the Swan.

In the finderscope: Three bits of light – Theta Cygni, the easily seen star Iota Cygni, and (if it's dark enough) a knot of light in the Milky Way – make a wedge shape in the finderscope. The star in the knot, at the southeast corner of this wedge, is 16 Cygni.

Alternately, start your finderscope at Kappa, and step from Kappa to Iota; 16 Cygni is one step farther in that direction, just east and a little north of Theta.

Once you have found 16 Cygni, the nebula NGC 6826 can be seen just to its east, in the same low-power telescope field.

NGC 6826 *in a star diagonal at low power*

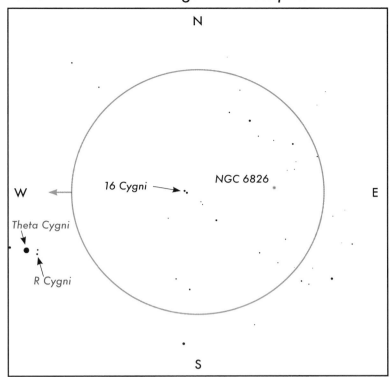

NGC 6826 *in a Dobsonian at medium/high power*

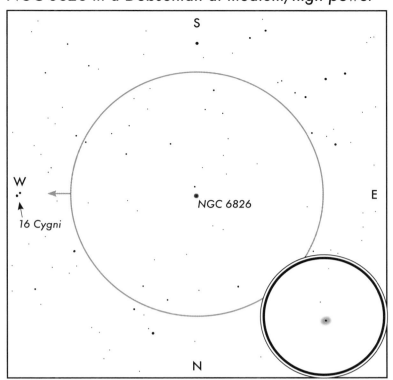

In a small telescope: You should see 16 Cygni, a fairly easy double star, in your telescope. Look less than half a degree due east from 16 Cygni (or just center on 16 Cygni and wait three minutes) for a rather faint "star" to center itself. But this "star" seems to become much dimmer whenever you stare straight at it. That's the nebula. On closer inspection (try medium power), with averted vision it looks like a tiny disk of faint light, smaller in size than the separation of the two stars in 16 Cygni; staring straight at it you see an 11th-magnitude star.

In a Dobsonian telescope: The double star 16 Cygni is easy, a widely separated pair of equal white stars. Move half a degree east and, with averted vision, look for a bright blue hazy circle; that's NGC 6826. As with a smaller telescope, when you look directly at it the haze will disappear and you'll only notice the central star. However, given the extra light-gathering power of a Dob, you can pass this object through a nebula filter and see both the hazy disk and the central star at the same time.

The **Blinking Planetary**, NGC 6826, puts out as much light as a ninth-magnitude star but this light is spread out over a small disk rather than concentrated at a point. Thus, the nebula shows clearly with averted vision, but as soon as you try to stare directly at it, it may disappear entirely. (When you stare straight at the nebula, the central part of your vision, which is more sensitive to detail than to dim light, picks up only the pinpoint center star, not the cloud surrounding it.) This "here again, gone again" effect gives this nebula its name.

This particular planetary nebula is located about 2,000 light years from us. The cloud of gas has expanded out to a diameter of about 15,000 AU, or about one quarter of a light year. Planetary nebulae (see page 99) are clouds of gas thrown off when a red-giant star destroys itself; here you can also see the white dwarf left behind after the explosion.

The **16 Cygni** stars are both G types, very similar to our own Sun, 70 light years distant. In fact, 16 Cygni B, has a spectrum virtually identical to our Sun's. It is often used as a *solar-analog* star; comparing the light from other bodies against this star is like comparing them against our Sun itself.

What's more, measurements of tiny shifts in its spectrum (as is done for spectroscopic binaries) reveals that 16 Cygni B itself has a planet, roughly twice as big as Jupiter, with an 802-day period. However, its very eccentric orbit carries it

from 0.7 AU (like Venus) to 2.7 AU (asteroid belt distance) from its star, sweeping out the temperate, habitable zone and thus making the presence of any Earth-like planets there very unlikely.

The primary star, 16 Cygni A, is itself a double star but its companion is too near and faint to be seen, even in a Dob.

Also in the neighborhood: Find Theta Cygni in your low-power telescope field. Immediately to the east you should see either one or two stars. The one that's there all the time is a run-of-the-mill ninth-magnitude star. Immediately to the southwest of it is **R Cygni**, *a long-period variable star. It changes in brightness from 7th magnitude to 14th magnitude, and back, over a fairly regular period of about 14 months. Thus, for a good part of the time it is too dim to be seen in anything less than a 10" telescope. It is a rare S-type star, red, cool and dim – just the opposite of the O- and B-type blue stars so prominent in open clusters.*

16 Cygni			
Star	**Magnitude**	**Color**	**Location**
A	6.0	Yellow	Primary star
B	6.0	Yellow	40" SE from A

find more at: www.cambridge.org/features/turnleft/seasonal_skies_july-september.htm

In Vulpecula: The *Dumbbell*, a planetary nebula, M27

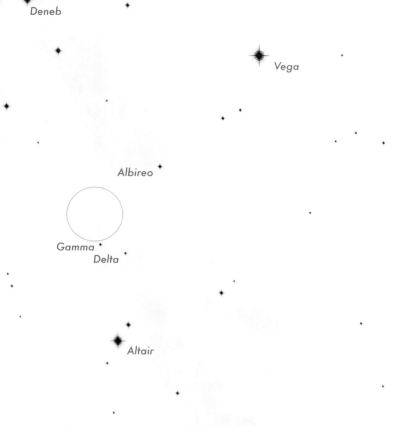

Star maps courtesy Starry Night Education by Simulation Curriculum

Dark sky
Medium power
Nebula filter
Best: June–Oct.

- Bright, big planetary nebula
- Spectacular when the night is dark
- Fascinating detail with aperture, filter

Where to look: Find Altair, the southernmost of the three stars that make up the Summer Triangle. Just to the north is a narrow group of four easily visible stars. This is the constellation of Sagitta, the Arrow. The leftmost star, Gamma Sagittae, is the point of the arrow, and the other three make a narrow triangle that are the arrow's feathers. Let the distance from the star in the middle of the arrow to the star at the point be one step; move your telescope that distance due north from the arrow's point. (That brings you into a different constellation, Vulpecula, the Little Fox.)

In the finderscope: Look for a sixth-magnitude star, 14 Vulpeculae; the nebula lies just south of that star. You may be able to see it as a tiny fuzzy spot in a rich field of faint Milky Way stars.

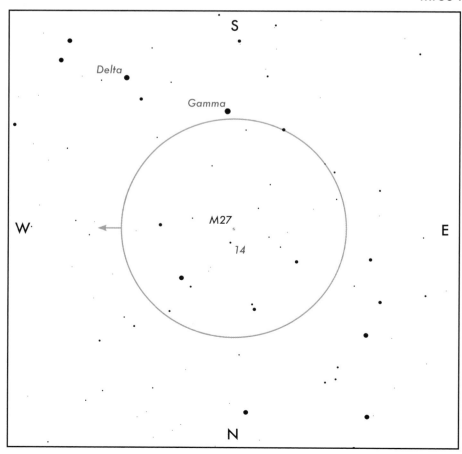

The Dumbbell in a star diagonal at low power

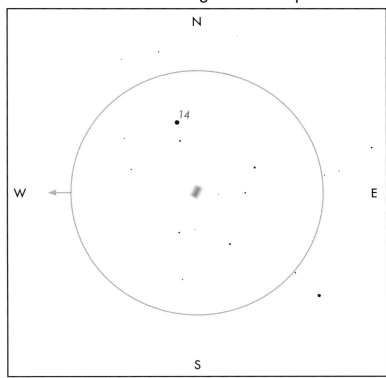

The Dumbbell in a Dobsonian at medium power

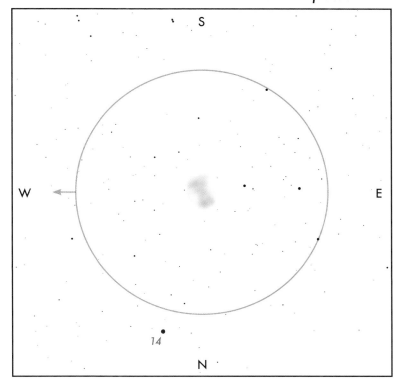

In a small telescope: In low power, aim for 14 Vulpeculae; the nebula should be visible just to its south. Center on it, then switch to medium power to observe it. The nebula looks like a bright but out-of-focus bow tie (or, more classically, a weight-lifter's dumbbell) in a field rich with stars.

In a Dobsonian telescope: Once you've found the nebula (put 14 Vulpeculae in a low-power field and move south), try looking with a higher magnification and a nebula filter. The dumbbell shape stands out very well, with a dimmer haze of nebular light filling in the area around it.

On a good dark night, the ethereal, extended glow of light from the **Dumbbell**, M27, seems to hang in space in among the surrounding stars. The contrast between the pinpoints of light from the stars and the diffuse glow from the nebula makes this one of the more striking, and prettier, sights in the sky. Let your eye relax, and take the time to look around the object as your vision adapts to the dim light. Take it slowly, and the nebula will reward you.

On a dark night, look for slight asymmetries in the shape and irregularities in brightness.

This planetary nebula is an irregular ball of thin, cold gas expanding out from a central star (much too faint to be seen in a small telescope) which provides the energy to make the gas glow. The gas is mostly hydrogen and helium, very cold and very thin.

The nebula lies 1,200 light years away from us, and extends more than 2 light years in diameter. It is expanding at about 30 kilometers per second, and so the cloud of light has been growing steadily at about one arc second per century. If it has been growing at this rate since it was formed, then it must have taken about 10,000 years to reach its present size.

For more on planetary nebulae, see page 99.

In Sagitta: A globular cluster, M71

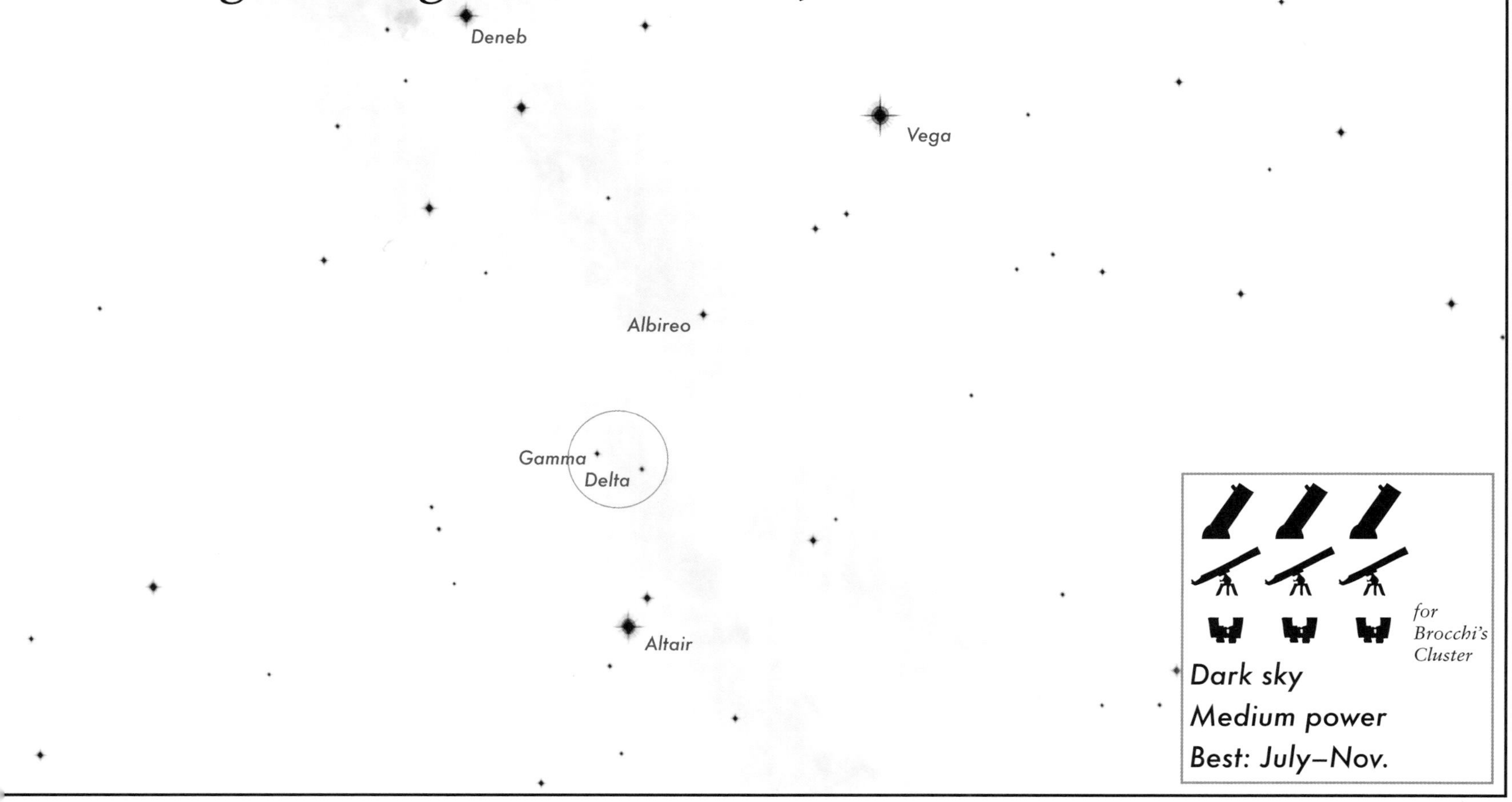

Deneb

Vega

Albireo

Gamma
Delta

Altair

Dark sky
Medium power
Best: July–Nov.

for Brocchi's Cluster

Star maps courtesy Starry Night Education by Simulation Curriculum

- Easy to find, tricker to see
- Unusually small and young globular cluster
- Nearby: Brocchi's Cluster, a fine binocular sight

Where to look: Find Altair, the southernmost of the three stars that make up the Summer Triangle. Just to the north find a narrow group of four moderately bright stars; this is the constellation called Sagitta, the Arrow. The leftmost star, Gamma Sagittae, is the point of the arrow. The other three make a narrow triangle that are the arrow's feathers. The star in the middle of the arrow, marking where the feathers meet the shaft of the arrow, is Delta Sagittae. Go to the point halfway between Gamma and Delta Sagittae.

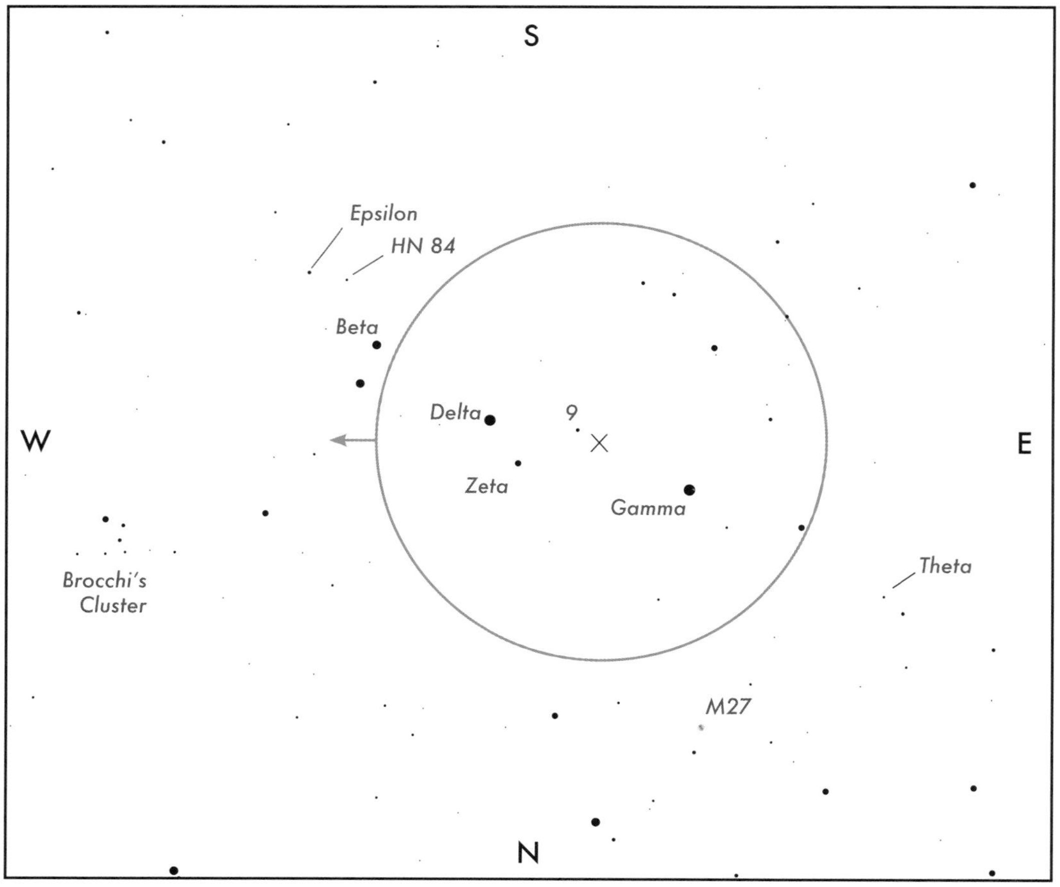

In the finderscope: The nebula probably will not be visible in the finderscope unless it's an exceptional night. Look instead for the four main stars of Sagitta, all of which should fit in your finderscope, and aim at a spot halfway down the shaft of the arrow between the middle star and the point. There's a seventh-magnitude star just to its west, 9 Sagittae.

S

Epsilon
HN 84

Beta

Delta   9
                    ×
W   Zeta   E
               Gamma

Brocchi's
Cluster   Theta

M27

N

M71 in a star diagonal at low power

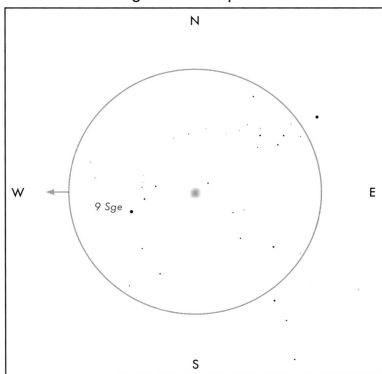

In a small telescope: Look for 9 Sagittae in low power; to the east, M71 looks much like a dim planetary nebula or a galaxy, hazy and irregular in shape. With medium power, a 3.5" telescope may just reveal some graininess to the light. The object is quite dim; you'll need a good dark night to see it well. Averted vision helps.

M71 in a Dobsonian at medium power

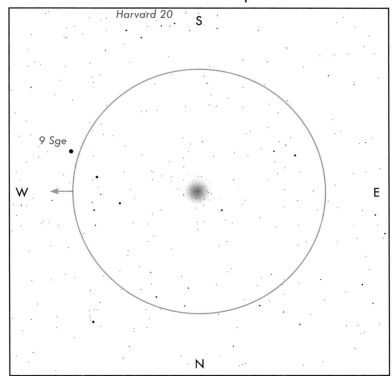

In a Dobsonian telescope: This is a rather odd cluster in a Dobsonian. At low power, it is quite grainy; at medium power you might be able to make out the constituent stars. It's probably not bright enough to look its best at high power, however, unless the night is very dark.

M71 is neither large nor bright, but graceful all the same. It looks different from other globular clusters; indeed, there was once debate as to whether it really was a globular cluster at all. The pattern of star colors matches what is seen in globular clusters (see page 115), but M71 looks much redder than a typical globular. We now understand why: the light from the cluster passes through an intervening cloud of gas and dust before we can see it, which scatters away blue light and turns the stars reddish, much like a setting Sun is turned red.

This cluster is estimated to be 30 light years in diameter, lying almost 20,000 light years away from us.

Also in the neighborhood: Due south of M71, in a Dob you can just make out a faint open cluster, Harvard 20, estimated to be 60 million years old, about 6000 light years away.

Just northeast of Delta Sagittae (the middle star in the arrow) is a fifth-magnitude star, Zeta Sagittae. It's a double star, with a ninth-magnitude companion 8.5" northwest of its primary. The primary is itself a binary star, but much too close to be split; there's also a fourth star in this system, far from the primary and too faint to be seen in a small telescope.

Another double star lies off the point of Sagitta. Step from Delta to Gamma Sagittae (from the middle of the arrow to

the point); one step farther brings you to Theta Sagittae. Theta consists of a sixth-magnitude star with a ninth-magnitude companion 12" to the northwest. This is a crowded star field; less than two arc minutes to the southwest is a seventh-magnitude star that looks like it might be part of this system, but in fact it is just a field star.

Step the other direction from Delta, to Beta; a step farther brings you to a pair of sixth-magnitude stars, Epsilon Sagittae and HN 84. Both have faint, distant companions. Epsilon B is magnitude 8.4, 87" to the east of its primary; two other, fainter stars some two arc minutes nearby may also be associated with that star. The companion of HN 84 is magnitude 9.5, 28" to the north-northwest of its primary.

Five degrees west and a bit north of M71 – roughly one finderscope field – is a pleasant collection of sixth- and seventh-magnitude stars called Brocchi's Cluster, or Collinder 399, or, more prosaically, the Coat Hanger. More than two degrees wide, it's too spread out for most telescopes but just right for binoculars or the finderscope. Though once thought to be an open cluster, data from the European Hipparcos satellite giving stellar distances and motions has revealed that, in fact, this is just a chance grouping of unrelated stars, lying from 200 to 1,000 light years away from us. Still, it's a pretty sight on a summer night.

find more at: www.cambridge.org/features/turnleft/seasonal_skies_july-september.htm

In Scutum: The *Wild Ducks*, an open cluster, M11

Tarazed
Altair
Alshain
Delta
Lambda
Antares

Star maps courtesy Starry Night Education by Simulation Curriculum

for V Aquilae

Dark sky
Low/medium power
Best: July–Oct.

- Elegant open cluster
- Rich location
- Awesome in a Dob!

Where to look: Find Altair, the southernmost of the stars in the Summer Triangle. It's a bright (first-magnitude) star. Altair is flanked by dimmer stars on either side, one north and a bit west (Tarazed), and a dimmer one south and a bit east (Alshain). These stars are the head of Aquila, the Eagle.

The body of the eagle is a line of stars running, from head to tail, northeast to southwest. In roughly equal steps, in a straight line, go from Tarazed (the flanking star north of Altair) down to Delta Aquilae; then to Lambda Aquilae, the tail star of Aquila. You'll see two fainter stars to the south and west of Lambda; all three should fit into your finderscope field.

In the finderscope: Put Lambda Aquilae and its neighbors in your finderscope, and imagine a clock face. The middle star is the hub of the clock; Lambda Aquilae, the brightest of the three, is at the 4 o'clock position; and the third star (called Eta Scuti) is at the 9 o'clock position. Step from the hub, to Eta Scuti, to one step further on. M11 is just a smidgen southwest of this point.

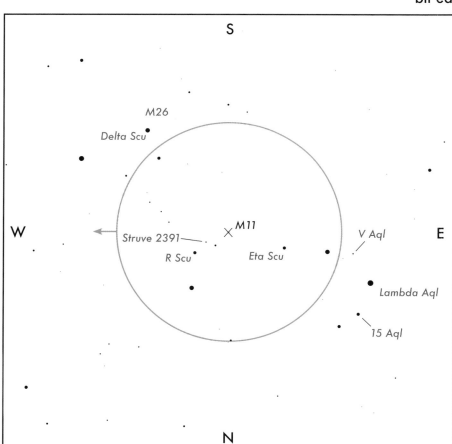

S

M26
Delta Scu
W
Struve 2391
R Scu
M11
Eta Scu
V Aql
E
Lambda Aql
15 Aql

N

M11 *in a star diagonal at low power*

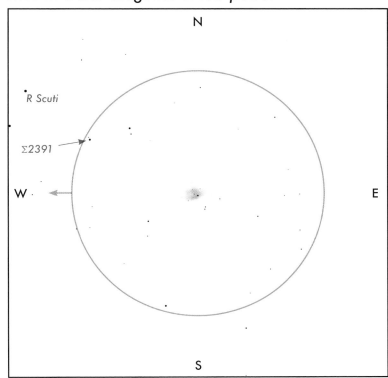

M11 *in a Dobsonian at medium power*

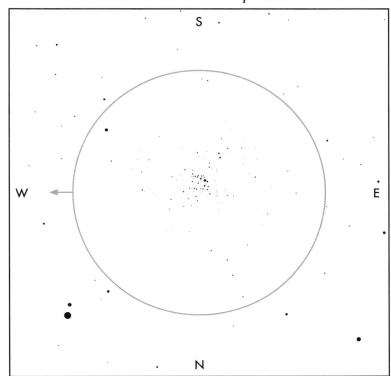

In a small telescope: Find in low power (shown here), observe in medium power. The cluster is a grainy wedge of light spreading out westward from a ninth-magnitude star. In moderate-sized telescopes (4" or larger), one can begin to see linear bare patches across each of the wings of the V.

In a Dobsonian telescope: Literally hundreds of stars are visible with a Dob in this cluster, while averted vision gives a hint of many more just beyond our ability to resolve them. With the added aperture, the shape no longer looks like a wedge but more like a spherical cluster with a gap in the western side of the sphere.

The **Wild Ducks**, M11, is a nearly circular group of more than 1,000 stars, but most of the brighter stars are in the wedge-shaped cluster of "flying ducks." In a small telescope, the bright star at the point of the wedge is quite visible and two or three others down the wedge may just barely be resolved within a background haze of light.

Small and not quite bright enough to be seen easily under higher power in a small telescope, it's still a pleasant sight. But in a Dobsonian on a good night, this cluster can be quite awesome, resolved into 100 or more individual stars against a background haze of grainy light. In these telescopes, it is one of the nicest of all the open clusters.

The cluster is a bit more than 50 light years in radius, lying some 6,000 light years away from us. The wedge of ducks is roughly 20 light years across. With so many stars in such a small area, the stars must on average be less than a light year apart from each other. Most of the brighter stars are hot young blue and white stars, spectral class A and F, but over a dozen stars have evolved into the red-giant stage. Because the O and B stars have had time to evolve into these giants, while the A stars have not yet reached that stage, one can estimate than the whole cluster is probably about 100 million years old. (For further information on open clusters, see page 71.)

*Also in the neighborhood: Half a degree to the northwest of M11 are two sixth-magnitude stars; the somewhat fainter one to the west (further from M11) is the double star **Struve 2391**. The primary star is magnitude 6.5, while its companion, much*

dimmer at magnitude 9.6, lies 38" to the north-northwest.

*Another half a degree northwest beyond these stars is the variable star **R Scuti**. Its brightness varies, irregularly, from magnitude 5.7 to as dim as magnitude 8.6; it can change through that range in a matter of a month or so.*

*A fifth-magnitude star just north of Lambda Aquilae is the double star **15 Aquilae**. The primary is yellow, magnitude 5.4, with a deep-blue seventh-magnitude companion 38 arc seconds southwest.*

*South of Lambda, about the same distance, is the deep-red variable (mags 6.6 – 8.4) **V Aquilae**, a pretty sight. It is a carbon star, similar to La Superba (page 113).*

*Southwest of M11, just outside the finderscope field of view, is the open cluster **M26**. Compared to other Messier objects, it's small and dim. But a Dobsonian can make out about two dozen stars, magnitude 10 and fainter, packed into a kite-shaped region only a couple of arc minutes across. Try medium power.*

And finally, note the fifth-magnitude star southwest of M11, Delta Scuti. It is 190 light years away and unimpressive now; but in 1.25 million years it'll only be 9 light years from Earth, and shine as bright as Sirius in our skies.

find more at: **www.cambridge.org/features/turnleft/seasonal_skies_july-september.htm**

In Sagittarius: The *Swan Nebula*, M17

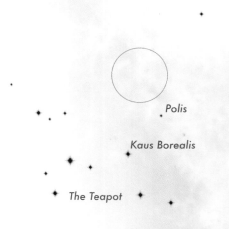

Tarazed
Altair
Alshain
Delta Aql
Lambda Aql
Polis
Kaus Borealis
The Teapot
Antares

for M24, MIlky Way

Dark sky
Low power
Nebula filter
Best: July–Oct.

Star maps courtesy Starry Night Education by Simulation Curriculum

- Beautiful swan shape of light
- Background nebulosity and cluster of stars
- In a region rich in open clusters

Where to look: Find the Teapot, low in the southern sky. From the top of the Teapot, Kaus Borealis, look north and west for a third-magnitude star, Polis. Aim your finderscope at Polis.

Alternately, if the Teapot is not up, find Altair, the southernmost of the stars in the Summer Triangle. Altair is flanked by dimmer stars on either side, Alshain and Tarazed, which stars make up the head of Aquila, the Eagle. In roughly equal steps, in a straight line, go from Tarazed (the flanking star north of Altair) down to Delta Aquilae; then to Lambda Aquilae, the tail star of Aquila. Two more steps should bring you to Polis.

In the finderscope: Start at Polis then move northward, through the M24 star cloud, until you see the fifth-magnitude star HR 6858 and the fourth-magnitude Gamma Scuti (with a faint neighbor and a distinctive triangle of stars to its east) in your finderscope. Aim at HR 6858; Gamma Scuti should sit at the northeast edge of the field of view.

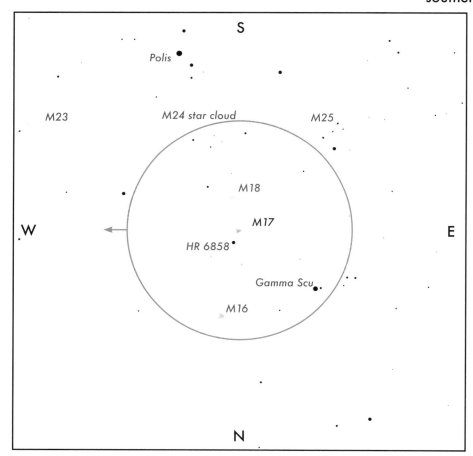

S
Polis
M23   M24 star cloud   M25
W   E
M18
M17
HR 6858
Gamma Scu
M16
N

M17 in a star diagonal at low power

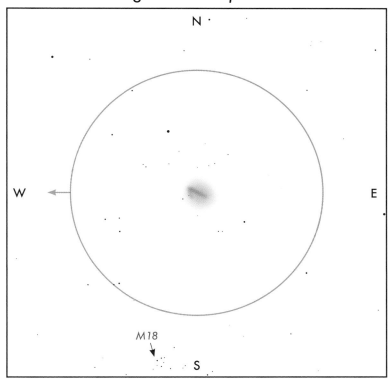

M17 in a Dobsonian at low power

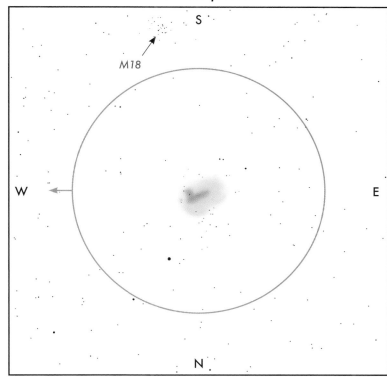

In a small telescope: M17 is a small bar of light with a tiny extension at one end, making it look like a mirror image check mark (√). At the opposite end, a faint lumpy haze of light envelops a loose open cluster of about a dozen stars, extending to the north and east.

With a bit of imagination, you can see it resemble a swan swimming towards you: the main bar of light is the swan's neck, and the short extension to the south from the westward end of the bar is the swan's head and beak, while the faint nebulosity surrounding the open cluster behind the neck is the swan's body.

In a Dobsonian telescope: M17 is an impressive, ghostly ball of light with an extension at one end, making it look like a check mark (√). At the long end of the check, a wisp of light extends up and around a loose open cluster of about a dozen stars. Dimmer outer parts of the nebula extend out to the north and east.

A nebula filter enhances the image of a swan swimming towards you.

The **Swan Nebula** will be difficult for observers in Canada or northern Europe to appreciate, but from the latitude of New York or Rome (and points south) it is a delight. A good, dark night can really bring out all the details of its nebulosity; the "check mark" bar of light is easily seen, while the swan's body extending up into the cluster can be more challenging.

M17 is a cloud of gas, mostly hydrogen and helium, and dust glowing from the energy of young stars embedded in the gas. This is a region where stars are formed; the cloud includes enough material to make many thousands of stars. The bright bar in the center of the nebula is about 10 light years long; the whole nebula extends across a region 40 light years wide. The nebula is located 5,000 light years away from us.

For more on diffuse nebulae, see page 55.

*Also in the neighborhood: The nebula and open cluster **M16** lies to the north of M17. Move your finderscope until Gamma Scuti sits just inside the east-southeast edge of the field of view (at about the 2 o'clock position, with south up). This should put M17 near the top (south) edge of the finderscope field; M16 should then be in the center of your telescope view. Known as the Eagle Nebula, the Hubble space telescope took a famous image of its pillars of dust where stars are being formed; but in a small telescope, M16 merely appears to be a loose open cluster. Binoculars show it as a dim, hazy patch of light. With a 3" telescope, you may be able to pick out about two dozen individual stars. In a Dobsonian telescope on a good night, you can begin to see the nebula among the stars. It's quite a young open cluster, estimated to be only about three million years old. Look for a conspicuous little double star near the edge of the cluster.*

*Just south-southwest of M17 is a small open cluster, **M18**. It's an inconspicuous clumping of about a dozen stars. Three other open clusters, **M23**, **M24**, and **M25**, are indicated in the finder field. (M24 is more properly a star cloud, a dense collection of otherwise unrelated stars, beautiful in binoculars.)*

This part of the Milky Way, near the center of our Galaxy, is rich with interesting clusters and nebulae, many featured in this book on their own pages. On a good dark night, some can even be visible to the naked eye as knots of light within the Milky Way. It can be fun to put yourself in Messier's shoes; just point your telescope (or binoculars!) in this general area and let yourself wander at will, discovering these wonderful objects on your own.

find more at: **www.cambridge.org/features/turnleft/seasonal_skies_july-september.htm**

In Sagittarius: Two open clusters, M23 and M25

Star maps courtesy Starry Night Education by Simulation Curriculum

Dark sky
Low/medium power
Best: July–Sept.

Where to look: Locate the Teapot, low in the south. Find the star that makes the knob at the top of the teapot, called Kaus Borealis. Aim above it and to the west at a fainter star, Polis.

In the finderscope: Both objects are found in the same way. First center on Polis, then move about one finder field diameter away.

M23: Put Polis at the east-southeast edge of your finderscope field of view, then look west-northwest. If Polis is at the 2 o'clock position of your finderscope, this puts the cluster at 8 o'clock. Center your telescope there.

M25: This cluster is just as far from Polis as M23, but to the east instead of the west. Put Polis at the west-southwest corner of the finderscope, the 10 o'clock position, and aim for the east-northeast corner, or 4 o'clock.

M23 looks prettiest in a small telescope at lowest power. Indeed, it is a large, hazy, conspicuous patch of light in 10×50 binoculars. It has over 100 stars in a cloud roughly 30 light years in diameter. They are located about 4,000 light years from Earth. This particular cluster is somewhat unusual in having so many stars of nearly the same brightness and appearance. Most of the stars are of spectral class B, with a few yellow giant stars but no obvious red or orange stars present.

M25 is about 2,000 light years away from us. It contains almost 100 stars, gathered in a loose ball about 20 light years across.

M23 in a star diagonal at low power

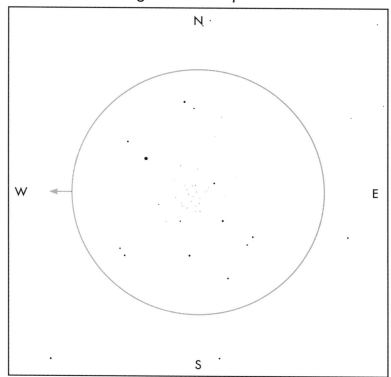

M23 in a Dobsonian at medium power

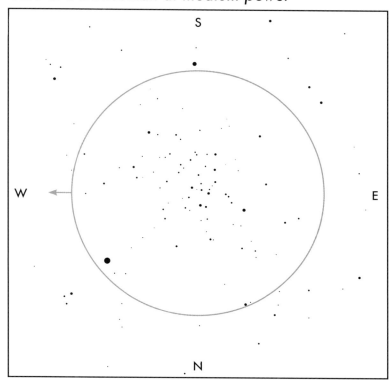

In a small telescope: M23 is a large, loose collection of stars; about 30 can be seen against a very grainy haze of light from other dimmer stars, too faint to be resolved in a small telescope. **M25** is a smaller, more compact cluster. Look for five or six conspicuous stars, with about two dozen much fainter stars visible around them.

In a Dobsonian telescope: M23 shows a nice wide sprin-gling of strings of stars, perhaps 50 or so visible with a Dob. **M25** is a very loose, widely spaced grouping of about 60 stars, half a dozen of them quite bright.

M25 in a star diagonal at low power

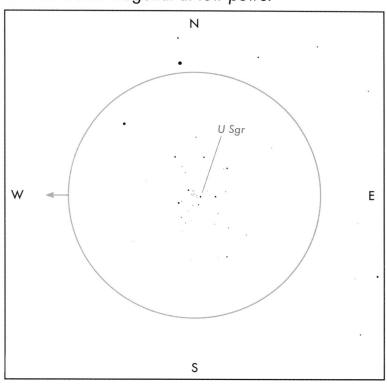

M25 in a Dobsonian at medium power

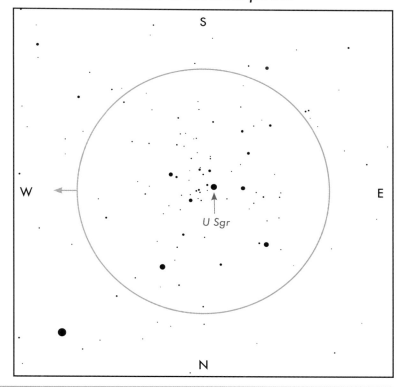

*Also **in the neighborhood:** M24, a loose cloud of stars visible to the naked eye as a patch of Milky Way north and a bit east of Polis, is just less than half of the way from M25 to M23. In binoculars or a low-power, wide-angle telescope it's lovely; but for most small scopes it's just too big to be seen to best*

*advantage. Just north of M24 is **M18**, a small open cluster, and M17, the Swan Nebula. They're described on page 142.*

*The middle of three conspicuous stars in the center of M25, **U Sagittarii**, is a Cepheid variable (see page 197), taking about a week to vary between magnitudes 6.3 and 7.1.*

find more at: www.cambridge.org/features/turnleft/seasonal_skies_july-september.htm

In Sagittarius: The *Lagoon Nebula*, M8, and an open cluster, NGC 6530

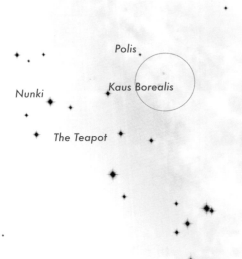

Polis

Nunki

Kaus Borealis

The Teapot

Antares

Star maps courtesy Starry Night Education by Simulation Curriculum

Dark sky
Any power
Nebula filter
Best: July–Sept.

- Nice open cluster for any sky
- Spectacular nebula/cluster mix in dark skies
- Beautiful in all apertures

Where to look: Locate the Teapot, low in the south. Find the handle of the Teapot, and step from the star at the top of the handle, Nunki, to the star at the top of the Teapot itself, Kaus Borealis. One step farther in this direction is where the nebula and cluster are located. On a dark, moonless night you can see a clump of Milky Way sitting near this spot even without a telescope.

In the finderscope: Quite a few stars should be visible in the finderscope, since you're looking towards the center of the Milky Way, but none of them will be particularly bright. The open cluster, NGC 6530, should be visible in the finderscope as a faint patch of light. Don't confuse this patch of light, however, for M20 or M21 just to its north.

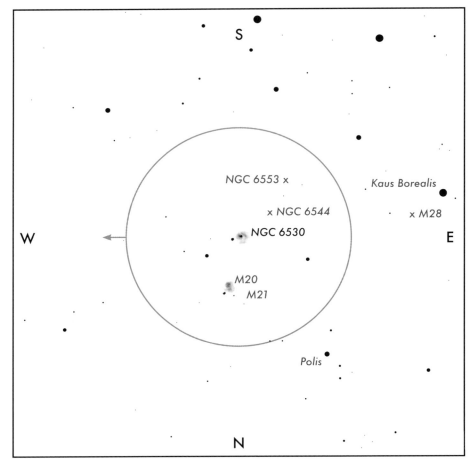

S

NGC 6553 x

x NGC 6544

Kaus Borealis

x M28

NGC 6530

W

E

M20

M21

Polis

N

M8 in a star diagonal at low power

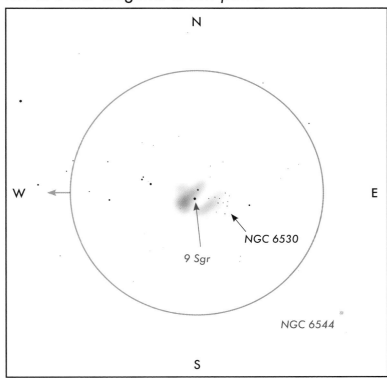

M8 in a Dobsonian at high power

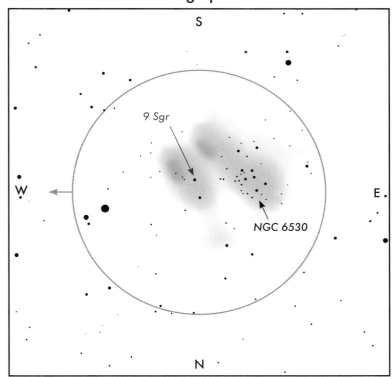

In a small telescope: Even on a poor night, you'll see an open cluster of a dozen or more stars (NGC 6530) near a conspicuous (seventh-magnitude) star. On a better night, you'll see a small patch of light (M8) just on the opposite side of that star. If the night is clear and dark, this patch becomes a bright irregular cloud of light; a dim tendril of the nebula extends around and into the open cluster, like a coral island encircling a lagoon. The nebulosity extends towards a fairly bright star (9 Sagittarii) on the opposite side of the cluster from the nebula.

In a Dobsonian telescope: More than two dozen stars can be seen in the open cluster. Among them, visible on a dark night, is the subtle glow of M8; it's more pronounced if the sky is particularly dark. Look for a dark lane in the nebulosity running northeast–southwest between the sixth-magnitude star 9 Sagittarii and the open cluster.

Even on a bad night, the open cluster will be visible, although most of the nebula may be lost in the sky brightness. This is especially true if you live in the north, since this object is always low in the southern sky. It will be difficult for most observers in Canada or Europe to appreciate it fully. However, if you are far enough south on a good dark night this object can be spectacular. Be sure to let your eyes dark-adapt, and try using averted vision to see the extension of the nebula around the open cluster.

The **Lagoon Nebula:** M8 is a huge cloud of ionized hydrogen gas roughly 50 light years in diameter and located about 5,000 light years away from us. The two stars we see in the nebula provide the energy to ionize the gas and make it glow. Like the Orion Nebula (see page 52) this is a region where young stars are forming; it is estimated that there is enough gas in this cloud to make at least 1,000 Suns.

NGC 6530: The open cluster is located near the nebula, also about 5,000 light years away from us. It's a fairly round group of about two dozen stars, some 15 light years in diameter. Most of the stars appear to be very young, still going through the last stages of birth. Several of the stars are of a type called *T-Tauri* stars, stars that have just started shining. T-Tauri stars emit a powerful *stellar wind* of hot plasma that pushes away the last bits of gas left over from the nebula from which the stars were formed. This particular cluster is thought to be one of the youngest open clusters known, only a few million years old.

We talk more about open clusters on page 71 and about diffuse nebulae on page 55.

Also in the neighborhood: There are two pleasant, if small, globular clusters nearby, worth a look.

NGC 6544 is within 1° to the southeast of M8 and NGC 6530. Notice its location in the bottom right corner of the low power star diagonal chart. A bit under 10,000 light years from us, it is currently crossing the galactic plane (as all globular clusters do, twice every orbit about the center of the Galaxy). It's small, only 2 arc minutes across, with most of the light inside the central one arc minute core; so its light is quite concentrated, making it as easy to see as an 8th magnitude star.

NGC 6553 is one more step away, and just a tad south. It is twice as far away from us (about 20,000 light years) and slightly fainter, at magnitude 8.1.

Don't expect to resolve either of these globulars into their component stars with a small telescope.

find more at: www.cambridge.org/features/turnleft/seasonal_skies_july-september.htm

In Sagittarius: The *Trifid Nebula*, M20, and an open cluster, M21

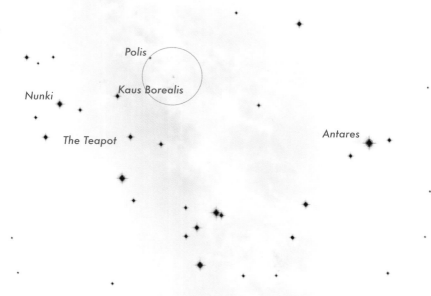

Star maps courtesy Starry Night Education by Simulation Curriculum

Dark sky
Medium power
Nebula filter
Best: July–Sept.

- Spectacular nebula/cluster pair on dark nights
- Open cluster visible in any sky
- Nice neighbor to M8

Where to look: Locate the Teapot, low in the south. Find the handle of the Teapot, and step from the star at the top of the handle, Nunki, to the star at the top of the Teapot itself, Kaus Borealis. One step farther in this direction brings you very close to where the nebula and cluster are located. On a dark, moonless night you can see a clump of Milky Way sitting near this spot even without a telescope.

In the finderscope: The nebula, M20, and open cluster, M21, should be visible in the finderscope as a faint patch of light. Don't confuse this patch of light, however, for M8 and NGC 6530 just to its south. If the night is not dark enough to see NGC 6530 in the finderscope, you may have trouble seeing M20 in the finder; instead, find Polis and move southwest towards the fifth magnitude star 4 Sagittarii.

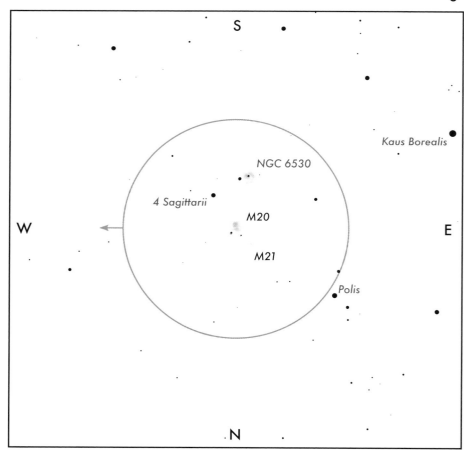

The Trifid Nebula, star diagonal, medium power

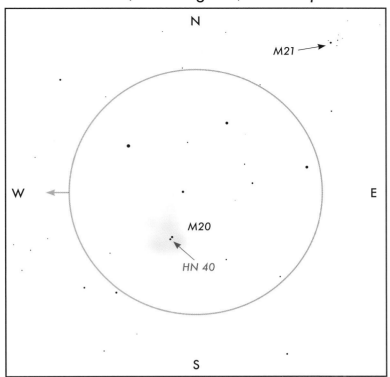

The Trifid Nebula in a Dobsonian at medium power

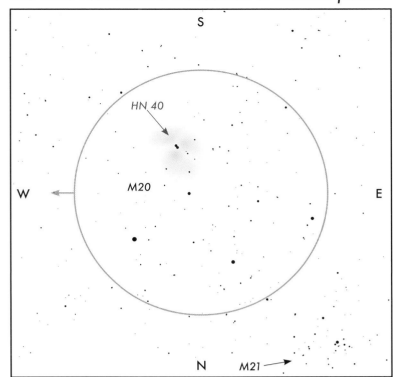

In a small telescope: It may be easier first to find M8 in the telescope itself (see page 146), and then move slowly north. After you move a full field of view (at low power) north, M20 should begin to come into view. M20 appears to envelop a faint double star, HN 40. The nebula will look like an irregular patch of light. Move northeast to find the open cluster M21.

In a Dobsonian telescope: The background field of stars of the Milky Way is so rich that the 40 or more stars of the open cluster itself do not necessarily stand out all that easily. But if the sky is dark, the nebula is unmistakable.

The extra light-gathering power of a Dob means you can use medium power and a nebula filter to enhance the contrast of M20 against the background sky. Look for the dark lanes splitting the nebula into three pieces; averted vision helps. Finally, note how M20 appears to envelop a faint double star, HN 40.

Move northeast to find the open cluster M21, a clump of stars in an already rich part of the Milky Way.

On good dark nights, the **Trifid Nebula** (M20) can be a glorious object. In a three-inch telescope, you might just begin to see the dark lanes that separate it into three patches (hence its name; "trifid" means "split into three"). In a Dobsonian you can see them clearly. The dark lanes go from the center out towards the west, northeast, and southeast. In addition, a dimmer patch of light extends some distance to the north of the main part of the nebula.

On poorer nights, if you have to contend with street lights or hazy air, you may need to use averted vision to see anything at all except the tiny double star.

Because the nebula is located so far to the south, it will not rise very far above the horizon when viewed from north of latitude 40°. It will be difficult for most observers in Canada or Europe to appreciate this object fully.

The open cluster, **M21**, is a loose, irregular group of about 50 stars, of which a dozen or so are visible in a small telescope. The cluster fills a region of space about 10 light years wide. It's thought to be a young cluster, about 10 or 15 million years old.

The nebula and the open cluster are located about 2,500 light years away from us.

The nebula, M20, is a cloud of ionized hydrogen gas, about 25 light years across, in which young stars are being formed.

It's estimated that there's enough gas in this cloud to make several hundred Suns. Like the Orion Nebula (see page 52), the gas glows because it is being irradiated by the light of the stars embedded in the gas. The dark lanes, visible on good nights, are caused by dusty clouds obscuring the light.

The main source of energy causing the Trifid Nebula to glow is **HN 40**, visible as a double in M20 near the center of the nebula. It is actually a multiple star. The brightest member is magnitude 7.6; to the southwest, 11 arc seconds away, is a magnitude 8.7 companion. Closer but fainter, there is a magnitude 10.4 star at 6.1 arc seconds to the north-northeast, while a more distant and fainter companion sits at 18 arc seconds in the opposite direction. You'll probably need a Dob to pick out the latter two stars. It has even fainter companions, at 13th and 14th magnitude, too dim even for a Dob.

We talk more about open clusters on page 71 and diffuse nebulae on page 55.

find more at: www.cambridge.org/features/turnleft/seasonal_skies_july-september.htm

Double stars around Scorpius

Omicron Ophiuchi: Look east of Antares and north of the Scorpion's sting for the third-magnitude star Theta Ophiuchi. In a finderscope you'll see it's the bottom of a V of stars. The star to the northwest is our fifth-magnitude double. The color contrast (blue–orange) and slight difference in brightness and separation make this a lovely pair.

36 Ophiuchi: The other star in the V, northeast of Theta Ophiuchi, is 44 Ophiuchi. Step from 44 to Theta; two steps fur-ther brings you to 36 Ophiuchi, a close pair of matched orange stars. There's also a distant companion, 12 arc minutes east-northeast of A.

24 Ophiuchi: Step northwest from 36 Ophiuchi to a pair of dim (magnitude 6.7) stars in an east–west line, 28 and 31 Ophiuchi. (Just to their southwest is the globular cluster M19; see page 154.) Further northwest is another slightly brighter pair (magnitudes 5.7 and 5.9) orient-ed northeast–southwest: 26 Ophiuchi and HR 6308. To the northwest of them is 24 Ophiuchi, a good Dob challenge.

for 36 Oph; Xi Sco/Σ1999

Steady sky
High power
Best: July–August

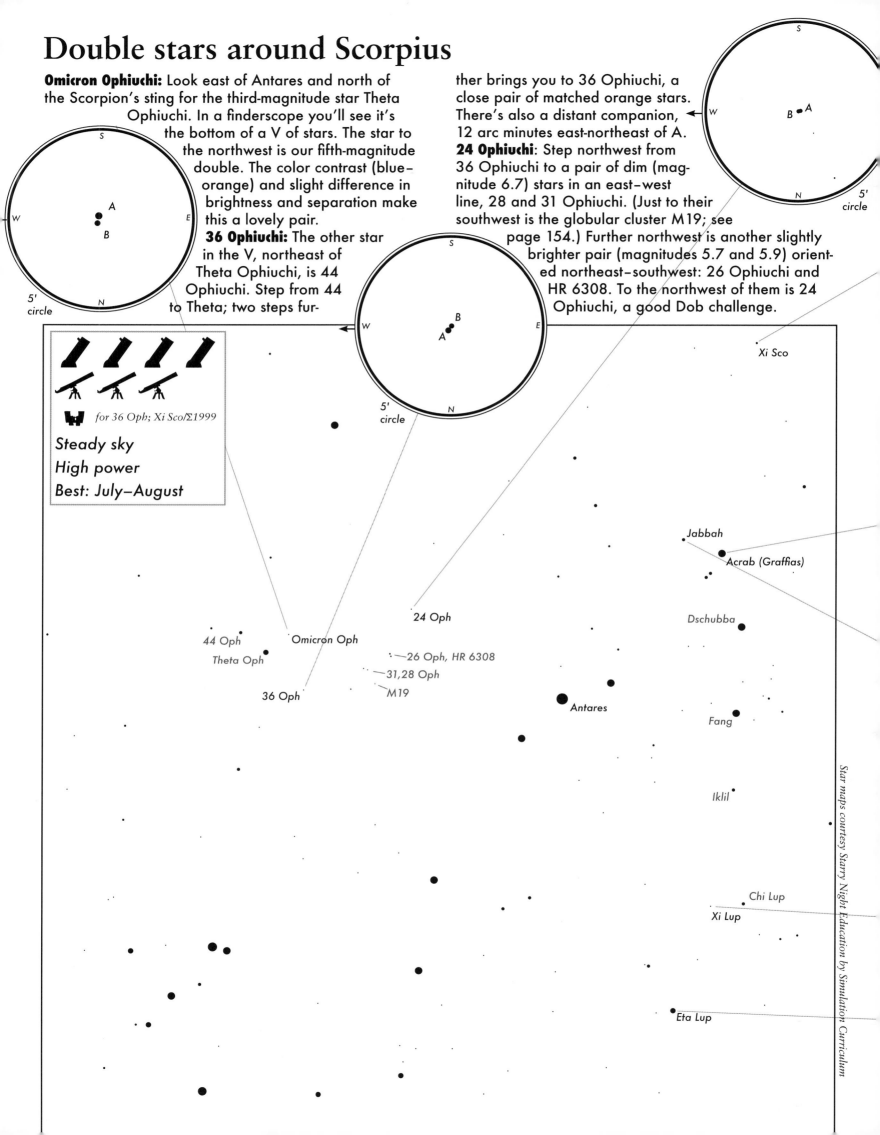

Xi Sco

Jabbah
Acrab (Graffias)

Dschubba

24 Oph

44 Oph
Theta Oph
Omicron Oph

26 Oph, HR 6308
31,28 Oph
M19

36 Oph

Antares

Fang

Iklil

Chi Lup
Xi Lup

Eta Lup

Star maps courtesy Starry Night Education by Simulation Curriculum

Find the claw of the Scorpion, three bright stars east of Antares: Acrab, Dschubba, and Fang. Step up from Fang to Dschubba and you'll be just west of Acrab. Two more steps in this direction brings you to a fourth-magnitude star, **Xi Scorpii.** It's a double-double, though each pair has its own name. Xi A and C are tough for a 3" because of the brightness difference; keep trying. In 1998 Xi B was too close to Xi A to split even in a Dob but by 2010 it was a full arc second north of A. The other half of this system, **Struve 1999**, lies four arc minutes south of Xi. That's a 10th of a light year; its period around Xi is about half a million years. Struve 1999 itself is just splittable in a 3".

Acrab (formerly known as Graffias) is actually a quadruple star system (at least) a bit over 500 light years away from us. The primary star and its closest companion are a pair closer together than Mercury and the Sun (too close to split), and they orbit each other in a week's time. Another star, B, orbits about 100 AU from this first pair, taking several hundred years to complete one orbit; it, too, is too close to split. Stars A and C are separated by a distance of more than 2,000 AU, and it takes more than 20,000 years for C to complete one orbit. C might itself be a double star, making this whole system a quintuple.

Jabbah is a fun double-double. A small telescope easily splits it into a widely separated north–south pair, A and C; C can be further split even with a 3" (on a good night) while a Dob can also split A. This system is about 450 light years from us; thus A and C are a 10th of a light year apart, while the C–D distance is 10 times the size of Pluto's orbit. Follow the stars of the claw south; past Fang, the next fourth-magnitude star, easy to see in the finder, is Iklil; then further south is Chi Lupi. Just to its east is a fifth-magnitude star, **Xi Lupi**. It's an evenly matched pair of blue stars, some 200 light years from us. One step further south (low in the haze for most northern observers) is **Eta Lupi,** whose strong brightness contrast is enhanced by its easy spacing. This star is estimated to be nearly 500 light years away.

Star	Magnitude	Color	Location
Omicron Ophiuchi			
A	5.2	Orange	Primary star
B	6.6	Blue	10" N from A
36 Ophiuchi			
A	5.1	Orange	Primary star
B	5.1	Orange	4.9" SE from A
C	6.5	Orange	730" ENE from A
24 Ophiuchi			
A	6.3	Blue	Primary star
B	6.3	Blue	1.0" WNW from A
Xi Scorpii/Struve 1999			
Xi Scorpii:			
A	5.2	Yellow	Primary star
B	4.9	Yellow	0.9" N from A
C	7.3	Orange	7.9" NE from A
Struve 1999:			
A	7.5	Yellow	4.7' S from Xi
B	8.1	Yellow	11" E from A
Acrab (Graffias) (Beta Scorpii)			
A	2.6	Yellow	Primary star
C	4.5	Blue	12" NNE from A
Jabbah (Nu Scorpii)			
A	4.4	Blue	Primary star
B	5.3	Blue	1.3" N from A
C	6.6	Blue	41" NNW of A
D	7.2	Blue	2.3" ENE from C
Antares (Alpha Scorpii)			
A	1.0	Red	Primary star
B	5.4	Green	2.5" W from A
Xi Lupi			
A	5.1	Blue	Primary star
B	5.6	Blue	10" NE from A
Eta Lupi			
A	3.4	Blue	Primary star
B	7.5	Blue	14" NNE from A
C	9.3	Blue	116" WSW from A

Antares is itself a double star, but the greenish secondary is close to, and much fainter than, the red primary. It's a challenge even for a Dob … but look for two "events" when it is occulted by the Moon.

Acrab, Jabbah, Antares, Sigma Scorpii, and Mimosa (Beta Crucis, see page 236) all move through the Galaxy at a similar rate. They may all have shared a common formation, coming from the same open cluster a few hundred million years ago.

In Scorpius: Two globular clusters, M4 and M80

Acrab (Graffias)

Dschubba

Antares   Alniyat

Star maps courtesy Starry Night Education by Simulation Curriculum

for M4

Dark sky
Medium power
Best: July–Aug.

- Nice contrast in globular-cluster views
- M4 is large; stars can be resolved
- M80 is small but bright

Where to look: Find Scorpius, low in the south. The three stars to the west of the very bright red star, Antares, are the claw of the scorpion. The top star is Acrab (a.k.a. Graffias); the one below it is Dschubba. M80 is halfway between Antares and Acrab. A quarter of the way from Antares to Dschubba is a fainter star, Alniyat.

In the finderscope, M4: Both Antares and Alniyat should be visible in the same finder field, making M4 an easy object to find. If Antares is the hub of a clock, Alniyat will sit at 8 o'clock; M4 is half as far away from Antares, at 9 o'clock. It will probably be visible as a smudge of light in the finderscope.
M80 lies a bit more than halfway along a line from Antares to Acrab.

S

M4

Antares

Alniyat

W   E

× M80

Dschubba

Acrab (Graffias)

Nu

N

M4: In a three-inch telescope under dark skies, you should just be able to make out individual stars towards the edge of this cluster. It is larger, brighter, and much easier to resolve into stars than M80, because it is much closer to us than M80, and the stars in M4 are much more loosely spaced.

M4 is one of the largest of the globular clusters near us. It is about 7,000 light years away, rather close by globular-cluster standards, and it's nearly 100 light years in diameter. Thousands of individual stars have been counted in photographs made by large telescopes, and fainter stars in this cluster must number in the hundreds of thousands.

M4 in a star diagonal at low power

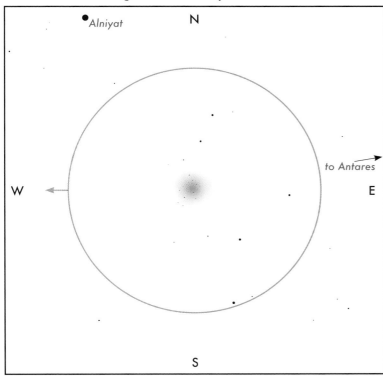

M4 in a Dobsonian at medium power

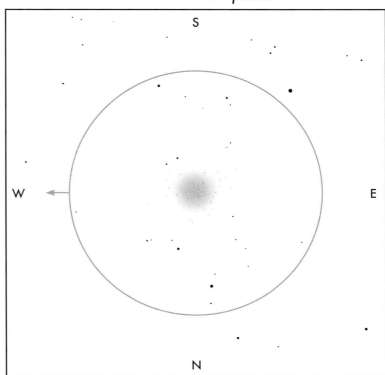

In a small telescope: Find in low, observe in medium power. **M4** is a smooth bright circle of light, milky in the center and grainy along the edges, but with a little brightening towards the center. **M80** looks almost like a star at low power; at medium power you see a hazy patch of light, concentrated at its center and fading out towards the edges.

In a Dobsonian telescope: M4 is not particularly concentrated, but well resolved, with several strings of stars visible, adding to the appeal of its appearance. In **M80** notice two levels to the light, with a compact grainy core surrounded by a hazy halo of light.

M80 in a star diagonal at low power

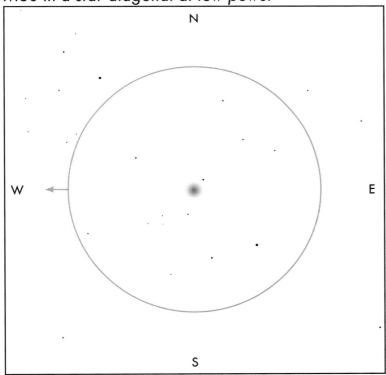

M80 in a Dobsonian at medium power

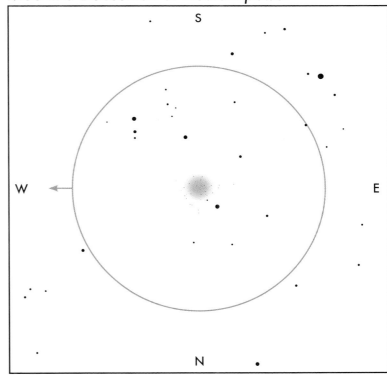

M80: Its main attraction is that this cluster, though small, is quite bright. In a small telescope, it looks like a small, bright featureless ball, almost like a bright planetary nebula. But in a Dob under excellent conditions and high power, one can begin to resolve a sort of mottled lumpiness in its central core, though even then there is no real resolution into individual stars.

M80 consists of hundreds of thousands of stars in a ball some 50 light years across, 30,000 light years away from us. One historical curiosity about this cluster is that, in 1860, one of its stars was observed to go nova, becoming for a few days as bright as the whole rest of the cluster.

For more on globular clusters, see page 115.

find more at: www.cambridge.org/features/turnleft/seasonal_skies_july-september.htm

In Ophiuchus: Two globular clusters, M19 and M62

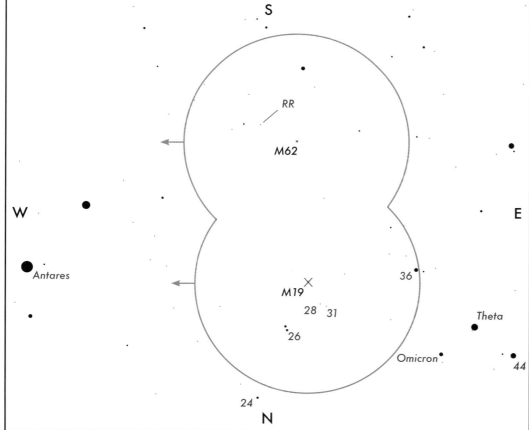

Dark sky
Medium power
Best: July–Aug.

Star maps courtesy Starry Night Education by Simulation Curriculum

Where to look: Find Antares, the bright red star of Scorpius. Look east from Antares towards Kaus Borealis at the top of the Teapot for a third-magnitude star, Theta Ophiuchi. Theta is the brightest star between Antares and Kaus.

In the finderscope, M19: Theta Ophiuchi is the middle star of three in a northeast–southwest line. Aim for the southwestern star, 36 Ophiuchi, and then move back towards Antares. As 36 Ophiuchi leaves the finder field, look for a fuzzy dot of light. That's M19. It sits just south of two distinctive star pairs, sixth-magnitude 26 and seventh-magnitude 28/31 Ophiuchi.

M62: Find 44 Ophiuchi, northeast of Theta and 36. Step southwest from 44 to 36 (hopping over Theta); one step further, and the nebula should be visible in the finderscope as a fuzzy spot of light.

M19 looks oval-shaped, oriented north to south. The bright central region seems disproportionately large compared to other globular clusters. The outer regions look grainy; and on a good night the central part may, too. This globular is located close to the center of our Galaxy, deep in the heart of the Milky Way. Because there is a lot of dust in this region of space, it is hard to pin down its precise location or the number of stars present. Most estimates place it about 30,000 light years from us, some 30 light years in diameter, and with probably around 100,000 stars.

M62 is obscured by dust, and so the cluster looks off-centered. That dust obscures its brightness, making it look

M19 in a star diagonal at low power

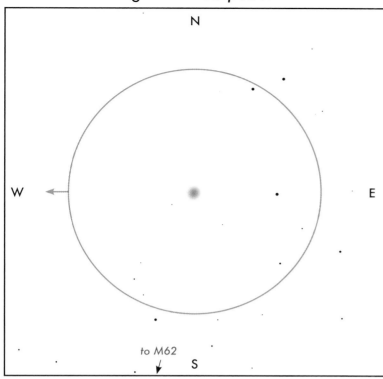

In a small telescope: Find in low, observe in medium power. **M19** looks like a fairly bright disk of light, somewhat oval-shaped. **M62** is an elongated ball of light; it gives the impression of being a bit lopsided.

M62 in a star diagonal at low power

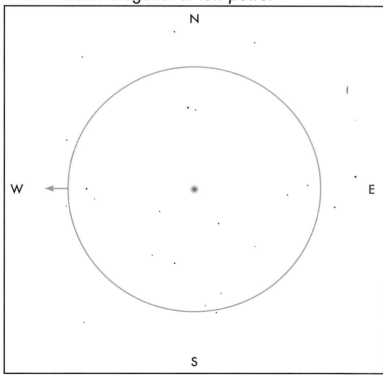

about 10 times dimmer than it would without the dust. The cluster is located about 25,000 light years from us, about two thirds of the distance from us to the center of our Galaxy. It is a collection of several hundred thousand stars, in a ball about 40 light years in diameter. Though its light appears grainy, a small telescope can't resolve the individual stars in the cluster.

One final curiosity: M62 happens to lie on the (arbitrary) boundary between Scorpius and Ophiuchus.

M19 in a Dobsonian at medium power

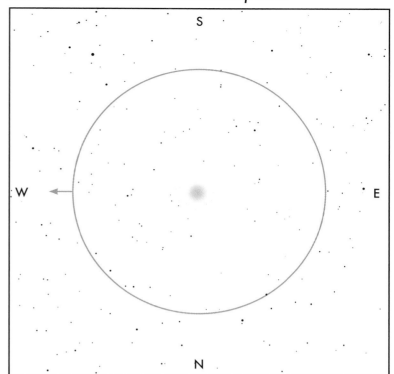

In a Dobsonian telescope: M19 is a uniform, moderately bright haze with a touch of graininess. Look for distinct stars along its edges. **M62** is a compact, hazy ball of light, just on the edge of graininess.

M62 in a Dobsonian at medium power

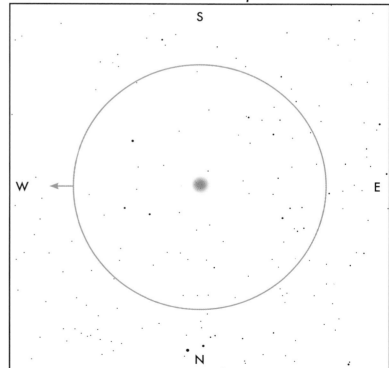

Also in the neighborhood: In the M62 finder field, **RR Scorpii** *is a red Mira-type variable, changing from a maximum that is usually a bit brighter than sixth magnitude (easy) to twelfth (too faint to see) over a period of about nine months. Thus, every four years its maximum occurs during this season, the northern hemisphere summer. Give it a try!*

Omicron, 24, *and* **36 Ophiuchi** *are double stars, described in more detail on page 150.*

find more at: www.cambridge.org/features/turnleft/seasonal_skies_july-september.htm

In Scorpius: Two open clusters, M6 and M7

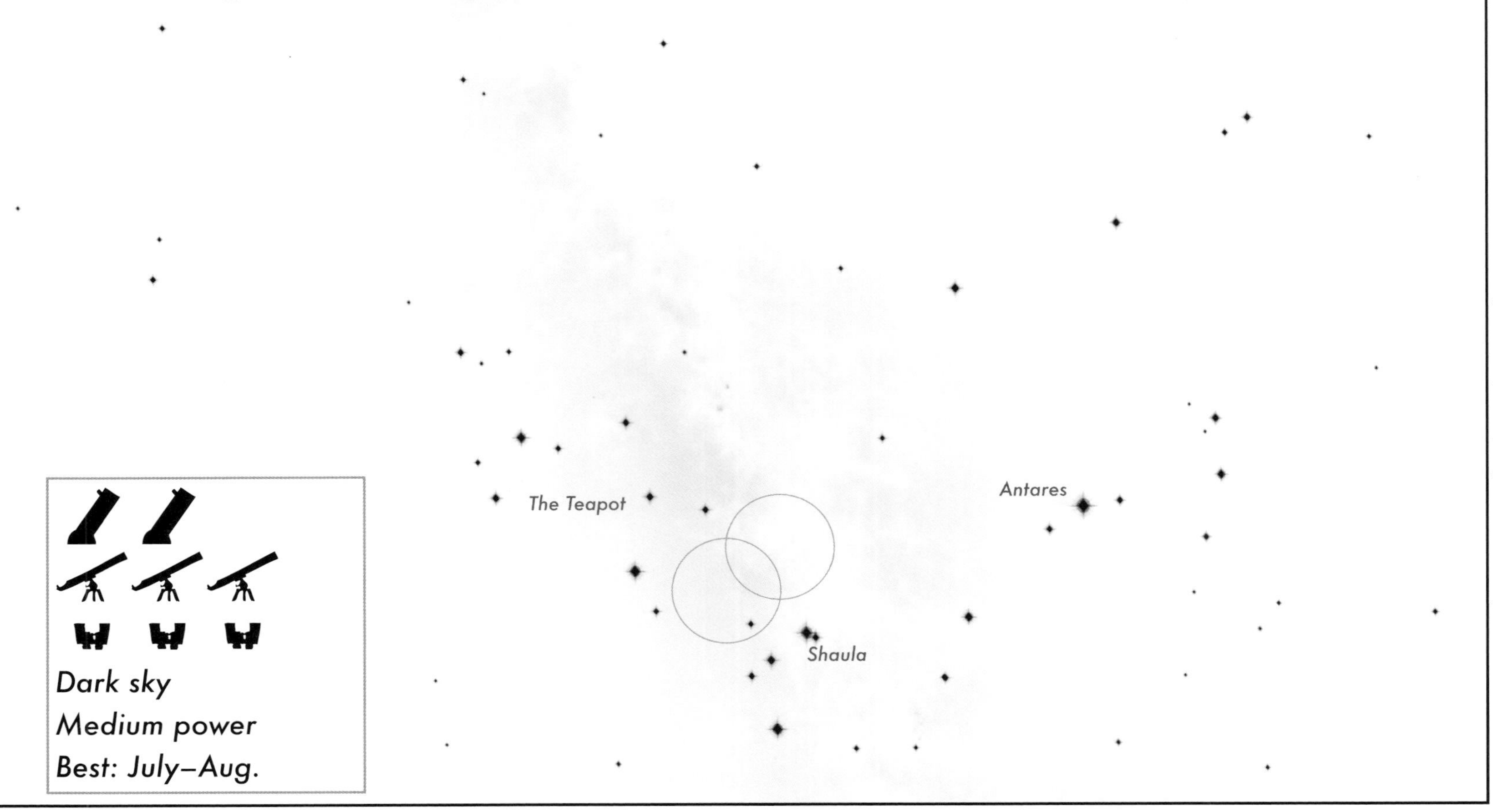

Star maps courtesy Starry Night Education by Simulation Curriculum

Dark sky
Medium power
Best: July–Aug.

The Teapot

Antares

Shaula

Where to look: Find Scorpius, low in the south. Follow the curve of stars from Antares down and around to the south and east, until you arrive at Shaula, a second-magnitude star at the end of the curl, the sting of the Scorpion.

In the finderscope: M7: Aim first at Shaula. In the finderscope you should see it (and its neighbor) at the lower left corner of an equilateral triangle of stars, with Kappa Scorpii and Fuyue. Step northeast from Kappa, to Fuyue, to a fuzzy patch of light: M7.

M6: Once centered on M7, look in the finderscope for another fuzzy patch of light about a finderscope field to the north and west (down and to the left in the finder, looking south). That patch is M6.

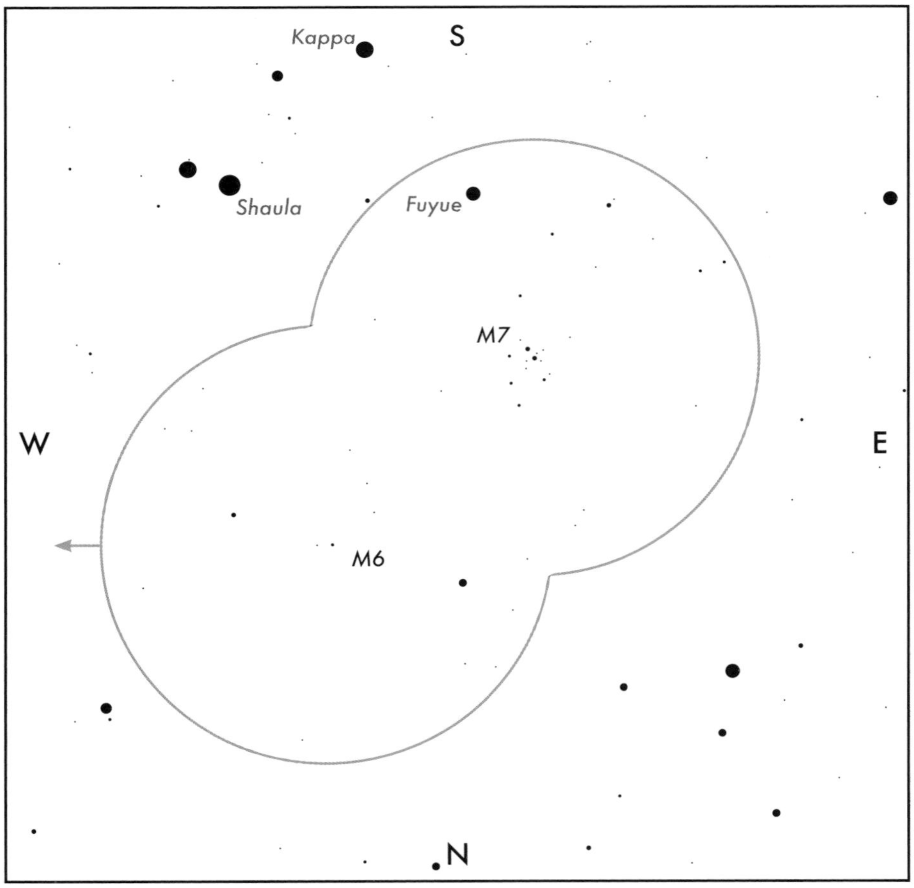

Kappa   S

Shaula   *Fuyue*

W   M7   E

M6

N

M7 consists of about 80 stars spattered across a region about 20 light years in diameter. It lies roughly 800 light years away from us. **M6** is made up of about 80 stars in a region 13 light years across, located roughly 1,500 light years away from us.

The majority of stars in **M6** are spectral types B and A (see page 71) but some of them have already evolved into their red-giant phase, and so it's estimated that this cluster is about 100 million years old. Judging from the number of bright-orangish stars, stars that have evolved through their main-sequence phase and into giants, one can estimate that M7 is probably older than M6.

In among the bright stars of M7 are many more dim ones, visible on a good dark clear night. The longer you look, the more you can begin to spot them; in a small telescope, many will appear to be just on the verge of visibility. One does not see a hazy fuzz of light in the background (unlike some

M7 in a star diagonal at medium power

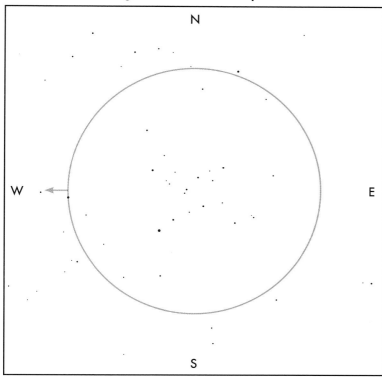

M7 in a Dobsonian at medium power

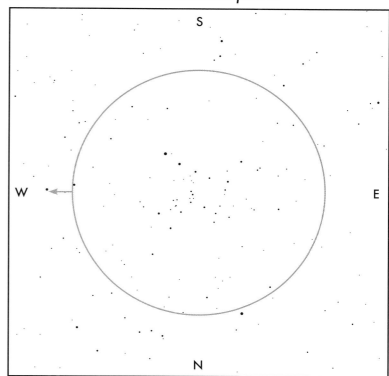

In a small telescope: M7 is a loose spread of about 20 stars, some quite bright, filling the telescope field.
M6 is looser and smaller than M7. Its brightest star, BM Scorpii, is distinctly orange and varies in brightness from sixth to eighth magnitude over a period of about 28 months, in no regular pattern.

In a Dobsonian telescope: M7 is big and loose, threaded with several strings of stars. Note the subtle yellow of the brightest star, southwest of the other, white–blue stars.
M6 has about 75 stars visible in a half-degree circle; the distinct orange star BM Sco stands out at the eastern edge of the cluster.

M6 in a star diagonal at medium power

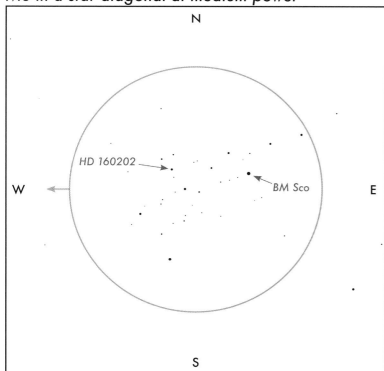

M6 in a Dobsonian at medium power

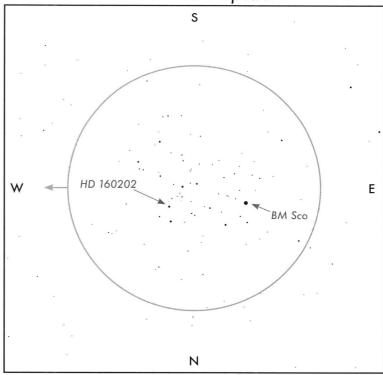

other open clusters); but it has a relatively large number of faint stars, spread out over a large area. By contrast, because M6 has fewer faint stars there's less of that feeling that, "with a bigger telescope, we could see more stars."

In M6, note the three brightest stars near the western edge of the cluster. One of them, HD 160202, is normally a seventh-magnitude star, though on several occasions it has dropped in brightness by more than two magnitudes. However, on July 3, 1965, it was observed to flare suddenly from eighth to as high as first magnitude! Within 40 minutes, it had returned to its normal brightness. These sudden changes are poorly understood.

find more at: www.cambridge.org/features/turnleft/seasonal_skies_july-september.htm

In Sagittarius: Two globular clusters, M22 and M28

Star maps courtesy Starry Night Education by Simulation Curriculum

Kaus Borealis

The Teapot

Antares

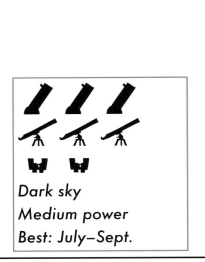

Dark sky
Medium power
Best: July–Sept.

Where to look: Locate the Teapot, low in the south, to the east of Antares. (The five brighter stars look like the outline of a house; with other fainter stars nearby one can make out a handle and spout.) This is the constellation Sagittarius. Find the star at the top of the Teapot.

In the finderscope, M22: Put the topmost star of the Teapot, Kaus Borealis, in the upper left corner of the finderscope. Look to the northeast for a small triangle of rather faint stars (the brighest, 24 Sgr, is notably orange-red). The globular cluster is located just to the east (right) of these stars. It should be visible in the finderscope, especially on a dark night. Under excellent conditions it can even be seen with the naked eye.
M28: Go roughly half a finderscope field northwest from Kaus Borealis. The globular cluster may be visible as a fuzzy patch of light.

S

Kaus Borealis

M28 ✕

24 Sgr

W

M22 ✕

E

N.

If **M22** were further to the north, and thus easier for more people to see, it would be even better known than M13. As it is, if you live south of latitude 40° N (but north of the equator) on a good dark night you may well think it is the nicest globular cluster in the sky, especially for a small telescope. (Of course, the southern hemisphere globular clusters Omega Centauri and 47 Tucanae put both to shame; see pages 238 and 212.) M22 is not the brightest of the globular clusters visible in the northern hemisphere, but it is certainly the easiest to resolve into stars; even in a 3" telescope it will look grainy.

By contrast, **M28** almost looks like an out-of-focus star. It is difficult to resolve individual stars in the cluster, even at high power, but you may get an impression of graininess in the light.

M22 in a star diagonal at low power

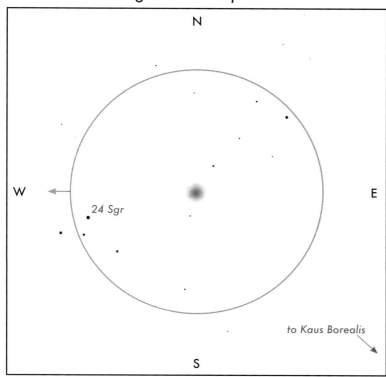

In a small telescope: Find in low, observe in medium power. **M22** appears ghostly and large, almost half as big across as a full Moon: a slightly flattened, evenly lit ball of light that fades out gradually at the edges. On a really crisp night, you can see the brightest of the individual stars in this cluster. **M28** looks like a much smaller, but still reasonably bright, concentrated ball of light.

M28 in a star diagonal at low power

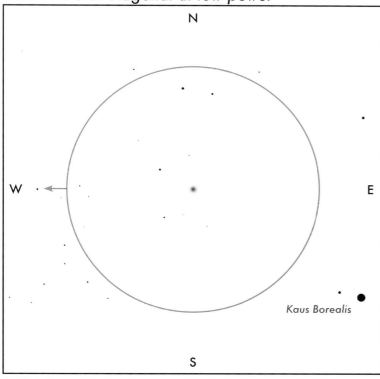

M22 in a Dobsonian at medium power

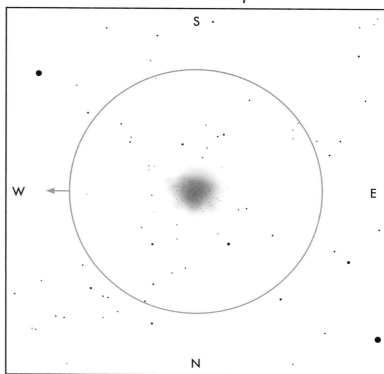

In a Dobsonian telescope: M22 is wonderfully resolved, an asymmetric ball of light threaded with strings of stars. **M28** shows a strong central concentration, but it is only marginally resolved; averted vision helps.

M28 in a Dobsonian at medium power

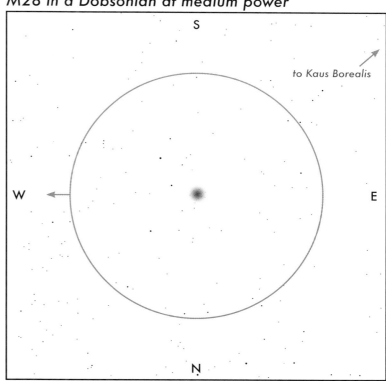

M22 consists of some half a million stars arrayed in a ball 50 light years in diameter, located 10,000 light years away from us. It's relatively close to the center of the Galaxy, and in fact there is a cloud of dust between it and us; without this obscuring dust, it would appear to be five times as bright as we actually see it. M28 contains a hundred thousand stars in a ball 65 light years in diameter. It is located 15,000 light years from us, about one and a half times as far away as M22. Like M22, it is positioned near the plane of the Milky Way Galaxy; it sits about halfway between us and the galactic center.

find more at: www.cambridge.org/features/turnleft/seasonal_skies_july-september.htm

In Sagittarius: Two globular clusters, M54 and M55

Star maps courtesy Starry Night Education by Simulation Curriculum

Dark sky
Low/medium power
Best: Aug.–Sept.

Where to look: Locate the Teapot, low in the south. Four stars make up the handle of the teapot: Nunki, Phi and Tau Sagittarii, and Ascella. Aim at Ascella.

In the finderscope, M54: From Ascella move about two degrees west-southwest, and you should be centered on M54. It may not be visible in the finderscope itself.

M55: Step to the southeast from Nunki to Tau, and continue two more steps in this direction. This should carry your telescope to a fairly sparse region of Sagittarius. If you reach the Terebellum/59/60 or Theta stars, you've gone too far. M55 should be visible (it's sixth magnitude) in your finderscope, like a fuzzy star.

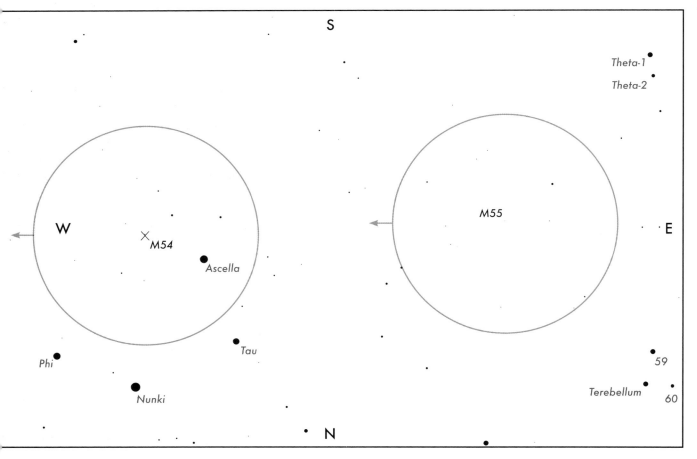

Also in the neighborhood: Ascella is a close double star with a 21-year period. Given our perspective on its orbit, their separation reaches 0.4" in 2028, then shrinks for a while before reaching 0.6" in the mid 2030s. Because both stars are of similar brightness, a Dob at that time under perfect conditions might just split them, or at least see them as one star elongated east–west.

In its orbit around the Galaxy, however, 1.2 million years ago you'd have seen a pair of stars each brighter than Sirius – separated by over 2 arc seconds!

M54 in a star diagonal at low power

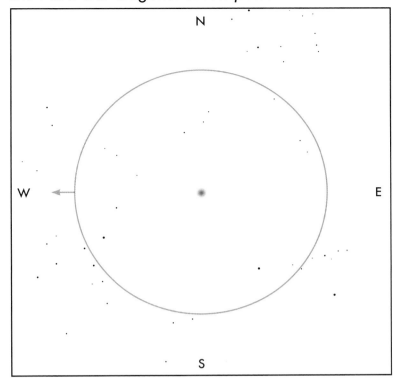

M54 in a Dobsonian at medium power

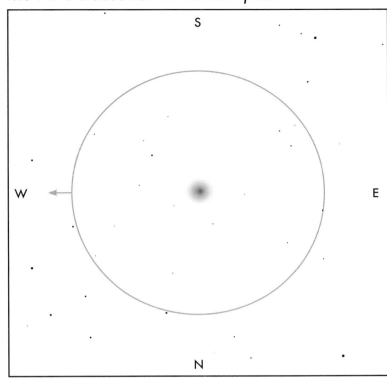

In a small telescope: M54 is very small and very round, a rather dim ball of light reminiscent of a planetary nebula. **M55** is larger and grainier, like a patch of smoke.

In a Dobsonian telescope: M54 is a ball of light with a visible central concentration, but not resolved into stars. **M55** is remarkably grainy, with dozens of resolved stars.

M55 in a star diagonal at low power

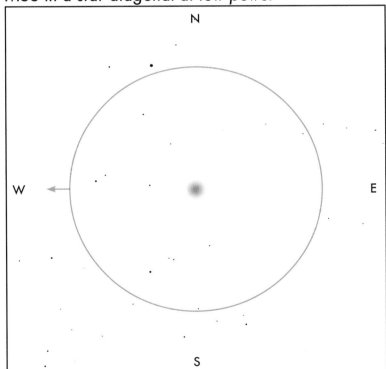

M55 in a Dobsonian at medium power

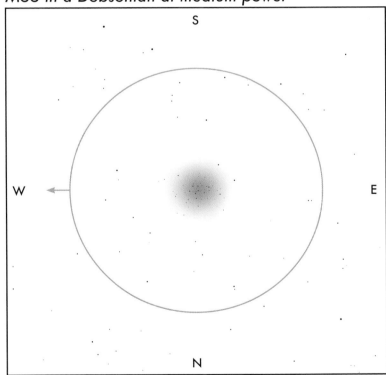

M54 and **M55** are difficult to see from Europe because they lie so close to the southern horizon. From the southern USA, and further south, both objects look more impressive. M55, in particular, begins to appear brighter in the center; with a Dobsonian you begin to resolve individual stars.

Globular clusters (see page 115) are rated according to how concentrated the stars are in its center: a class 1 cluster is extremely concentrated, while a class 12 is very loose and sparse. The contrast between M54 and M55 illustrates this scale. M54 is a compact class 3 globular cluster, 60 light years wide of about 100,000 stars, 50,000 light years away. It looks smaller than M55 to us because most of its stars (and hence most of its light) is concentrated in a small region at its core. M55 is a class 11 cluster, one of the least tightly packed of all globulars; its several hundred thousand stars are more uniformly distributed than those of M54. In addition, it is both closer to us (only 30,000 light years away) and larger (about 75 light years in diameter).

find more at: www.cambridge.org/features/turnleft/seasonal_skies_july-september.htm

Seasonal skies: October–December

The nights are often surprisingly chilly at this time of year (autumn in the north, spring in the south), and so it is important to dress appropriately and let your telescope come to the same temperature as the outdoors before you use it. In many areas these nights also tend to be damp and humid, and so dew is a common problem. Keep the telescope lenses covered, or at least pointed down rather than radiating to the sky, until you're ready to use it. We talk more about this on page 15.

The center of the Milky Way has set by now. There are no prominent diffuse nebulae among the objects in this chapter – though Orion will be well up by midnight, if you're willing to wait. Instead, the most prominent objects now are galaxies, especially the Great Galaxy in Andromeda (best seen in the northern hemisphere, though it can peep above the northern horizon even as far south as Sydney) or, if you are well south of the equator, the Magellanic Clouds.

If you're in the northern hemisphere during these months, now is the best season to prowl through the open clusters and double stars of Perseus and Cassiopeia (pages 187–195). However, beware of the harvest Moon! During summer or

Looking west

Finding your way:
October–December sky guideposts

Look first to the south and west. There you will see the three bright stars of the *Summer Triangle* slowly setting. **Altair** is to the south; the brilliant blue one further north, nearest the horizon, is **Vega**; and a bit higher overhead is **Deneb**. Deneb is at the top of a collection of stars in the form of a cross, standing almost upright between Vega and Altair during this time of the year.

High overhead are the four stars sf the *Great Square*. Although they are not particularly brilliant, they stand out because they are brighter than any other stars near them. The square is lined up so that its sides are aligned north–south and east–west.

South of the Square, the skies are nearly barren of stars. There is one very bright star, called **Fomalhaut** (pronounced something like "foam a lot"), low to the south. If you see any other bright object to its north, it is probably a planet.

North of the Great Square, you'll see the five main stars of *Cassiopeia* making a bright "W" shape (or an "M" depending on your point of view). It will be visible even as far south as Brazil.

A string of second-magnitude stars runs from the northeast corner of the Great Square in a line south of Cassiopeia; that's the constellation of Andromeda.

winter, every day past full Moon gives you about an hour or so more moonless sky before the Moon rises. But in autumn, you only get about 20 extra minutes of dark sky for each day past full Moon. This effect is due to the angle the Moon's orbit makes with the horizon; the farther from the equator you live, the worse the effect becomes. It's great for harvesters, hunters, and romantics, but terrible for those of us who want to look at deep-sky objects.

(Southern hemisphere observers, of course, have this problem during their autumn, April through June.)

For southern hemisphere observers, this is the best season to explore the wonders of the Small Magellanic Cloud and its nearby globular clusters, 47 Tucanae and NGC 362 (see pages 210–215).

To the west: Many excellent objects from July–September are still up at this time. Northern hemisphere observers have the added advantage that nightfall comes earlier, so they'll have an extra chance to catch them above the western horizon before they set:

Object	Constellation	Type	Page
Epsilon Lyrae	Lyra	Quadruple star	128
Ring Nebula	Lyra	Planetary nebula	130
Albireo	Cygnus	Double star	132
Blinking	Cygnus	Planetary nebula	134
Dumbbell	Vulpecula	Planetary nebula	136
M11	Scutum	Open cluster	140

Looking east

The *Milky Way* runs through Cassiopeia and continues across the northern horizon. It's out of reach for observers in southern climes; but some of its best objects are visible now in the north, especially around Cassiopeia – see the following chapter for northern skies. It's nice just to scan your telescope through Cassiopeia and Perseus at random, not looking for any object in particular.

To the east, you'll see the (northern hemisphere) "winter" stars begin to make their appearance. Look for a little cluster of stars called the *Pleiades*.

East and south of the Pleiades is the reddish star **Aldebaran**; north of the Pleiades, the brilliant star **Capella** rises. It is a magnitude-zero star, a match for Vega in the west.

Along the horizon, the bright stars of Orion, **Betelegeuse** and **Rigel**, and the Twins will be just rising. It is worth staying up a bit later in the evening if you can, and try to get an early look at some of the best of the January–March objects to be found there (see pages 53 ff).

During this time of year, the *Big Dipper* can be hard to find even in northern climes because it lies right along the northern horizon during the evening. Any trees or buildings there, or even just haze in the sky, may obscure it from view until the wee hours of the morning. If you are observing from south of 30° N – Mexico, Egypt, India, and points south – you'll have a hard time seeing it at any time of the night during these months.

In Pegasus: A globular cluster, M15

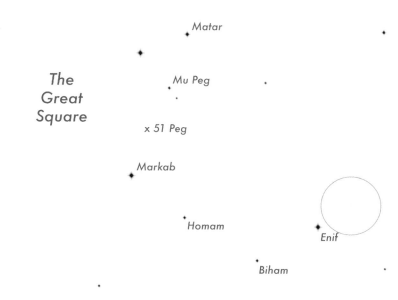

51 Peg

Dark sky
Medium power
Best: Aug.–Oct.

Star maps courtesy Starry Night Education by Simulation Curriculum

- Pleasant, easy to see
- Improves at high power
- Nearby, historic planet discovery

Where to look: Locate the Great Square, high overhead. Find the star in the southwest corner, Markab.

Head west-southwest from Markab past a third-magnitude star (Homam) to another third-magnitude star, Biham, all in a rough line. Northwest from Biham is a second-magnitude star called Enif. It's about as bright as the stars in the Square. Step northwest from Biham to Enif; half a step farther is the cluster.

In the finderscope: The object will appear like a fuzzy spot in the finderscope, right next to a star of similar brightness. Enif will probably be just outside the field of view (it depends on your finderscope) and there are few other bright stars nearby.

M15 in a star diagonal at medium power

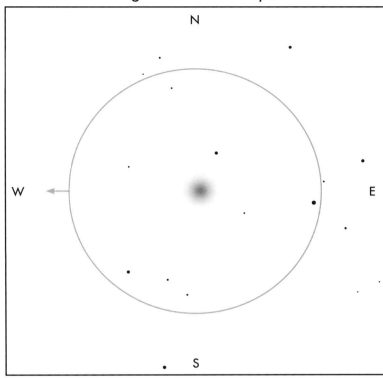

M15 in a Dobsonian at medium power

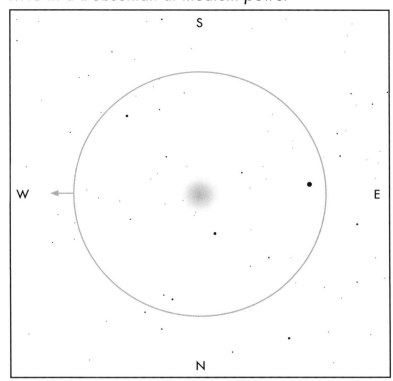

In a small telescope: The cluster seems to have a small bright central nucleus, surrounded by a much larger, uniformly dim outer region. It looks like a star surrounded by an evenly lit round ball of light.

In a Dobsonian telescope: The core of the cluster appears something like a dull star that fades away quickly as you move from the center. The rest of the cluster has the feel of a super-sized open cluster, with many individual stars resolved. Look at high power for lanes where the graininess is missing, giving the appearance of streamers of light.

M15 is bright enough that, once you've found it, you can try using a higher power eyepiece to try picking out detail (especially if the sky is very dark) but in general it looks nicer under low power. A 4" to 6" aperture is needed to begin to see individual stars in the cluster.

This globular cluster is a gathering of a few hundred thousand stars, in a ball about 125 light years in diameter, located some 40,000 light years from Earth. It is unusual in having such a large concentration of stars at its center, producing the bright starlike nucleus we see in our small telescopes.

M15 has a number of other interesting features. It is a source of X-rays, which implies that one of the members of this cluster is a neutron star or possibly a black hole. And large telescopes have found a faint planetary nebula within this cluster.

For more about globular clusters, see page 115.

Also in the neighborhood: *With binoculars or your finderscope, look for **51 Pegasi**: a featureless magnitude 5.5 star in a remarkably bleak part of the sky. It's hard to find, and once you do there's nothing to see – it's just another star in a small telescope. So why bother? Purely for its notoriety: it was the first ordinary star (after our Sun) to be shown to have planets. In 1995 Michel Mayor and Didier Queloz of the Geneva Ob-*

servatory looked for tiny shifts in the spectrum of this star due to the pull of an orbiting companion (much as spectroscopic binaries are detected) and found that 51 Peg, a G-type star much like our Sun, only 50 light years away, had a companion a bit smaller than Jupiter. This planet orbits remarkably close to the star, 0.051 AU or roughly five star diameters from its surface, and completes one "year" every 4.2 Earth days.

To find 51 Peg, start again at the southwest corner of the Great Square, Markab. The northwest corner is Scheat; to its right, and a bit north, is an equally bright star, Matar. From Matar, head back down towards Markab. You'll step past a fourth-magnitude star, Mu Pegasi (it's just northeast of another fourth-magnitude star, Lambda Pegasi); then a gap; then Markab. Aim your finderscope at the gap, halfway between Mu and Markab, and you'll see a modestly bright star. That's 51 Pegasi. (To confirm you've got it, look for an eighth-magnitude star immediately to its east.)

find more at: www.cambridge.org/features/turnleft/seasonal_skies_october-december.htm

In Aquarius: A globular cluster, M2

Dark sky
Medium power
Best: Aug.–Oct.

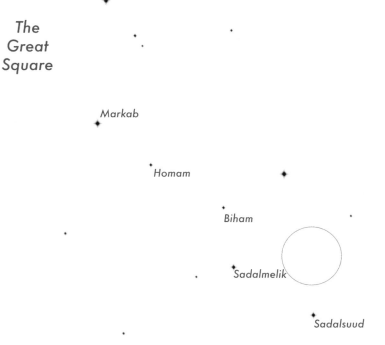

The
Great
Square

Markab

Homam

Biham

Sadalmelik

Sadalsuud

Altair

Star maps courtesy Starry Night Education by Simulation Curriculum

- Tricky to find
- Conspicuous in a sparse field
- Granular appearance in larger telescope

Where to look: Locate the Great Square, high overhead. Go to the star at the southwest corner, Markab. Head west-southwest from Markab past a third-magnitude star (Homam) to another third-magnitude star, Biham, all in a rough line. South and west of Biham, find Sadalmelik; southwest from there is Sadalsuud.

In the finderscope: From Sadalsuud, move the telescope due north. On a really good night, you may see the cluster itself in the finderscope as a faint fuzzy star in a field of about half a dozen equally faint stars. More typically, you won't see anything until you come to a pair of sixth-magnitude stars in a northwest–southeast line, 25 and 26 Aquarii, northeast of the cluster itself. In that case, aim at the pair and notice a third such star, 11 Aquarii, to their northeast. Step back (southwest) from 11 Aquarii, to 26 Aquarii, and one step further should bring you into the neighborhood of M2.

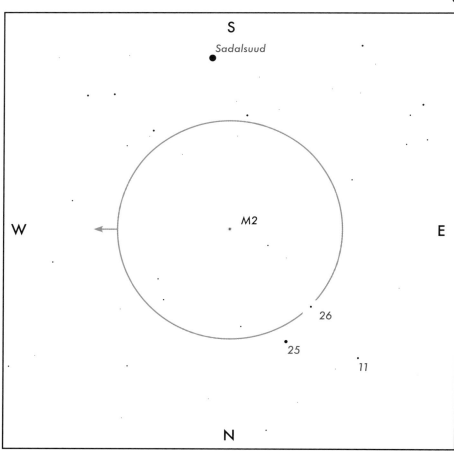

S

Sadalsuud

W

M2

E

26

25

11

N

M2 in a star diagonal at low power

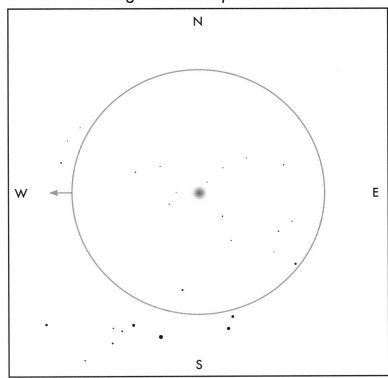

M2 in a Dobsonian at medium power

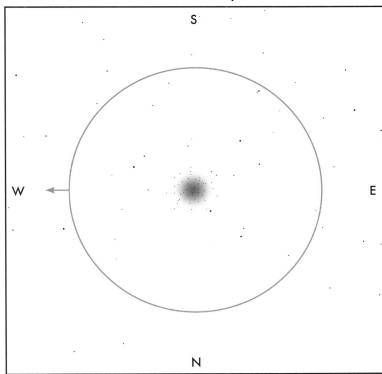

In a small telescope: Find with low power (as illustrated here), observe at medium power. M2 stands out as a conspicuous small disk of light among a scattering of faint stars; the object is round, uniformly bright, and featureless at low power. Note the four stars south of the object, which you will encounter as you move up from Sadalsuud; they're a sign that the cluster is about to come into view.

The cluster will be brighter and so easier to find at low power, but once you've found it, raising your power to medium or even high power allows you to see the cluster as something more than a featureless disk.

In a Dobsonian telescope: You will see the object looking mottled or granular as you start to make out individual stars. In a Dob it is visibly elongated, northwest–southeast. There's a modest concentration of light in the center, but it is grainy across the entire disk of the cluster, with tens of stars distinctly visible. Using averted vision, you can begin to see streamers of light in all directions. At high power, crazy spurs of grain and darker lanes become visible in the disk of light.

M2 is a globular cluster of several hundred thousand stars, gathered in a ball about 175 light years across, making it one of the largest globular clusters known. It is located about 40,000 light years away from us. As can be gathered from its low number, it was one of the first objects catalogued by Charles Messier, who first recorded it in 1760, though he noted that in fact another comet hunter had previously seen this nebula in 1746. Thirteen years after Messier, the brother–sister team of William and Caroline Herschel were the first to be able to make out individual stars within the cluster.

Some measurements of M2's motions indicate that it may be following a path in orbit around the center of the Galaxy that is both highly tilted with respect to the plane of the Galaxy, and very eccentric; at times its orbit (which takes a few hundred million years) it may reach a distance of more than half that to the Magellanic Clouds. It is also quite old even by globular cluster standards, perhaps 13 billion years old; it may be one of the objects around which our Milky Way Galaxy formed itself.

For more on globular clusters, see page 115.

In Aquarius: A globular cluster, M72, and the *Saturn Planetary Nebula,* NGC 7009

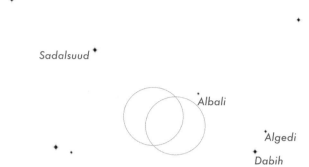

Dark sky
Medium/high power
Nebula filter (Saturn)
Best: Sept.–Oct.

Altair

Sadalmelik

Sadalsuud

Albali

Algedi

Dabih

Fomalhaut

The Teapot

Star maps courtesy Starry Night Education by Simulation Curriculum

- NGC 7009: unusual shape
- Challenging to find; best in a Dob
- Fun tour of an obscure part of the sky

Where to look: Aim halfway between the first-magnitude stars Altair and Fomalhaut. From the midpoint between them, look southwest (toward the Teapot) for a pair of third-magnitude stars, Algedi and Dabih, in a north-north-west–south-southeast line. Start there.

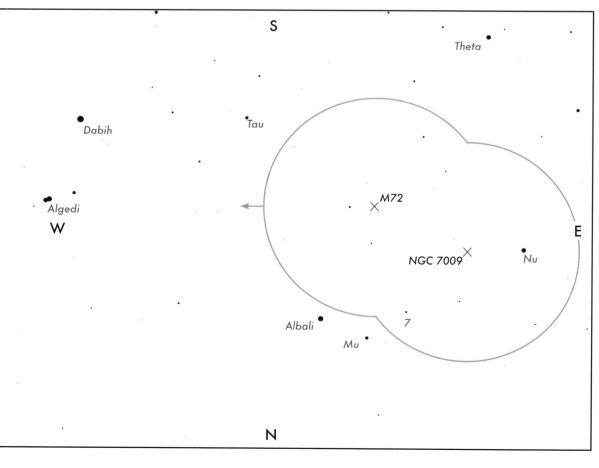

S

Theta

Tau

Dabih

Algedi

W

M72

NGC 7009

Nu

Albali

Mu

7

E

N

In the finderscope: Both Algedi and Dabih should fit in the finder field. Notice that Algedi is actually a distinctive pair of stars. Move it to the southern edge of your finder field, and move due east about two finder fields until you see three stars, the fourth-magnitude Albali and Mu Aquarii and fifth-magnitude 7 Aquarii. Center on Mu and then move due south, looking for three sixth-magnitude stars in a thin triangle oriented north–south. **M72** is due east of the middle star, just past a line connecting the other two. **NGC 7009** is half a finder field away; move north and east from M72 until you spot Nu Aquarii, and aim for a spot two degrees due west of Nu. Alternately, step southeast from Mu to 7 Aqr; two more steps, towards Nu, should bring the nebula into your low-power field.

M72 in a star diagonal at medium power

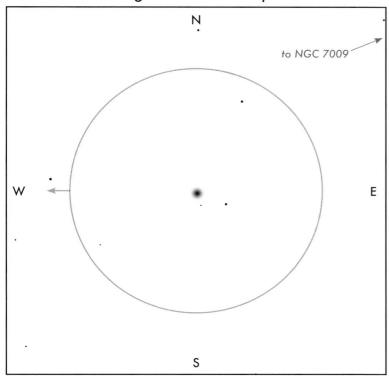

M72 (and M73) in a Dobsonian at medium power

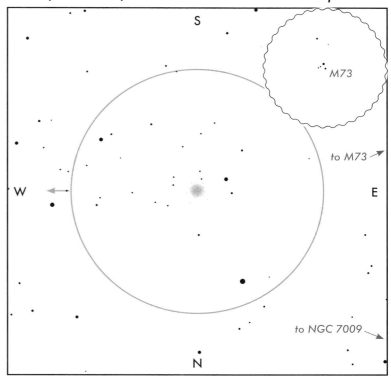

In a small telescope: The globular cluster **M72** is faint, barely visible in a small telescope, making this a good challenge for a dark night. The fun is in the finding, especially since this is a pretty sparse region of the sky.

The **Saturn Nebula** is a different sort of challenge, small but relatively bright. If you think you've found it, switch to a higher power to reveal that it is more than just a point of light. A nebula filter will enhance the disk against the sky compared to the background stars. The inset shows a high-power (20') field of view.

In a Dobsonian telescope: M72 has no sharp core; it appears as a disk of light, with a slight hint of graininess especially at the edges. Averted vision brings out a hint of a few stars. Look to the east for an apparent cluster of stars that are actually unrelated; this is Messier's M73.

The **Saturn Nebula** is almost starlike at low power, but with a distinctive pale blue–gray color. At higher power one sees an almost round, uniform disk with a hint of handles at the ends, reminiscent of its namesake. The inset shows a very high-power (5') field of view.

The Saturn Nebula, star diagonal, low/high power

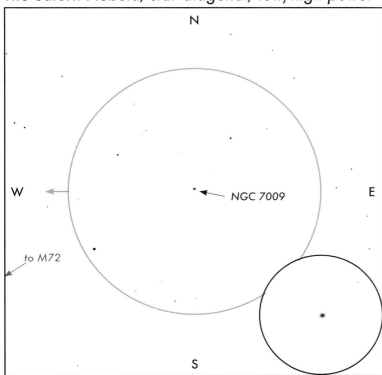

The Saturn Nebula in a Dob at medium/high power

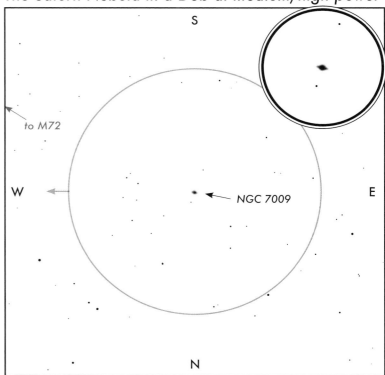

M72 is relatively young as globular clusters go, with an unusual number of blue stars. It is about 40 light years in diameter, located 50,000 light years away. The **Saturn Nebula** is ten times closer, at about 5,000 light years distance. Its green color is fluorescence in the nebular gas excited by the ultraviolet light of its hot central star.

find more at: www.cambridge.org/features/turnleft/seasonal_skies_october-december.htm

In Capricornus: A globular cluster, M30

Dark sky
Low, medium power
Best: Sept.–Oct.

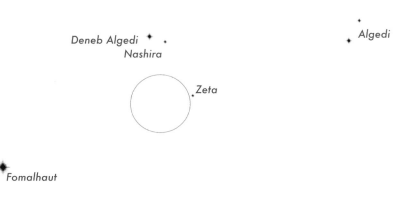

Altair

Sadalmelik

Sadalsuud

Deneb Algedi
Nashira

Zeta

Algedi

Fomalhaut

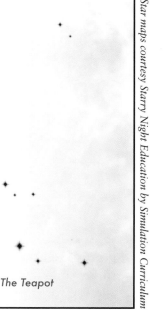

The Teapot

Star maps courtesy Starry Night Education by Simulation Curriculum

- Easy to get close to, harder to see
- Strong central concentration
- Messier Marathon challenge

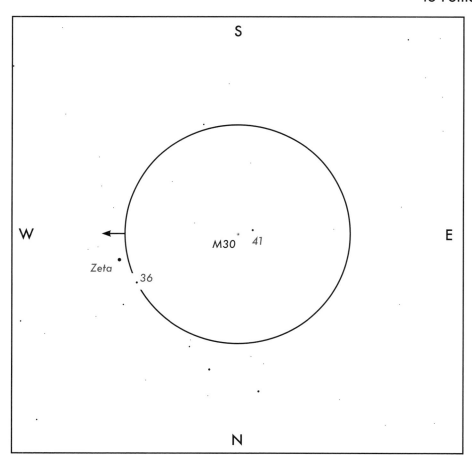

Where to look: Find Altair, the head of the Eagle and southernmost of the Summer Triangle stars; and look southeast to Fomalhaut, the only first-magnitude star in its region of the sky. Two thirds of the way from Altair to Fomalhaut are a pair of stars, Nashira and Deneb Algedi. (Deneb Algedi is not to be confused with a star to the west, named Algedi; nor with the more famous Summer Triangle star named Deneb, for that matter!) Starting from this pair, look south and a bit west for a third-magnitude star, Zeta Capricorni.

In the finderscope: Aim at Zeta Capricorni; it is distinctive for having a number of fourth- and fifth-magnitude stars nearby, most notably 36 Capricorni just to its northeast. From Zeta, move eastwards and a bit south until you see a fifth-magnitude star, 41 Capricorni; aim there. The globular cluster is just to its west.

M30 in a star diagonal at low power

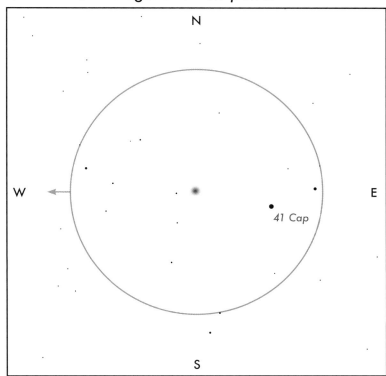

M30 in a Dobsonian at medium power

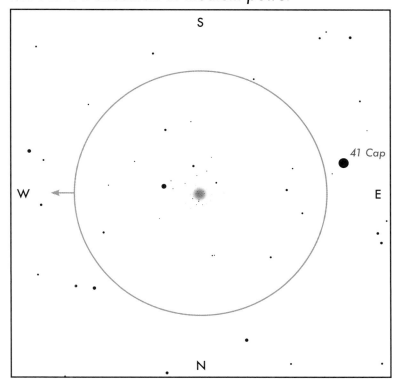

In a small telescope: Find at low power (illustrated here), observe at medium power. Finding the field of view where the cluster should be visible is relatively easy, given the presence nearby of the fifth-magnitude star 41 Capricorni. But seeing the cluster itself can be a challenge if you're observing from suburban skies, or from a location north of latitude 35° N (where it will always be low on the horizon). It's a small and rather dim disk of light, not one to be shared with anyone who doesn't appreciate the thrill of the hunt.

In a Dobsonian telescope: This is a classic case of an object to be found under low power, but observed at higher power. Under low power, M30 and the star that we used to find it, 41 Capricorni, can fit in the same field of view, making the object easy to locate. But the cluster itself is only about ten arc minutes across, and it is hard to see any detail in it at this scale. Higher power reveals graininess, however, and one can begin to resolve individual stars.

The globular **M30** is a dense cluster of stars located about 25,000 light years from us. Its structure is significantly concentrated at the center: the whole cluster extends about 90 light years across, but the stars in its center appear to have collapsed into a core of only about a light year in diameter. In fact, half the mass of the cluster is confined to the inner 17 light years of its diameter. That's the equivalent of having several hundred thousand stars packed between us and our near-neighbor star, Sirius.

For amateur astronomers, M30 is particularly infamous because of its location, far from other Messier objects. A favorite test of one's observing skill among advanced amateurs is the "Messier Marathon," where you try to observe every object with a Messier number in one night. That's just possible in the northern hemisphere during a moonless night in late March; but at that time, this is the Messier object closest to the Sun and thus the most difficult to catch, rising low in the east right at sunrise.

For more about globular clusters, see page 115.

find more at: www.cambridge.org/features/turnleft/seasonal_skies_october-december.htm

In Aquarius:
The *Helix*, a planetary nebula, NGC 7293

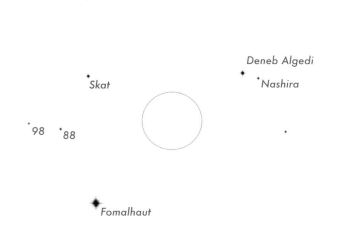

Dark sky
Medium power
Nebula filter
Best: Oct.–Nov.

Star maps courtesy Starry Night Education by Simulation Curriculum

- Challenge in a small telescope
- Famous in photographs
- Large but dim

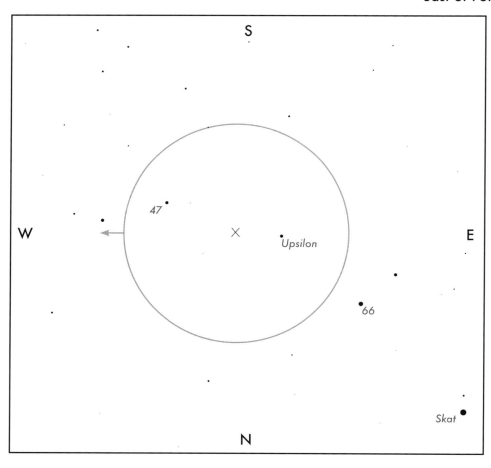

Where to look: Due south of the Great Square, the brightest star in the south, all by itself, is Fomalhaut. To the north and west of Fomalhaut are a pair of third-magnitude stars, Deneb Algedi and Nashira; to the north and east of Fomalhaut is the third-magnitude star Skat. Start by moving your finderscope from Fomalhaut to Skat, past a pair of fourth-magnitude stars (98 and 88 Aquarii).

In the finderscope: From Skat, you will be moving southwest, following a trail of fifth-magnitude stars. First find 66 Aquarii in the southwest edge of your finder field; move there, then look further southwest for Upsilon Aquarii; center there. You should be about halfway between Fomalhaut and Deneb Algedi at this point.

Once you have centered on Upsilon, look further west and a bit south for 47 Aquarii. The nebula is found about a third of the way from Upsilon to 47.

The Helix in a star diagonal at low power

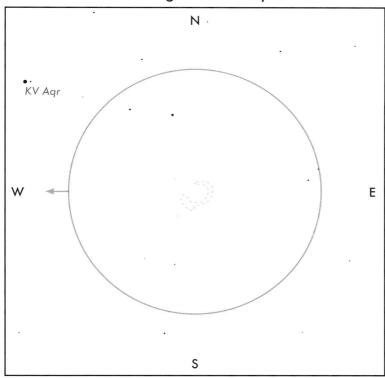

The Helix in a Dobsonian at medium power

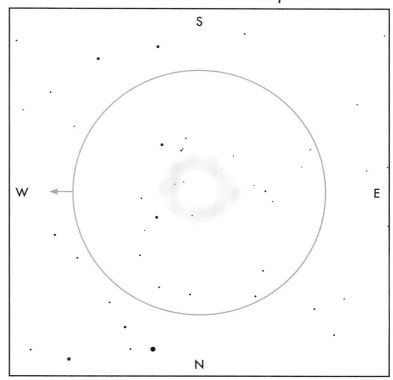

In a small telescope: The Helix is a large nebula but quite dim. For a small telescope, it is a challenge except on the darkest night. If you're in a wonderfully dark site, and you have a nebula filter, look where we have indicated with the dotted lines for a dim ring of light.

In a Dobsonian telescope: The Helix is a large nebula but quite dim. On a dark night, with a nebula filter, you'll see a large, fat, dim ring of light, looming ghost-like behind the other stars in the field of view.

Given its size, the Helix is a favorite target of astrophotographers. With a long-enough exposure, its red ring and greenish interior become visible. Alas, the red light that shows up so well in photographs or digital images is at a wavelength just beyond the reach of the typical human eye.

Located about 700 light years from us, the **Helix Nebula** is one of the closest planetary nebulae; hence its large size in an amateur telescope. At that distance, we can calculate that the nebula is about 2.5 light years across. Best estimates suggest that the gas cloud has been expanding for about 10,000 years, making it a relatively new feature in the Galaxy.

For more about planetary nebulae, see page 99.

Also in the neighborhood: *One degree northwest of the nebula (just outside the low-power field of view, as indicated in the star diagonal view illustrated above) is the variable and double star* **KV Aquarii.** *As a variable it's not particularly interesting (occasionally brightening by 0.2 magnitudes), but as a double it's a nice challenge. The primary is magnitude 7.1, and its companion is magnitude 8.0, located 6.9" to the south-southeast. There is a fainter, 10th-magnitude star three full arc minutes further east (as shown in the star diagonal view) but that is a field star, not part of the system.*

find more at: www.cambridge.org/features/turnleft/seasonal_skies_october-december.htm

In Cetus: a galaxy, NGC 247
In Sculptor: a galaxy, NGC 253, and a globular cluster, NGC 288

to The Great Square

Diphda

Deneb Algedi

Fomalhaut

Star maps courtesy Starry Night Education by Simulation Curriculum

Dark sky
Medium power
Best: Nov.–Dec.

- Three nice objects in the same neighborhood
- Good challenges for a Dob
- Beyond the reach of most smaller telescopes

Where to look: Look due south from the Great Square. The brightest star in the south, all by itself, is Fomalhaut. Northeast of Fomalhaut, the sky is almost completely empty of bright stars; the one exception is the second-magnitude star, Diphda, northeast of Fomalhaut. Aim there.

In the finderscope: From Diphda move due south toward a triangle of fifth- and sixth-magnitude stars. Aim at the northeastern star, HR 220, and then move north and slightly west for **NGC 247**.

Past the faint triangle, you'll see a diamond pattern of stars. Step south from HR 232 to HR 228 to the galaxy **NGC 253**.

From NGC 253, move the finder southeast until the fourth-magnitude star Alpha Sculptoris enters your finder field; then slowly return northwest, looking in your telescope at low power for the globular cluster **NGC 288**.

S

Alpha Scl

✕ NGC 288

NGC 253

228 —

232 —

W   E

— 220

✕ NGC 247

Diphda ● — N

NGC 247 in a Dobsonian at medium power

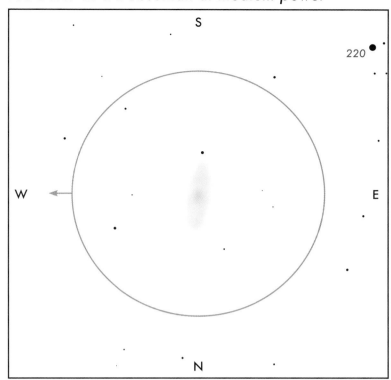

In a Dobsonian telescope: NGC 247 is dim, the hardest of these three objects to see. It is a faint lumpy dash of light oriented almost north–south. To orient yourself, look for the ninth-magnitude star (a part of our own Milky Way, not a member of this galaxy) at the southeast tip of the galaxy; at the opposite end, notice a bit of lumpiness and a faint concentration of light.

NGC 253 in a Dobsonian at medium power

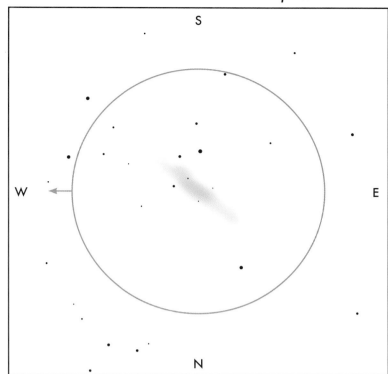

In a Dobsonian telescope: This galaxy is a large asymmetric bar of lumpy light. The edges look almost squared off, as if one were looking at a box of light. A relatively bright gash across the field of view, it is much easier to spot than its neighbor to the north, NGC 247.

It's not just a coincidence that we find these two galaxies so close to each other in the sky. **NGC 247** and **NGC 253** are, in fact, the brightest members of a cluster of galaxies known as the Sculptor Group, and they really are about as close to each other as they appear here. Both of them lie about 12 million light years from us, making this group one of the closest galaxy clusters to our own Local Group of galaxies (which includes the Milky Way, the Andromeda Galaxy, and the Magellanic Clouds).

NGC 253 is the central galaxy of this group, about which the others revolve. It is a galaxy currently undergoing a period of star formation: what's called a *starburst* galaxy. (For more about galaxies, see page 109.)

On the other hand, the globular cluster **NGC 288** is located within our own Galaxy and so its position in the sky so close to these two galaxies is, indeed, merely a coincidence. Compared to the millions of light years between us and the Sculptor Group, this cluster is a mere 30,000 light years away. (Learn more about globular clusters on page 115.)

NGC 288 in a Dobsonian at medium power

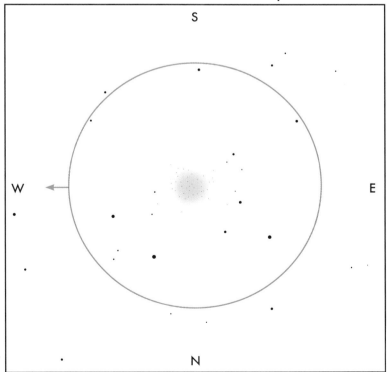

In a Dobsonian telescope: NGC 288 is brighter than NGC 247, and easier to spot. A globular cluster, not a galaxy, it's a round, lumpy ball of stars with poorly resolved edges. It appears grainy across the entire disk, with no clear concentration at its center.

find more at: www.cambridge.org/features/turnleft/seasonal_skies_october-december.htm

In Andromeda:
The *Andromeda Galaxy*, M31, with its companions, M32 and M110

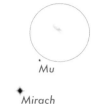

Cassiopeia

Mu

Mirach

The Great Square

Star maps courtesy Starry Night Education by Simulation Curriculum

Dark sky
Low power
Best: Sept.–Jan.

- Large, bright, spectacular galaxy
- Easy to find and see
- Something to see in any scope, any night

Where to look: Locate the Great Square, almost straight overhead. From the northeast corner, find three bright stars in a long line, arcing across the sky west to east, just south of Cassiopeia, the big W. From the middle of these three stars (called Mirach) go north towards Cassiopeia to Mu Andromedae. On a dark night, you may see a second star, Nu Andromedae, north of Mu in a slightly curving line. The galaxy is just barely visible to the naked eye on a good dark night, just to the west of Nu Andromedae.

In the finderscope: Start at Mu Andromedae and move north until you can see Nu Andromedae in the finder. Aim at Nu, and the galaxy should be easily visible in the finderscope.

S

Mirach

Mu

Nu

W

E

N

M31 *in a star diagonal at low power*

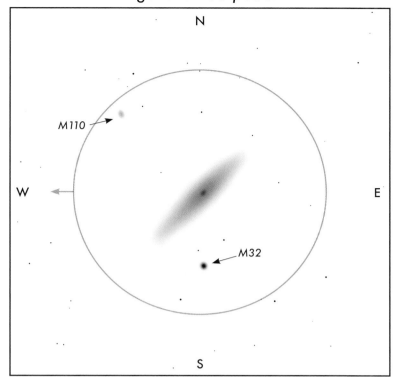

M31 *in a Dobsonian at low power*

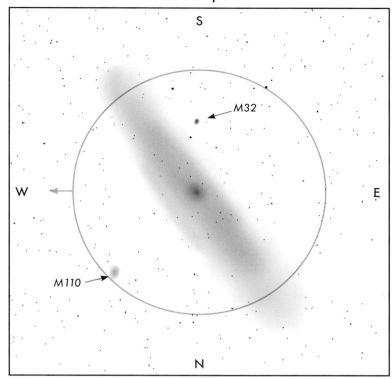

In a small telescope: The galaxy M31 looks like a bright oval embedded in the center of a long swath of light, which extends almost completely across the field of view.

Off to the south, and a bit east, is what looks like an oversized star making a right triangle with two faint stars. This is a companion galaxy, M32. Increasing magnification, you can see it is an egg-shaped cloud of light.

With M31 in the center of your low-power field, M110 (more properly called NGC 205) is just inside the field of view, to the northwest. It's on the opposite side of M31 from the other little companion, M32. It's dimmer, and spread over a larger area than M32, making it harder to see. It has an oval shape, elongated north to south.

In a Dobsonian telescope: The Andromeda Galaxy is one of the wonders of the sky, worth looking at even when the Moon is up but especially marvelous on a clear, dark night. The central core of M31 draws your eye, but the strong concentration of M32 will also be easily visible as a small disk of light. Let your eye wander and eventually averted vision will bring out the broader sweep of the galaxy, extending well beyond the field of view of your lowest-power eyepiece. As you become accustomed to seeing the extended disk of the galaxy, you should then be able to make out M110 as a similarly bright disk of light on the opposite side of Andromeda from the smaller but brighter M32.

The **Andromeda Galaxy** is the largest of 20 or so galaxies, including our own Milky Way Galaxy, the Triangulum Galaxy (page 178), and the Magellanic Clouds (pages 210 and 216), that make up the *Local Group*. With 300 billion stars, and a diameter of 150,000 light years, Andromeda is considerably larger than the Milky Way Galaxy. It is a spiral galaxy, which we're seeing nearly edge-on.

You can just make out M31 by the naked eye on a good dark night; it is the most distant object the human eye can see without a telescope. The bright nucleus of M31 is visible in a telescope or binoculars; the darker the night, the more of the surrounding galaxy you will see. The bright nucleus seems to sit a bit off from the center of a dimmer streak of light, barely visible at first; the longer you look, the more it seems to extend far beyond the field of view of even the lowest-power eyepiece.

Best estimates locate this galaxy at 2.5 million light years from Earth. Its close companion, **M32**, is 2,000 light years across, and lies 20,000 light years south of M31. **M110**, the other companion visible here, is more than twice as large as M32. These companions are elliptical galaxies. The Andromeda Galaxy also has at least a dozen other faint companion galaxies, most too dim to be seen with a small telescope.

Large telescopes can begin to resolve this galaxy into individual stars, and in the 1920s Edwin Hubble recognized that some of these stars seemed to be pulsating variable stars called Cepheid variables (named for the variable star Delta Cephei; see page 196). He knew that the rate of pulsation was related to the intrinsic brightness of such a star, so he could calculate how bright such stars should be for a given distance from us. By measuring how faint these stars were in Andromeda, he was the first to measure the enormous distance to this galaxy.

One fascinating area of recent research concerns the future evolution and merging of the Milky Way with the Andromeda and Triangulum (M33, page 178) galaxies. By measuring the relative speeds of these galaxies today, our best estimate is that the Milky Way and M31 will merge about 6 billion years from now, with M33 orbiting them both. The resulting galaxy will probably resemble a giant elliptical galaxy. The Sun most likely will end up relatively far from the center of this merged galaxy; indeed, some models suggest that sometime in the next 10 billion years our Sun (and its solar system!) may find itself moving through M33, while it is still bound gravitationally to the Milky Way–Andromeda merged galaxy.

For more on galaxies, see page 109.

find more at: www.cambridge.org/features/turnleft/seasonal_skies_october-december.htm

In Triangulum:
The *Triangulum Galaxy*, M33

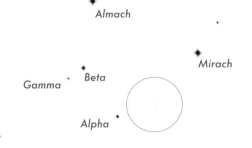

Cassiopeia

Almach

Mirach

Gamma   Beta

Alpha

The
Pleiades

The
Great
Square

Star maps courtesy Starry Night Education by Simulation Curriculum

- Beautiful view if the night is dark
- Spiral arms visible under good condition
- Aperture helps; dark skies help more

Where to look: Locate the Great Square, high overhead. From the northeast corner, find four bright stars in a long line, arcing across the sky west to east, just below (south of) Cassiopeia, the big W. Down and to the left (southeast) of the middle two stars (Mirach and Almach) you'll find three stars forming a narrow triangle, pointing roughly towards the southwest. This is the constellation Triangulum. Aim for Alpha, the star at the point.

In the finderscope: Use the distance from the northernmost star of this triangle, Beta Trianguli, to the point of the triangle, Alpha Trianguli, as a yardstick. From Alpha, look about a third this distance up and to the right (to the northwest, back towards Mirach) for a faint star called CBS 485. (If the sky isn't dark enough for you to see CBS 485, you probably won't be able to see the galaxy, either – see below!)

Step from Alpha Trianguli, to CBS 485, to half a step further. Look for a vague fuzz of light. That's the galaxy.

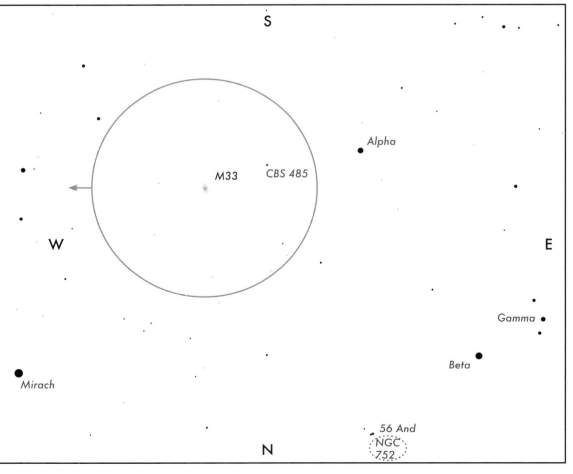

S

Alpha

M33   CBS 485

W

E

Gamma

Mirach

Beta

56 And
NGC
752

N

M33 in a star diagonal at low power

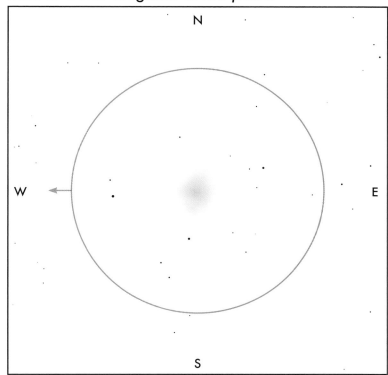

M33 in a Dobsonian at low power

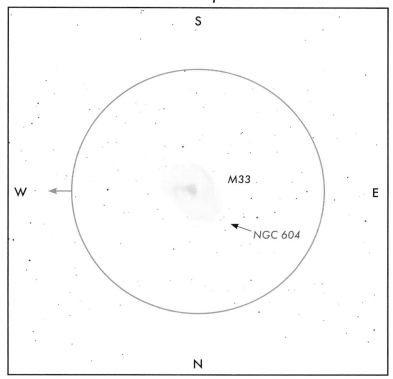

In a small telescope: Four stars in the shape of a "kite" should be visible in the telescope. The galaxy will look like a large but very faint patch of light in this kite. Be sure to use your lowest power.

The biggest challenge for this object is its relative faintness. A dark sky is an absolute necessity; if you do have such a sky, though, this can be one of the most impressive objects to see in a small telescope or even a pair of binoculars. It's worth staying up late to look for, if you've brought your telescope with you on a summer camping trip far from city lights.

In a Dobsonian telescope: The galaxy looks like a large, lumpy disk of light elongated in the northeast–southwest direction. The darker the night, the more detail you can see; if it is very dark, you should be able to make out two spiral arms running west to north, and east to south.

Within the Triangulum Galaxy is the star-formation region, NGC 604.

A light pollution filter that cuts out the yellow light from sodium vapor lamps can increase the contrast between the galaxy and the surrounding sky. Some people find that a nebula filter helps bring out NGC 604.

The **Triangulum Galaxy** is very large but it can be hard to see. Because it is so large, its total brightness is spread thinly over a relatively big area of the sky. There's little contrast brightness from the edge to the center, unlike the galaxy in Andromeda, so you may even have a hard time recognizing it when you've got it.

Bigger telescopes don't necessarily help. It can be hard to find even in a 14" telescope if the sky isn't particularly good. But on a crisp dark night, it can look lovely in a pair of binoculars or a small, wide-field telescope. It all depends on sky darkness; if the Moon is up, don't even waste your time trying!

On the other hand, if the sky is really dark and your eyes are truly dark-adapted, a Dobsonian opens up a view that can take your breath away. Averted vision clearly shows a big S shape – a lumpy, mottled pair of spiral arms. Three knots of light stand out in particular. Two can be found about 10 arc minutes south and southwest of the center of the galaxy; and the other one is 10 arc minutes to the northeast, on the upper-right edge of the S. This latter region has its own NGC catalog number, **NGC 604**. It is a cloud of hot hydrogen gas similar to the Orion and Tarantula nebulae where stars are being formed. Try looking for it with a nebula filter!

Another member of our Local Group along with the galaxies in Andromeda, the Triangulum Galaxy is located about three million light years from us, not much farther from us than the Andromeda Galaxy. Indeed, they're less than a million light years from each other, and may be orbiting each other. Stargazers in the Andromeda Galaxy should have a lovely view of the Triangulum Galaxy, and vice versa.

This galaxy has the mass of nearly 10 billion Suns. It is a classic example of a spiral galaxy, except that it is curiously lacking a bright central core. That is one of the reasons it is so difficult to pick out in a small telescope (unlike the Andromeda Galaxy, whose bright core shows where to start looking).

For more on galaxies, see page 109.

Also in the neighborhood: For those nights that just aren't good enough to find the Triangulum Galaxy, you can take solace from your frustration in a much easier object, a nearby open cluster. Note the location of **NGC 752**, *marked on the finder chart. To find it, call the distance from Gamma to Beta Trianguli one step; go from Gamma, to Beta, and continue two more steps. Though not visible to the naked eye, it should be visible as a fuzzy dot in the finder and jump out at you, a loose open cluster, in a low-powered telescope, near what looks like a very wide double (the optical pair* **56 Andromedae** *– see page 180). It holds about 100 stars, located some three thousand light years from us.*

find more at: www.cambridge.org/features/turnleft/seasonal_skies_october-december.htm

Andromeda doubles

Cassiopeia

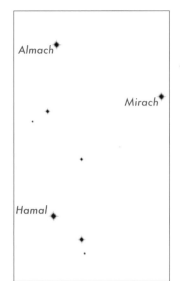

Almach

Mirach

Hamal

The Pleiades

The Great Square

Star maps courtesy Starry Night Education by Simulation Curriculum

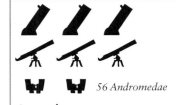

56 Andromedae

Any sky
High power
Best: Sept.–Feb.

Star	Magnitude	Color	Location
Almach (Gamma Andromedae)			
A	2.3	Yellow	Primary star
B	5.1	Blue	9.8" ENE from A
59 Andromedae			
A	6.1	Blue	Primary star
B	6.7	Blue	17" NE from A
56 Andromedae			
A	5.8	Yellow	Primary star
B	6.0	Yellow	200" WNW from A
C	11.9	Dim	18" ESE from A
6 Trianguli (also known as Iota Trianguli)			
A	5.3	Yellow	Primary star
B	6.7	Blue	3.7" ENE from A
Lambda Arietis			
A	4.8	Yellow	Primary star
B	6.6	Blue	37" NE from A
Mesarthim (Gamma Arietis)			
A	4.5	White	Primary star
B	4.6	White	7.5" N from A

S
W
E
N
5' circle

Almach: Locate the Great Square, high overhead. From the northeast corner, find three bright stars in a long line, arcing across the sky west to east, below (south of) the big W of Cassiopeia. Almach is the third and eastern-most of these three stars. The pair shows a distinct color contrast: the primary is orange–yellowish, the secondary is blue.

This is actually a quadruple star system, about 200 light years from us. Almach A is a giant K star, larger but cooler than our Sun. The others are all dwarfs. Star B, the main companion, orbits A at about 600 AU. The others, C and D, orbit B. Star C is in a 64-year orbit around B, too close to split in a Dob, while B and C together take thousand of years to orbit A.

Almach C's orbit is quite elliptical. In 2002, C reached its closest approach to B, less than 0.1" separation, beyond the reach of any amateur telescope. But by 2024, it reaches its greatest separation when B and C are almost two thirds of an arc second apart; with excellent conditions, by that time you might just be able to resolve them in a Dob.

The B/D system is a *spectroscopic binary*. From shifts in the lines of star B's spectrum, we calculate that there must be a star D only a few million miles from B (a few percent of the distance from the Sun to the Earth) and orbiting B in less than three days.

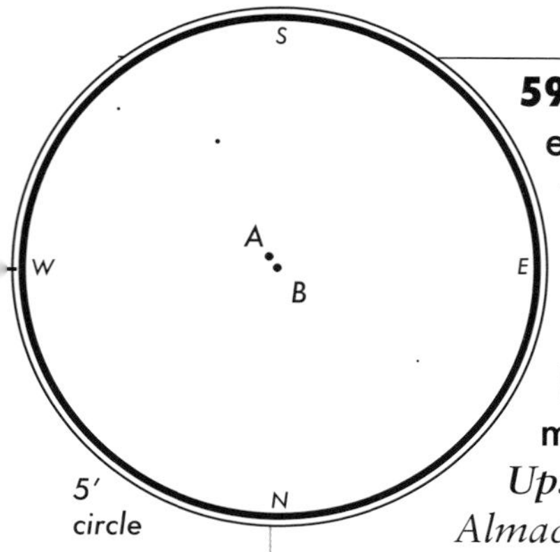

59 Andromedae: At the southern edge of the Almach finderscope field is a sixth-magnitude star, 59 Andromedae. (Don't mistake it for a brighter star, 58 Andromedae, further south.) It's a pleasant pair of closely matched white stars.

Upsilon Andromedae, 3° west of Almach, is a Sun-like star 44 light years from us orbited by a whole system of planets. One of them is a two-Jupiter-mass planet with a period of eight months, in an elliptical orbit at a distance comparable to Earth and Mars. If it has large moons (like Jupiter does), they could be habitable…

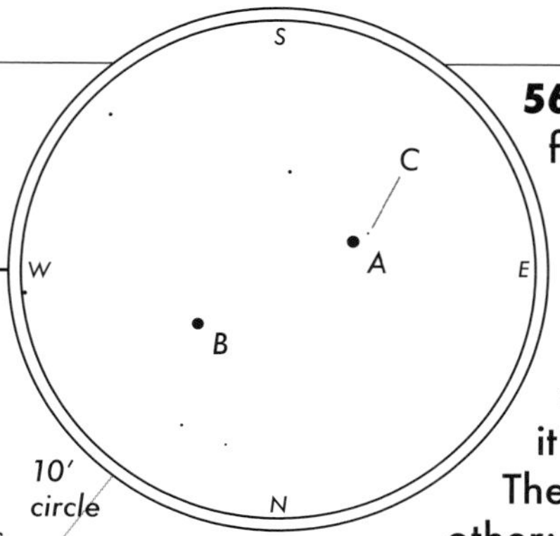

56 Andromedae: South of Almach find the second-magnitude star, Beta Trianguli. Aim at Beta, and notice a pair of sixth-magnitude stars about two degrees northwest. Widely separated, and near an open cluster (NGC 752), it's a fun object for binoculars. The stars don't actually orbit each other; one is 320 light years from us, the other over 1,000 light years away. Star A does have a true companion, a dim (magnitude 12) star 18" to its east-southeast: a nice Dob challenge.

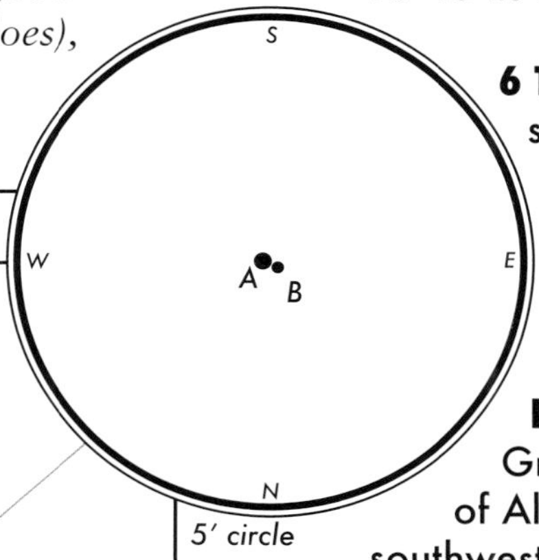

6 Trianguli: South of Almach find the second-magnitude stars, Beta and Alpha Triangula. Aim your finder at Alpha and look at the eastern edge of the field for a fifth-magnitude star, 6 Trianguli: another Dob challenge.

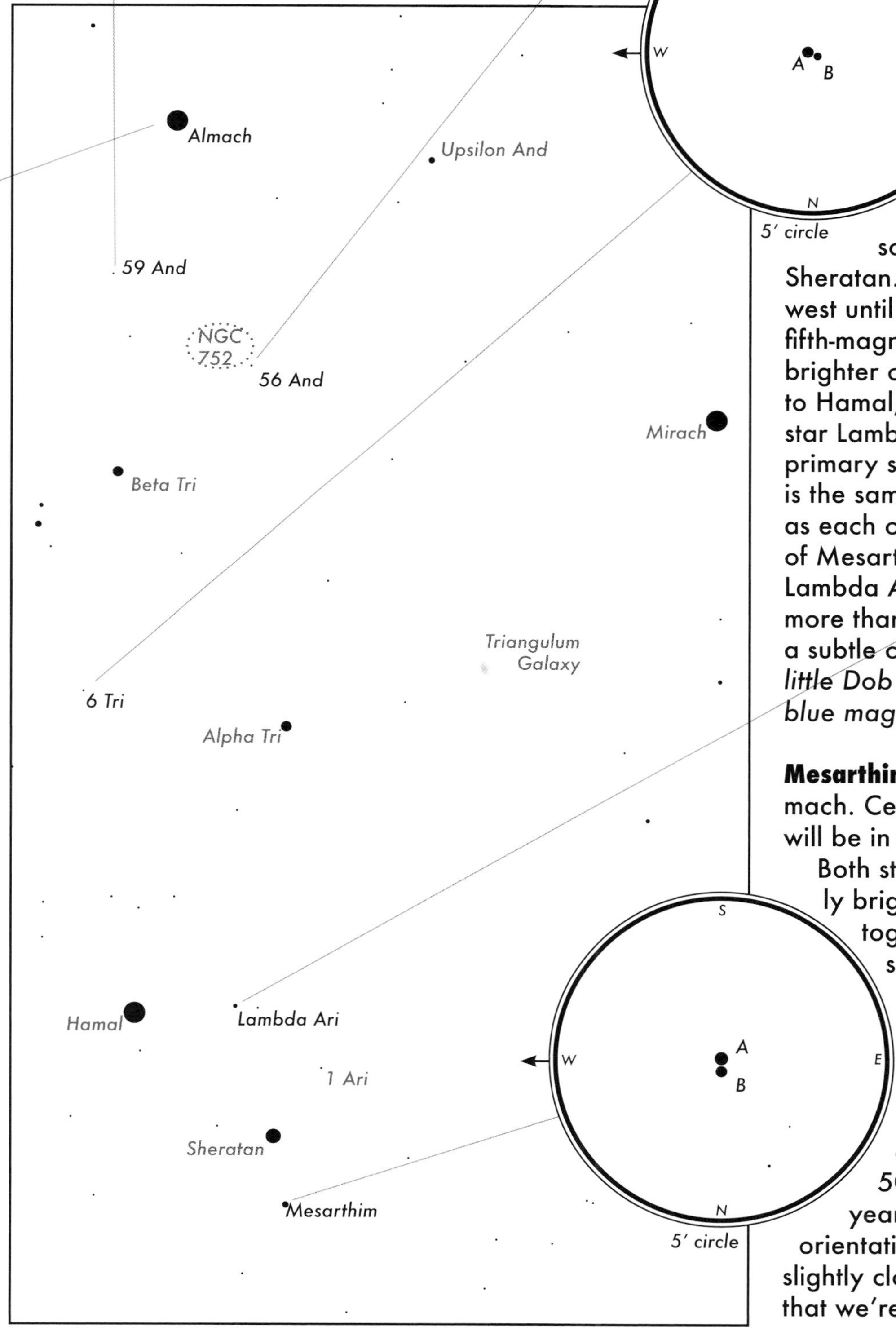

Lambda Arietis: Halfway between the Great Square and the Pleiades, south of Almach, are two stars in a northeast–southwest line, Hamal and Sheratan. From Hamal, move west until you see a pair of fifth-magnitude stars; the brighter of the two, closer to Hamal, is the double star Lambda Arietis. The primary star of this pair is the same brightness as each of the members of Mesarthim (below) but Lambda Arietis' companion is more than 10 times fainter. Note also a subtle color contrast. *(Nearby, 1 Arietis is a lovely little Dob double: an orange mag. 6.3 primary with a blue mag. 7.3 companion 2.9" to the south-southeast.)*

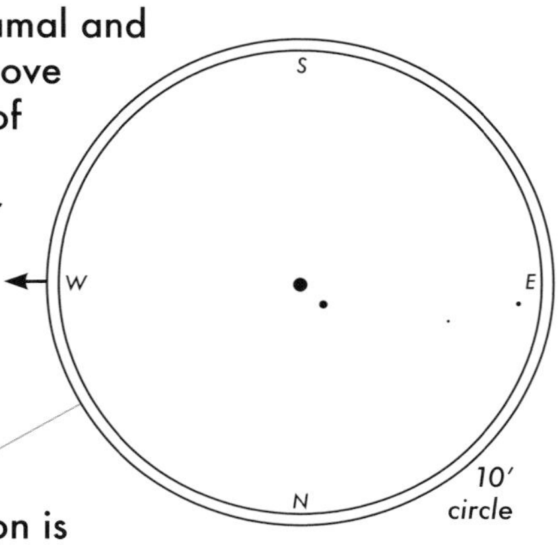

Mesarthim: Find Hamal and Sheratan, south of Almach. Center Sheratan in the finderscope; Mesarthim will be in the same finder field of view, to the south. Both stars in Mesarthim appear to be about equally bright, and blue–white. They're fairly close together, so you'll want your highest power to separate them. These stars are quite fun because they are so evenly matched in brightness, like cat's eyes staring back at you. Each star is between three and four times as massive as our Sun. They are about 200 light years from us, and orbit each other slowly, separated by at least 500 AU over a period of about of 3,000 years. Over the past 230 years their relative orientation appears unchanged but they have moved slightly closer together. From this we can conclude that we're probably seeing their orbit edge-on.

In Perseus:
An open cluster, M34

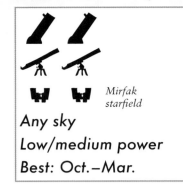

Star maps courtesy Starry Night Education by Simulation Curriculum

Cassiopeia

Capella

Delta

Mirfak

Algol

Almach

The Pleiades

The Great Square

Mirfak starfield

Any sky
Low/medium power
Best: Oct.–Mar.

- Box of jewels
- Rich part of the Milky Way
- Nearby: famous variable star Algol

Where to look: The brightest star between Cassiopeia and the brilliant star Capella is a fairly bright (second-magnitude) star called Mirfak. Just to the south of Mirfak is another star, of nearly equal brightness, called Algol. (However, Algol is a variable star – see below.) The next bright star to the west from Algol is Almach. Draw a line between Algol and Almach; M34 is located just to the north of this line.

In the finderscope: Look not quite halfway along the line from Algol to Almach, then move a bit to the north. The cluster may be visible in the finderscope as a grainy patch of light.

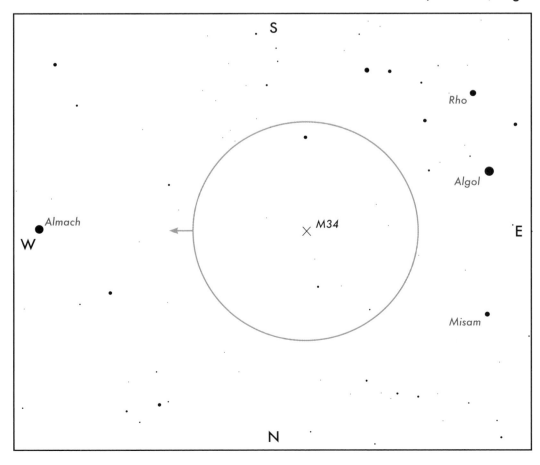

M34 in a star diagonal at low power

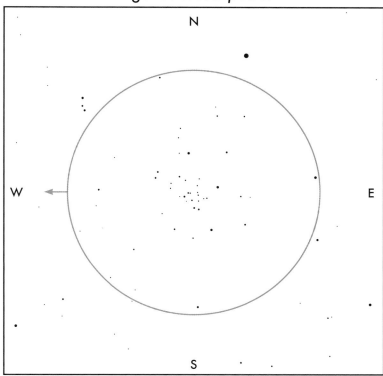

M34 in a Dobsonian at medium power

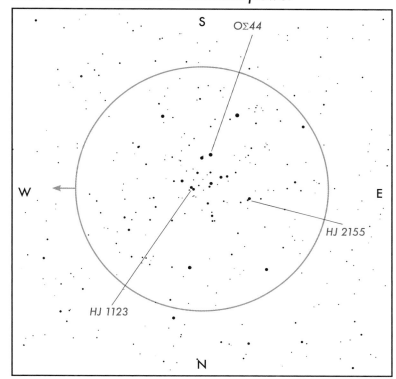

In a small telescope: The cluster is conspicuous at low power. You'll see a dozen or so moderately faint stars in a fairly small area, with another dozen spread further out, rather evenly distributed. The bright star south-south-east of the cluster's center has a distinctly orange color.

In a Dobsonian telescope: Find in low power, but observe at higher power. In among the myriad of stars in this cluster are a number of nice double stars. **Otto Struve 44** is a Dob challenge, magnitude 8.3 and 9.0 stars, separated by only 1.4" to the northeast; but the bright star an arc-minute-and-a-half to its west-northwest is also a member of the group. **HJ 2155** is a magnitude 8.3/10.3 pair separated by 17" (secondary to the northwest). **HJ 1123** is easier, magnitudes 8.4 and 8.5 with a 20" separation (secondary to the southwest).

When looking for **M34**, don't be fooled by the background stars of the Milky Way. This can be a problem, especially with bigger telescopes or higher-power eyepieces.

The cluster is quite distinct and pretty, looking like a box of jewels. There are about 80 stars in M34, filling a region in space about five light years across. Current estimates place this cluster at 1,500 light years from us. It's a fairly middle-aged open cluster, a bit more than 100 million years old. For more on open clusters, see page 71.

Also in the neighborhood: Algol is perhaps the most famous variable star in the sky. It is an eclipsing binary, changing brightness when its dim companion orbits in front of the brighter primary star. Normally a second-magnitude star, it drops to magnitude 3.5, less than a third of its usual brightness, for 10 hours out of a period of a little less than three days.

The effect is quite noticeable, and fun to look for – but it can throw you off if you're looking for it as a guidepost star when it's at its dimmest. Compare it to Delta Persei, a magnitude 3.1 star just to the southeast of Mirfak. When Algol is deep in eclipse, it is noticeably dimmer; the rest of the time, it is considerably brighter than Delta.

Just south of Algol is another variable, **Rho Persei.** *A red-giant star, the fusion processes that power this star are*

unstable, causing it to fluctuate in brightness. It varies from magnitude 3.3 to 4.0, with a period ranging from five to eight weeks. Compare this star with Kappa Persei, a constant magnitude 4.0 star. It's located half again as far north of Algol as Rho is to the south.

The starfield near **Mirfak,** *with a nice sprinkling of fairly bright stars, is a particularly pretty one for binoculars or a finderscope. However, the stars are too widely spaced for a telescope to show them to best effect.*

Due west of M34 is **Almach,** *a double star described on page 180.*

find more at: www.cambridge.org/features/turnleft/seasonal_skies_october-december.htm

Northern skies

For most of the readers of this book, those who live north of 35°N, the objects in this chapter will be visible on any clear night of the year at some time during the night. For observers from the southern hemisphere, these objects represent sights you'll have been missing until you find yourself in the northern hemisphere. (If you do get to travel north, be sure to bring a small telescope…or at least a pair of binoculars!)

However, even though these objects are indeed always above the horizon, obviously some objects will be easier to see in December, and others in June, unless you're willing to get up in the wee hours of the morning or look low in the sky (where the viewing conditions are never at their best).

Note that for these objects, you'll always be looking northwards (of course) and so your usual sense of "up" and "down" in the telescope will be upside down compared to when you look southwards. (Notice in the pictures below that "north" is at the bottom of the picture.) Thus for the objects in this chapter we have reversed the orientation of north and south

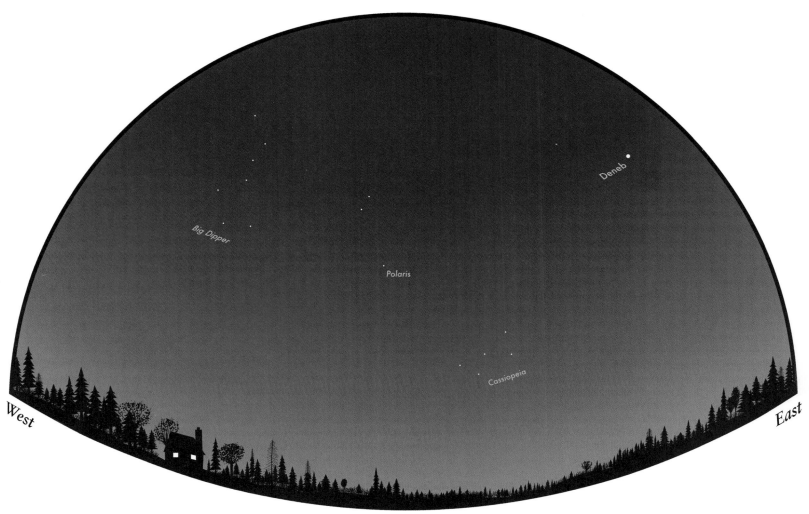

Looking north, June

Finding your way:
Northern sky guideposts

The key constellations to find in the north are the *Big Dipper* (known variously as The Plough, The Carriage, The Saucepan…and, technically, not a constellation itself but just the brightest part of Ursa Major, the Big Bear) and *Cassiopeia*, with five stars in the form of a big W.

From January through June, the Big Dipper will be higher in the sky during the evening hours, and then for the second half of the year Cassiopeia holds the upper sky. Both are distinct, bright, and unmistakable.

Between these two star groups is the second-magnitude star **Polaris**, which marks (to within one degree) the location of the north celestial pole. Face Polaris and you are always facing north. All the other stars in the north move counterclockwise around Polaris, which itself appears to stand still in the sky through the night and throughout the year. Note in particular the two second-magnitude stars between Polaris and the end of the Dipper's handle; they are called *The Guardians* as they appear to march around Polaris during the course of a night.

The Big Dipper is the location of one of the easiest double stars (**Mizar**) and a number of galaxies, and its shape lends itself to use as a guidepost in a number of directions.

The most famous indicators are the *Pointer Stars*, the endmost two stars of the dipper's bowl, which point almost directly towards Polaris. These stars lie quite close to the 11-hour line of right ascension – the astronomical measure

in the finder and telescope views compared to how they are oriented in the other chapters. For instance, the finderscope view still assumes that your finder inverts what you see from a naked-eye view; but when you are looking northward, the uninverted "top" of the view would have been south so our doubly inverted view is north.

If that's too complicated to follow, just trust us: the orientations we have used in this chapter are what feels most natural for most telescope users when you have the book with you at the telescope.

We keep north to the top for the naked-eye views in order to connect up these charts with the naked-eye views of the other seasons.

To the south: *If you are visiting from the southern hemisphere, many objects that you've only been able to see low in the murk of the northern horizon will take on a whole new aspect. Be sure to take a look at these, some of our northern favorites:*

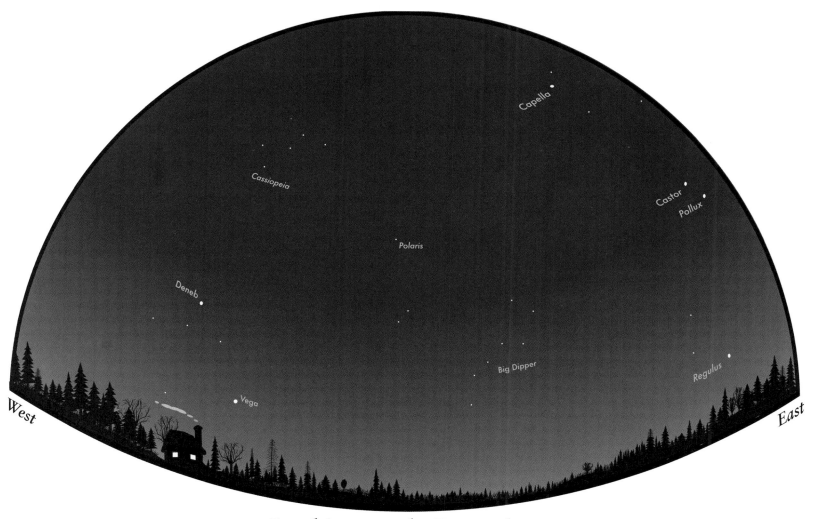

Looking north, December

of longitude on the celestial sphere, used to describe the precise location of stars in tables such as can be found on pages 246–251. The other two stars of the bowl point across the sky towards **Deneb**, one of the bright stars that make up the *Summer Triangle*. And the curve of the handle directs you towards the April–June guidepost stars Arcturus and Spica – "arc to Arcturus and spike to Spica." South of the Big Dipper you'll find *Leo* with its bright star **Regulus**.

Cassiopeia is home to a number of nice doubles and, most famously, a large number of fascinating open clusters. Only three of its stars have proper names, but all have Greek-letter designations; we use just the Greek letters here for simplicity.

The various segments of the W will point to the neighborhood of a number of interesting objects. The important thing is

to keep track of which side of the constellation you're looking. That's trickier than it sounds when you're looking at it upside down or at an odd angle, especially when it is nearly overhead.

Think of it being made of two V shapes. The V to the left is a soft, open V while the one to the right (which has a few extra third-magnitude stars not shown here) is more of a sharp V with a 90° angle. The open V is situated near Perseus; the sharp V is near Cepheus. South of the W are the three second-magnitude stars of the constellation Andromeda.

The rightmost star of the W, the end of the sharp V, lies near the zero-hour line of right ascension. Draw a line from Polaris to it, and think of that as the hour hand of a 24-hour clock…that runs backwards…only keeps standard time… and gains two hours a month for every month past March 21.

In Perseus:
The *Double Cluster,*
two open clusters,
NGC 869 and NGC 884

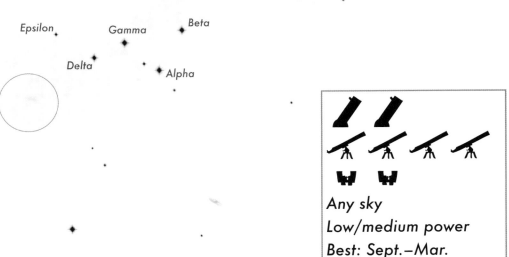

Star maps courtesy Starry Night Education by Simulation Curriculum

Any sky
Low/medium power
Best: Sept.–Mar.

- Two great open clusters in the same field of view
- Easy to find and see
- Nearby colorful double star

Where to look: Find Cassiopeia high overhead, looking like a large W. Call the distance from Gamma to Delta Cassiopeiae one step; step down a line from Gamma, through Delta, to a spot two steps beyond Delta.

In the finderscope: You should see a fuzzy patch visible in the finderscope. In fact, it can be seen with the naked eye even in suburban skies, and it is quite distinct in binoculars. That's the double cluster.

Also in the neighborhood: To the southeast of the double cluster is a triangle of third- and fourth-magnitude stars that stretch across the finderscope field, pointing to the north. The orange fourth-magnitude star at the northern point of the triangle, **Miram**, is a double star. (Miram should not be confused with another star in Perseus, Misam, mentioned on page 182.) Its blue companion, magnitude 8.6, is 28 arc seconds to the east-northeast. Though the companion is dim, the color contrast is lovely.

The Double Cluster in a star diagonal at low power

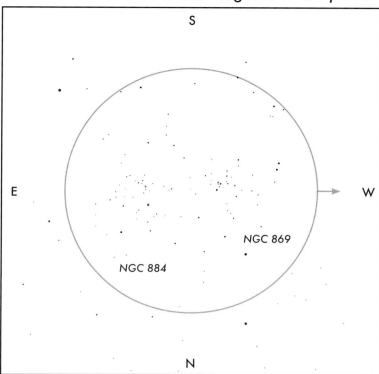

The Double Cluster in a Dobsonian, medium power

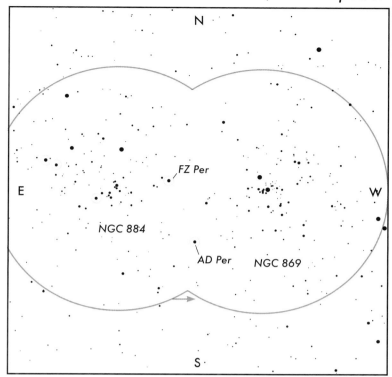

In a small telescope: The cluster closer to Cassiopeia is NGC 869; the farther one is NGC 884. In NGC 869 you should see about three dozen stars, most of them confined to a circular region half the size of the full Moon, with two stars distinctly brighter than the rest. In the center of this cluster, you'll see a patch of grainy light, the light from stars too faint and close together to be resolved individually. The other cluster, NGC 884, shows about 30 stars, somewhat more widely spread out than NGC 869. There seems to be a hole in the center of this cluster, where few stars are visible.

In a Dobsonian telescope: Low power helps you find the cluster and see both at once, but medium power lets you enjoy each cluster individually.

Notice the reddish stars circled in this figure; most of them are in NGC 884. They are M2-class supergiants and all are variable.

We label two of them, easily distinguishable between the clusters. FZ Persei varies from magnitude 9.8–10.8 over a period of about 6 months. AD Persei varies from magnitude 7.7–8.4, with a period of about a year.

The **Double Cluster** is much prettier in a smaller telescope than in a larger one. The haze of light behind the bright stars of NGC 869 is especially nice. You'll want to use your lowest power to get both clusters in the field of view at the same time.

Several stars in these clusters are distinctly reddish variable stars. Look straight at them; the edge of the eye may be best for dim objects, but the center of your eye picks up color better. One of these stars is between the central parts of the two clusters, a bit to the NGC 884 side (dead center on the telescope-view drawing). Another is to the southwest, almost half the way out to the edge. There are two others next to each other, two thirds of the way out towards the eastern edge of the telescope view, at the eastern edge of NGC 884. There is a fainter one just to the southeast of that cluster's center.

NGC 869, also known as *h Persei*, is a compact grouping 70 light years in diameter, located about 7,500 light years away from us. NGC 884 is also known as *Chi Persei*. (That's the Greek letter χ; the "ch" is pronounced as a hard c, as in "Christmas".) It is of similar size and distance as its near neighbor, about 70 light years across and again perhaps 7,500 light years away. The brightest of these stars are supergiants some 50,000 times as bright as the Sun.

Despite the presence of more red stars visible in NGC 884 than in NGC 869, it is apparently not older; the latest estimates

put the age of both at about 13 to 14 million years old. There is one suggestion that NGC 869 may be one or two hundred light years closer to us than NGC 884; even if that's the case, they're still very close to each other.

How many stars are there in these clusters? Exact counts are somewhat uncertain because there are dark clouds of dust between us and them. However, recent estimates suggest that there are at least 5000 stars in the core of each cluster (with NGC 869 the slightly more massive of the two); the entire region may have anywhere from 13,000 to 20,000 stars, with a total mass of at least 20,000 times the mass of our Sun. That would make them, combined, about 10 times as massive as either the Pleiades or the stars associated with the Orion Nebula, M42.

The view from a planet in one of the clusters would be spectacular: perhaps a hundred stars in the home cluster would look brighter than the brightest star in Earth's sky, while the other cluster would be far more impressive than any open cluster in our sky. Any such planet, though, would be very young (if indeed it would have had time to be formed at all)…on the order of only five million years old.

For more about open clusters, see page 71.

find more at: www.cambridge.org/features/turnleft/northern_skies.htm

Cassiopeia double stars

Iota Cassiopeiae is a quadruple star, of which only three components are visible in a small telescope. To find it, start at the eastern stroke of the big W, Epsilon Cassiopeiae and Delta Cassiopeiae. Step from Delta to Epsilon; one step farther in this direction brings you to Iota Cassiopeiae.

The A/C pair are easily split but C is much fainter than A, making it a challenge in a small telescope; it's easier in a Dob. The challenge for a Dob is splitting A/B, since B is so close to the primary. On a steady night it can look like a bulge on the brighter star. The stars in this complex are located 140 light years away from us. No colors are visible in its components.

Half a step farther northeast of Iota Cassiopeiae are two sixth-magnitude stars, SU and RZ Cassiopeiae. (SU is the one to the south, with a faint neighbor to its southwest.) RZ is perhaps the most dramatic variable star in the sky. An eclipsing binary, roughly every two days it fades within two hours' time from magnitude 6.4 to 7.8; two hours later, it will be back to its original state. SU Cassiopeiae also varies, going from magnitude 5.9 to 6.3 over two days' time.

Star	Magnitude	Color	Location
Iota Cassiopeiae			
A	4.6	White	Primary star
B	6.9	White	2.9" SW from A
C	9.0	White	7.0" ESE from A
Struve 163			
A	6.8	Orange	Primary star
B	9.1	Blue	34" NE from A
C	10.7	Blue	114" WSW from A
Achird (Eta Cassiopeiae)			
A	3.4	Yellow	Primary star
B	7.4	Red	13" NW from A
Burnham 1			
A	8.6	White	Primary star
B	9.3	White	1.1" E from A
C	8.9	White	3.7" SE from A
D	9.7	White	8.5" SSW from A
Struve 3053			
A	6.0	Orange	Primary star
B	7.3	Blue	15" ENE from A
Sigma Cassiopeiae			
A	5.0	Blue	Primary star
B	7.2	Blue	3.2" NNW from A

Struve 163 is a multiple star near Epsilon Cassiopeiae. From Epsilon, look north-northeast (away from the W) for a small triangle of fifth- and sixth-magnitude stars. This triangle points to the west, towards Struve 163. If the triangle is one step to the north-northeast of Epsilon, then Struve 163 is one step to the north-northwest. (Another way is to put Epsilon in your low-power telescope view, and move north-northeast. As it leaves your field of view, Struve 163 should just be entering it.) Struve 163 is a dim but colorful double, consisting of a seventh-magnitude orange primary star, with a ninth-magnitude blue companion located 35" to the north-northeast. With a Dob you can also spot the fainter, more distant companion C.

Achird (Eta Cassiopeiae) is located between the middle and fourth stars of the W, Gamma and Alpha, closer to Alpha. It is a fairly close double, and the primary star is almost four magnitudes (about 40 times) brighter than its companion, so steady skies and high power are needed to separate the two. It shows a sharp color contrast; the primary star is yellow (perhaps with a touch of green) while its companion is reddish. The colors stand out better with more aperture (i.e. in a Dob) under high power and a twilight sky; in a three-inch under high power, the faint red companion is just barely visible, with an almost ghostly appearance.

The brighter of these stars, star A, is a G-type star very similar to our Sun, hence its yellow color. It has about 10% more mass than our Sun and is about 25% brighter. The smaller star, B, has half the mass of our Sun packed into a quarter the volume and it is about 25 times dimmer than the Sun. From its reddish color one can infer that it's a much cooler star (spectral type M).

The stars are located just 19.4 light years from us, making them close neighbors. They are, on average, about 80 AU apart and take 500 years to orbit each other. Back in 1890 they were at their closest, and they have been moving apart since that time. In a telescope, this means that their separation can be seen to vary over the centuries from a minimum of 5 arc seconds to a maximum of 16 arc seconds. For the next few decades, they will be separated by about 13 to 14 arc seconds.

They also visibly move around each other, at roughly one degree per year. Back in the 1940s the dimmer star was west of the brighter star; nowadays it's more to the northwest.

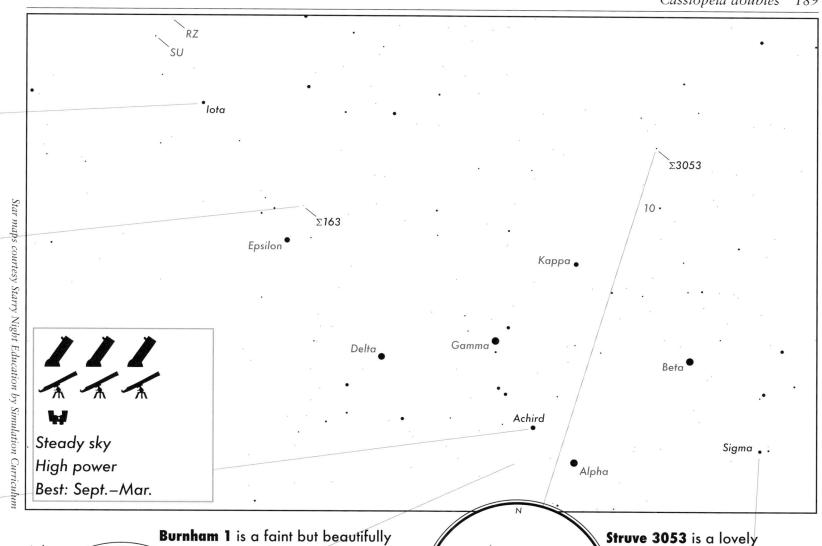

Steady sky
High power
Best: Sept.–Mar.

Burnham 1 is a faint but beautifully complex collection of four stars, eighth to tenth magnitude, set in faint nebulosity (NGC 281, popularly known as the "Pacman Nebula" – it looks like it is about to gobble up Burnham 1). To find them, imagine an equal-sided triangle of Achird and Alpha Cassiopeiae and an empty point southeast of these two stars. Start from Eta in your lowest power, and move down the leg of that triangle towards the empty point. You'll pass a red magnitude 7 star, then see a collection of eighth-magnitude stars including Burnham 1. If you reach another red seventh-magnitude star, you've gone too far.

circle: 10'

Struve 3053 is a lovely double star with remarkable orange–blue colors, but a challenge to find. Start at the middle star of the W and step northwest to the fourth-magnitude star, Kappa; one step farther brings you to a fifth-magnitude star, 10 Cassiopeiae. From there, look northward for three sixth-magnitude stars in a north–south line. The southernmost of these, slightly farther from the other two, is Struve 3053. An easy split, in a 3" the fun comes merely from the challenge of finding it; with the greater aperture of a Dobsonian, the colors leap out. The pair are located roughly 2,000 light years from us, and are separated by nearly 10,000 AU.

5' circle

Sigma Cassiopeiae is a fifth-magnitude star southwest of the easternmost star in the W, Beta Cassiopeiae. From Beta move south and west, looking for a pair of stars running east–west, coming into view as Beta moves out. Sigma is the eastern, slightly brighter star of that pair. It's a tight pair, seen better in a Dob, about 1,500 light years away.

5' circle

Achird

red m7 star

Burnham 1

m7 star – you went too far

Alpha

In Cassiopeia: The Cassiopeia open clusters

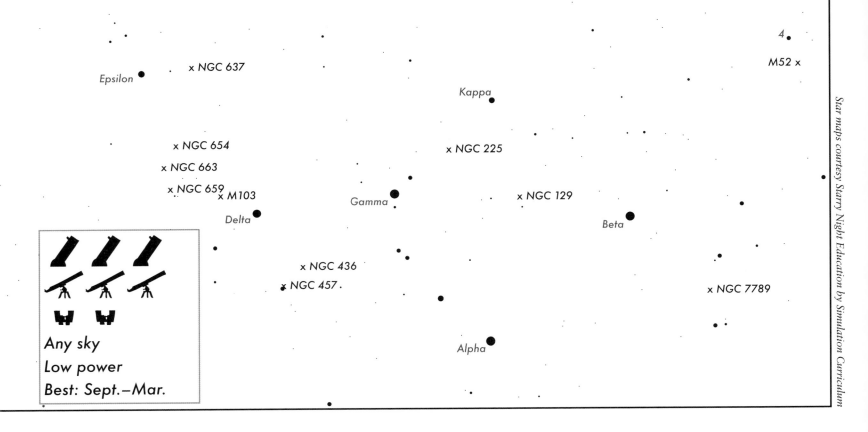

Star maps courtesy Starry Night Education by Simulation Curriculum

Any sky
Low power
Best: Sept.–Mar.

- A dozen different opens in one constellation
- Fascinating mixture of sizes and styles
- Rich region of Milky Way

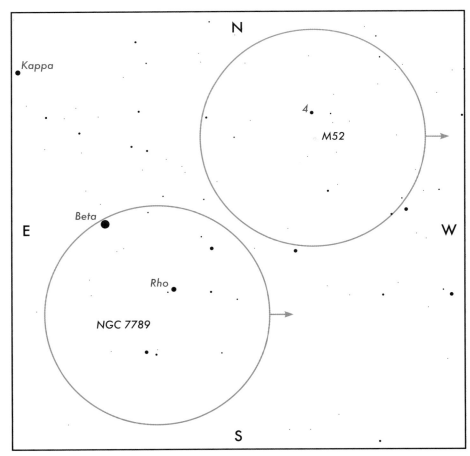

Western Cassiopeia in the finderscope: The three western stars of Cassiopeia's W shape, Alpha, Beta, and Gamma, make a box with a fainter fourth star, Kappa Cassiopeiae. **NGC 7789:** Step from Kappa to Beta, and continue half a step farther. In the finderscope you'll see two bright stars, each with a nearby dimmer companion. Aim for a spot halfway between these two pairs. **M52:** Step from Alpha to Beta; one step farther in this direction places you near a dim star, 4 Cassiopeiae. In the finderscope, find this star and aim for a fuzzy spot of light just to the south of it.

NGC 7789 is a very large and unusually old open cluster of stars. It's estimated that nearly 1,000 stars are in this cluster, filling a region of space 35 light years wide. Most of the stars in this cluster have evolved into red giants or supergiant stars, indicating that the cluster may be well over a billion years old. The cluster lies more than 7,000 light years away. M52 consists of about 200 stars in a region 15 light years in diameter. It lies about 5,000 light years away from us.

Note **Rho Cassiopeiae.** *It is an unstable supergiant star, one of the brightest stars in our Galaxy but several thousand light years distant from us. It is a variable star, apparently shedding large quantities of material now and again. This sort of behavior cannot continue indefinitely; this star may be on its way to becoming a supernova.*

NGC 7789 *in a star diagonal at low power*

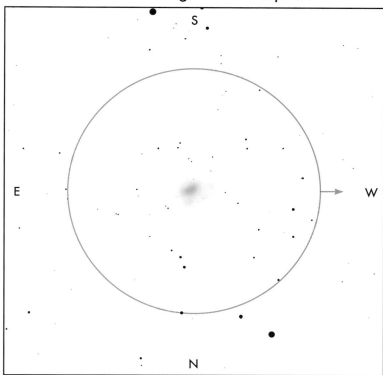

In a small telescope: The cluster looks like a large, round, dim disk of light. The disk does not appear to be smooth, but grainy, giving a hint that with just a little more resolution one would begin to see individual stars in this cluster: many of the stars are around magnitude 11, near the limit of a 3" telescope.

NGC 7789 *in a Dobsonian at medium power*

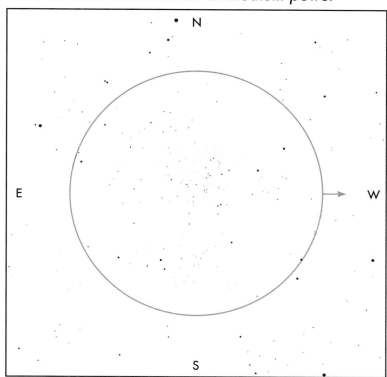

In a Dobsonian telescope: At low power, the cluster retains a sense of unresolved mystery; higher power resolves much of that mystery into a beautiful sparkling of dots against the last remaining bit of background haze.

M52 *in a star diagonal at low power*

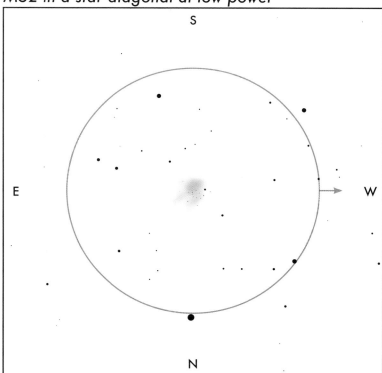

In a small telescope: You'll see a handful of individual stars set in a faint hazy field of light, seen best with averted vision: lumpy and rich, as if scores of stars are almost but not quite resolved. The cluster is dominated by an eighth-magnitude star on the western edge, with several dimmer stars resolved to its east.

M52 *in a Dobsonian at medium power*

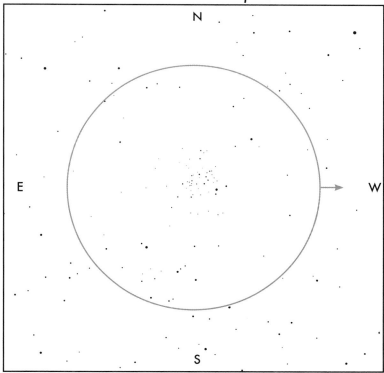

In a Dobsonian telescope: In medium power, you should be able to make out about 30 clearly resolved stars, with a hint of more in the background haze, including a number of strings of stars. Note the eighth-magnitude star at the southwest edge of the field of view which dominates the group.

find more at: www.cambridge.org/features/turnleft/northern_skies.htm

Central Cassiopeia in the finderscope: From Gamma, move northwest towards Kappa; look for **NGC 225** halfway from Gamma to Kappa.

Starting again from Gamma, find the star Phi in the direction opposite from Kappa; **NGC 457** stretches out to the northwest from this star.

Finally, find **NGC 129** halfway between Gamma and Beta. It lies near a sixth-magnitude star, HR 113.

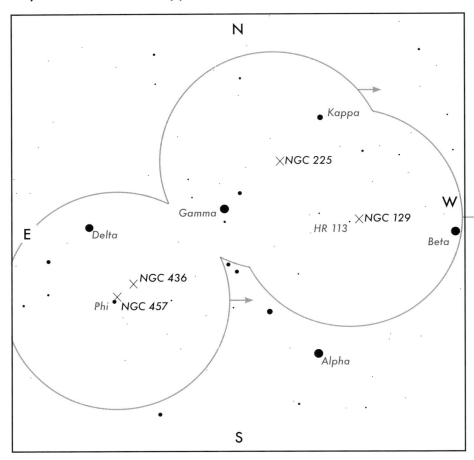

NGC 225 has about 20 bright members, and an estimated 50 or so dimmer ones. The cluster is about five light years wide, and it is located just under 2,000 light years away.

NGC 129 is a cluster of 50 stars in a region 20 light years across, located about 5,000 light years away from us.

NGC 457 is a bit over 5,000 light years away from us, some 200 stars, in a region about 20 light years in diameter. Is Phi is actually a member of this cluster? It is much brighter than the other stars in the cluster, but it appears to be moving in space at the same rate as the other stars. It is a yellow super giant, consistent with what you'd expect for a star in the cluster that had evolved through its red-giant phase. If it really is part of the cluster, and as far from us as the other cluster stars, then it would have to be one of the brightest stars in the Galaxy.

Nearby **NGC 436** is a small cluster of about 40 stars spread over four light years, about 4,000 light years away from us.

NGC 225 in a star diagonal at low power

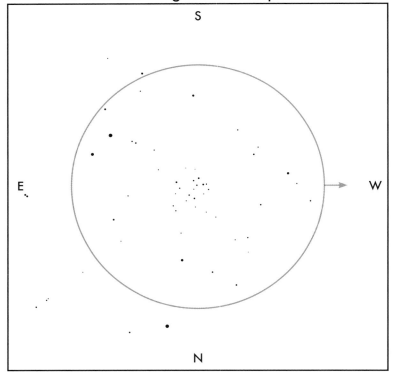

NGC 225 in a Dobsonian at medium power

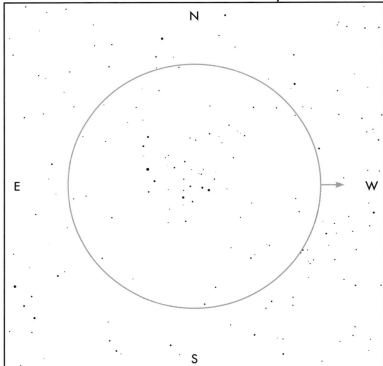

In a small telescope: NGC 225 looks like roughly a dozen stars in an elegant little half-circle, with a pair of stars inside the circlet. Using averted vision, you'll see hints of many more stars, including a hazy patch of light at the western side of the circlet.

In a Dobsonian telescope: About twenty stars are visible, in a very loose cluster even at low power. As a result, this cluster is not particularly well suited for a Dob; try medium power to bring out the cluster against the background fainter Milky Way stars.

NGC 436/457 in a star diagonal at low power

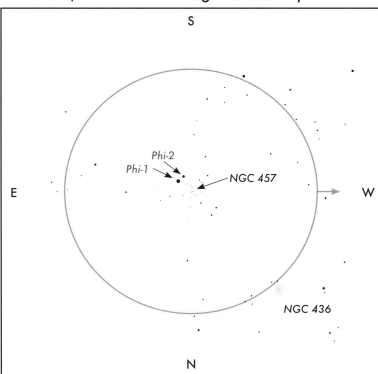

In a small telescope: NGC 457 is a group of perhaps two dozen stars of greatly varying brightness, spread out like wings. It's one of the nicer open clusters in the northern sky, one of the best of the ones that Messier missed. Nearby, NGC 436 is a small, faint haze of light.

NGC 436/457 in a Dobsonian at low power

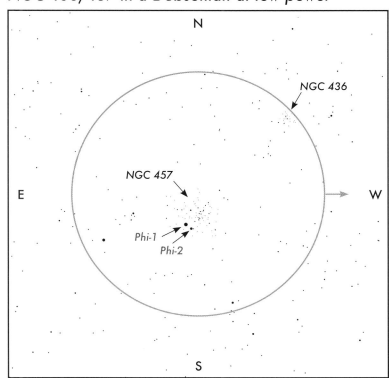

In a Dobsonian telescope: NGC 457 is remarkably rich, perhaps 80 stars visible, dominated by Phi-1 and Phi-2 at the southwest edge of the group (magnitudes 5 and 7, respectively). NGC 436 is a little clump of about two dozen stars, all but six or so quite dim, in the northwest corner of the field of view.

NGC 129 in a star diagonal at low power

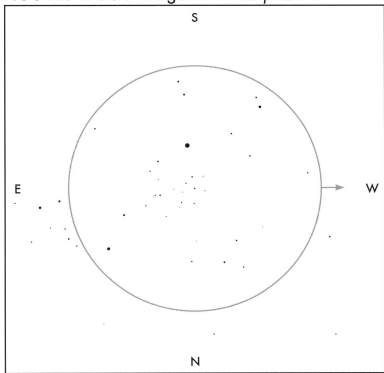

In a small telescope: NGC 129 is large, but not particularly dense. About 15 stars are visible in a small telescope, all of ninth-magnitude or fainter. There's a noticeable lack of stars in the center of the cluster. It's best in a small telescope, suppressing the background Milky Way.

NGC 129 in a Dobsonian at low power

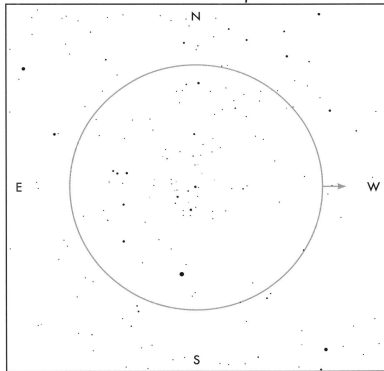

In a Dobsonian telescope: The cluster is not particularly obvious in a Dob. You can make out two loose groups with a gap between them, about 25 stars in all, including eight relatively bright ones; the rest fade into the background of the Milky Way.

find more at: www.cambridge.org/features/turnleft/northern_skies.htm

Eastern Cassiopeia in the finderscope: Aim at Delta and slowly move northeast, towards Epsilon, looking in the telescope itself for **M103**. Center on M103, and note your position again in the finderscope. Call the distance from Delta to M103 one step; one step beyond takes you to a clump of sixth-magnitude stars, and a farther step brings you to **NGC 663**; the clusters NGC 654 and NGC 659 should be nearby in your low-power telescope field.

Starting at Epsilon, in the finderscope you'll see a magnitude 5.5 star just to the west of Epsilon, HR 511. Step from Epsilon to HR 511; one step farther brings you to **NGC 637**.

M103 is a loose collection of about 50 stars in a region 15 light years wide. It is located about 8,000 light years from us, and is about 15 million years old.

NGC 663 has about 100 stars spread out over a width of 35 light years, located roughly 5,000 light years away from us. **NGC 654** is a small cluster, five light years in diameter, about 4,000 light years from us, with 50 or so stars. Many of the stars are bright in the infrared wavelengths; this implies that they are enmeshed in a large cloud of interstellar dust which absorbs and re-radiates their light. **NGC 659** is a cluster of 30 brighter stars (and perhaps 100 or more very dim ones) in a region about 10 light years in diameter, located 7,000 light years away from us.

NGC 637 is a very small cluster of about 20 stars, less than five light years wide, 5,000 light years away.

M103 in a star diagonal at low power

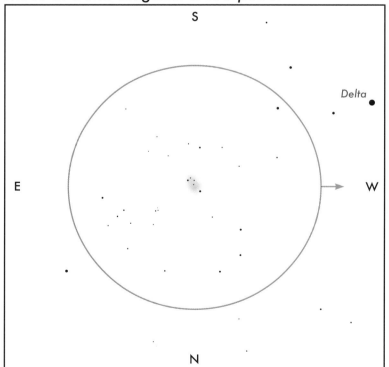

M103 in a Dobsonian at medium power

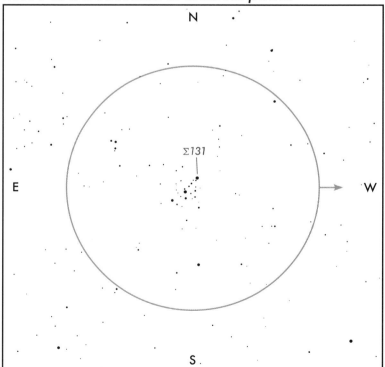

In a small telescope: M103 is a small, faint disk of light with only a few stars of the cluster easily visible. One of the stars is a distinct orange color. If you miss M103 when moving away from Delta, you may stumble on NGC 663.

In a Dobsonian telescope: Find in low, observe in medium power. See 30 or so stars in a little wedge, with the magnitude 7 double star Struve 131 (a field star, not a member of the cluster itself) visible at the northwest point of the wedge. (Struve 131 has a magnitude 10 companion 14" southeast of the primary star.)

NGC 659/663/654 in a star diagonal at low power

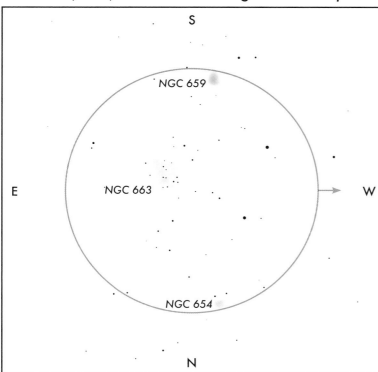

NGC 654/663/659 in a Dobsonian at low power

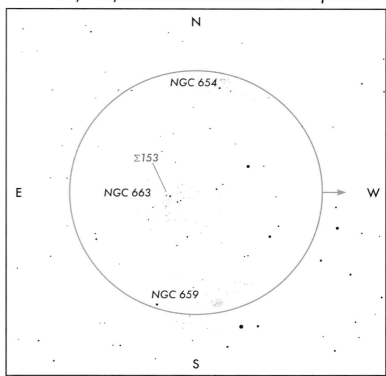

In a small telescope: NGC 663 is quite conspicuous, and quite possibly the nicest of the open clusters in this region. You'll see about 15 tiny yellow stars set like jewels in a background haze of light. NGC 654 is dim and hard to see in a 3" telescope; NGC 659 is a small, lumpy patch of light, also pretty challenging.

In a Dobsonian telescope: Find at low power, but explore at high; NGC 663, for one, has lots more to show, even if high power removes the last trace of mysterious unresolved haze. Look for some faint but fun doubles in the clusters; see in NGC 663, Struve 153 (magnitudes 9.4 and 10.4, with 7.7" separation to the east).

NGC 637 in a star diagonal at low power

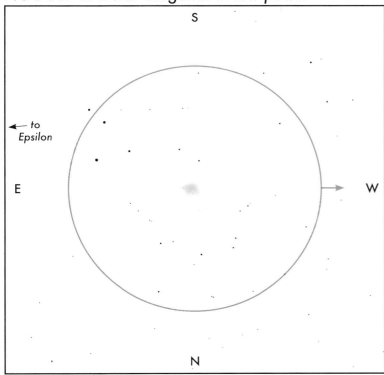

NGC 637 in a Dobsonian at medium power

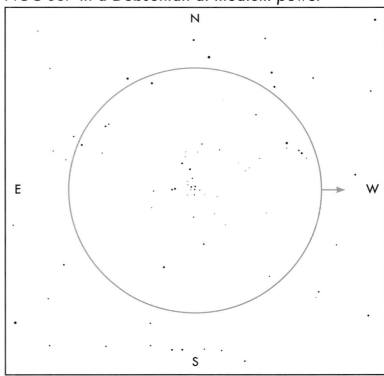

In a small telescope: NGC 637 is a dim little elongated patch of light. Within this patch, a few individual stars can be resolved, but one gets the impression that several more are just on the verge of being picked out.

To find it, put Epsilon in the scope and move west; but don't confuse this cluster with NGC 559 farther west.

In a Dobsonian telescope: A Dob resolves the cluster into a sweet collection of a dozen stars, loosely grouped against a background of Milky Way members. Try medium power.

find more at: www.cambridge.org/features/turnleft/northern_skies.htm

In Cepheus: The *Garnet Star,* Mu Cephei; and Delta Cephei and Struve 2816/2819, multiple stars

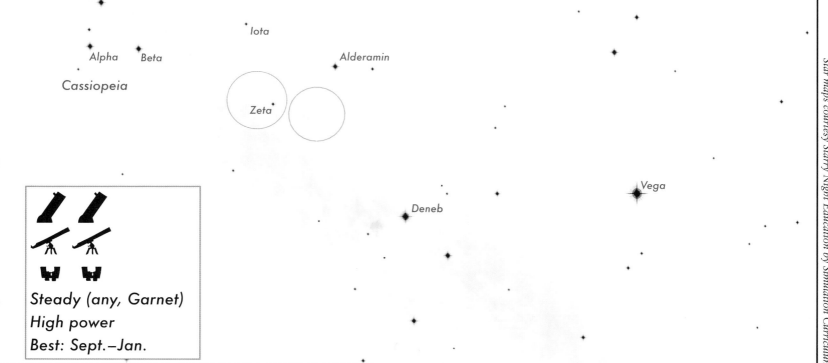

Star maps courtesy Starry Night Education by Simulation Curriculum

Steady (any, Garnet)
High power
Best: Sept.–Jan.

- Garnet star: glorious deep red color
- Struves: a double and a triple in a single field
- Delta Cephei: famous variable/double

Where to look: Locate Cassiopeia, the five stars which make a big W. Step from Alpha to Beta, the two stars making the rightmost stroke of the W, and continue in this direction until you see a triangle of third-magnitude stars: Alderamin, and Iota and Zeta Cephei. Alderamin is the brightest; start there.

In the finderscope: From Alderamin, move south and east towards Zeta. You'll pass two fifth-magnitude stars, 9 and Nu Cephei. From Nu, move straight south until you see a red fifth-magnitude star, the **Garnet Star**.

Struve 2816/2819: From the Garnet Star, follow a curl of fainter stars south and west, around to the multiple stars Struve 2816 and Struve 2819.

Delta Cephei: From the Garnet Star, look a finder field to the east for Zeta Cephei. It makes a nice triangle with a star to its south and Delta Cephei, to its east; Delta is the one at the narrow point of the triangle.

Struve 2819/2816 in a star diagonal at high power

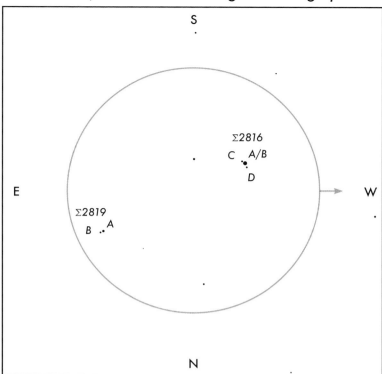

In a small telescope: The double stars are a fun split in a busy field of fainter Milky Way stars.

Struve 2819/2816 in a Dobsonian at high power

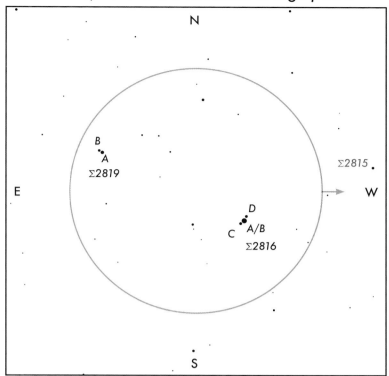

In a Dobsonian telescope: The doubles should be easily split in a Dob (the Struve 2816 A/B pair is too close and faint even for most Dobs). Note to the west, Struve 2815: primary, magnitude 8.6, and companions, magnitudes 10.0 (0.9" south-southeast) and 8.6 (7.5" east).

The **Garnet Star** is the name that Herschel gave to a star formally known as Mu Cephei. It is a red giant (hence its deep red color) and, as giants go, it is one of the largest. Recall the red giant Ras Algethi (page 124) whose diameter would, if placed where our Sun is, encompass all the terrestrial planets in our Solar System? The Garnet Star is more than five times bigger. It would enclose our Solar System out to the orbit of Saturn. The Garnet Star is 5,000 light years away from us.

Delta Cephei, besides being a pleasant double, is also a variable star. It's a Cepheid variable; in fact, it is *the* Cepheid variable, giving the name Cepheid to perhaps the most important class of variable stars. As part of its fusion process, a Cepheid pulsates with a regular period, usually a few days (5.3 days for Delta). But the period of the pulse is directly related to the luminosity of the star. That means when you observe a Cepheid's period, you can determine how intrinsically bright it is. That makes it a *standard candle*. Compute a Cepheid's intrinsic brightness from its period, observe how faint it looks to us, and (the fainter, the farther) you can calculate how far away it must be. Observing Cepheids in other galaxies is how Hubble (both the astronomer, Edwin Hubble, and later the space telescope named for him) measured how far away those galaxies are … and discovered that, the farther they are, the faster they appear to be moving away from us. This observation was the first

10' circle

confirmation of the *Big Bang* cosmology. From it we know how big and how old the Universe is.

Struve 2816 is 1,000 light years from us. Its companions orbit many thousands of AU from the primary star. The distance to **Struve 2819** is not well determined.

Surrounding the Struve doubles is a cloud of nebulosity known as IC 1396. Seeing this nebula is a challenge for a Dob on a very dark night. Photographs reveal that the nebulosity extends for nearly three degrees across the sky, bigger than six full Moons across, out to the Garnet Star. That's beyond the power of visual observers with an 8" telescope, however. The Garnet Star, Struve 2816, and presumably Struve 2819 are not connected with the nebula but only appear in the same line of sight; they're much farther away than the nebula.

Star	Magnitude	Color	Location
Delta Cephei			
A	3.5*	White	Primary star
B	6.1	White	41" S from A
	*variable: 3.5 – 4.4, 5.3 day period		
Struve 2816			
A/B	5.7	Yellow	Primary star
C	7.5	White	12" ESE from A
D	7.5	White	20" NW from A
Struve 2819			
A	7.4	White	Primary star
B	8.6	White	13" ENE from A

find more at: www.cambridge.org/features/turnleft/northern_skies.htm

In Draco: The *Cat's Eye*, a planetary nebula, NGC 6543

Dark sky
High power
Nebula filter
Best: Sept.–Jan.

Star maps courtesy Starry Night Education by Simulation Curriculum

Polaris

The Guardians

Altais

Aldhibah

Athebyne

The Big Dipper

- Fun, challenging hunt
- Elegant blue–green disk in high power
- Historical interest

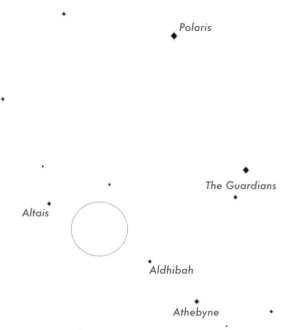

N

Dziban

Altais

28

27

E

× NGC 6543

W

42

Aldhibah

36

S

Where to look: Using the pointer stars in the bowl of the Big Dipper, find the North Star, Polaris. Look for the two Guardian stars that march around the pole between Polaris and the handle of the Big Dipper. Follow the line made by these two stars southward to the next reasonably bright (third-magnitude) star, Athebyne. Northeast of Athebyne you should see another third-magnitude star, Aldhibah. Step from Athebyne, to Aldhibah, to one step farther northeast. You should be about halfway between Aldhibah and another third-magnitude star, Altais.

In the finderscope: The view in the finderscope is pretty sparse. When you reach a pair of stars, 28 and 27 Draconis, turn south and east. Keep moving until this pair leaves your finder field and another pair, 42 and 36 Draconis, enter it. Aim for a spot halfway between the two pairs.

Alternately, aim for Aldhibah and wait four minutes until the planetary nebula drifts into the field of view.

The Cat's Eye in a star diagonal at low/high power

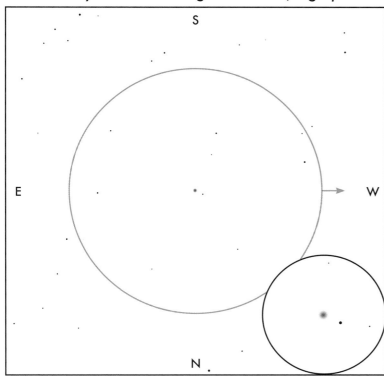

The Cat's Eye in a Dobsonian at high power

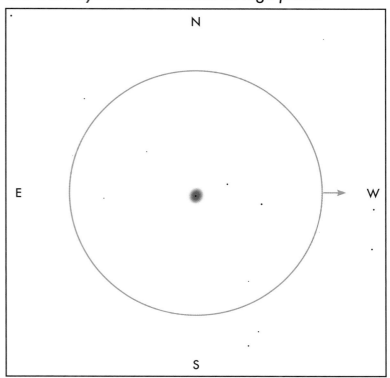

In a small telescope: Find with low power, confirm with high power. Look for a "double star" where the star to the east appears to be fuzzy, with a distinctive blue–green color. At high power (see the 20' inset) you should see it distinctly as a planetary nebula.

In a Dobsonian telescope: Find with low power, observe with high power. The planetary nebula has a distinctive electric-blue color. Note the stars to the west; when looking at the nebula straight on, the stars will appear brighter, but when looking directly at the stars your averted vision will make the nebula look brighter than the stars. Likewise a nebula filter will bring out the nebula and suppress the background sky (and other stars).

The **Cat's Eye** is a planetary nebula, the ember of a dying star (see page 99). It's a cloud of gas with an extended halo from its red giant phase about 9 light years across around a much smaller planetary nebula, maybe half a light year wide. That's the bit we can see here. All of that is eminating from a star located 5,300 light years from us that is going through its final stages of collapse.

Although planetary nebulae as such are relatively well understood, the evolution of this particular nebula is not particularly clear. Images from large telescopes show a remarkably complex set of clouds inside (giving it the peculiar "cat's eye" shape, unfortunately hard to see in an amateur telescope). One thought is that the central dying star may in fact be a double star, though that's yet to be confirmed.

This particular planetary nebula is one of the best, and longest studied, of its type. It was the first planetary nebula to have its spectrum of colors recorded, in the mid-nineteenth century, by the English amateur William Huggins; his results showed that the light from a planetary nebula is coming from hot gases and not from a star.

Also in the neighborhood: Dziban (Psi Draconis), a fourth-magnitude star just north of the 27/28 Draconis pair in the finderscope field, is an easy double star: a magnitude 4.6 primary with a magnitude 5.6 companion 30" to the north-northeast.

find more at: www.cambridge.org/features/turnleft/northern_skies.htm

In Ursa Major: Two galaxies, M81 and M82

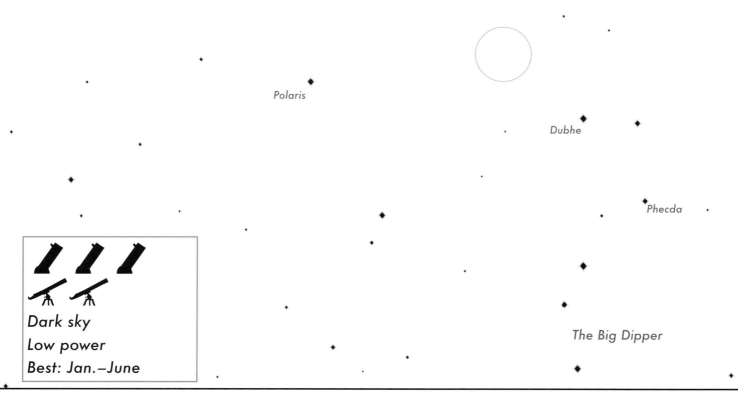

Polaris

Dubhe

Phecda

The Big Dipper

Dark sky
Low power
Best: Jan.–June

Star maps courtesy Starry Night Education by Simulation Curriculum

- Beautiful pair of galaxies
- Challenge to find, but relatively easy to see
- Interesting contrast of galaxy types

Where to look: Find the four stars that make up the bowl of the Big Dipper. Draw a line running diagonally across the bowl from Phecda (the one at the handle end that's not part of the handle) to Dubhe, the star at the opposite side of the bowl. Step from Phecda to Dubhe, and one step again along that line.

In the finderscope: The view in the finderscope is pretty sparse. Moving away from Dubhe, you'll first see a rough line of four fifth- and sixth-magnitude stars; the endpoints of this line, 38 and 32 Ursae Majoris, are indicated in the finder view. For most finders, these four stars appear just as Dubhe is moving out of the field of view. You're halfway there.

Keep moving away from Dubhe; as these four dim stars move out of the field, a fourth-magnitude star, 24 Ursae Majoris, will come into view. Aim at that star. Now put your lowest power eyepiece into your telescope, and start your last stage of star-hopping with the telescope itself.

N

M82

24

HD 3838

M81

E

W

38

32

Duhbe

S

M81 and M82 in a star diagonal at low power

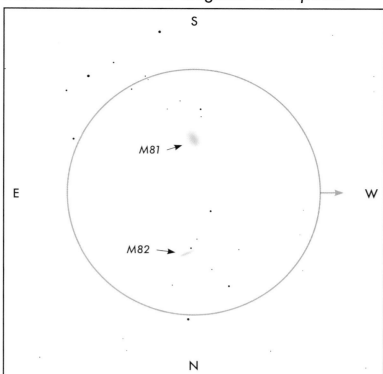

M81 and M82 in a Dobsonian at low power

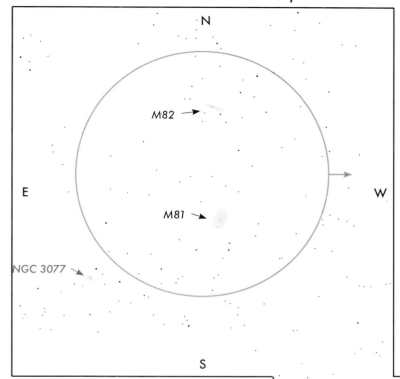

In a small telescope: The star 24 Ursae Majoris (see the finderscope view) is at the corner of a right triangle. It and two other stars make up the long leg of the right angle, and a single star marks the other leg. Aim for the end of the long leg, the sixth-magnitude star HD 3838, and then slowly move the telescope to the east. Keep an eye out for the galaxies. If you reach another faint triangle, you've gone too far.

The two galaxies look like two fuzzy spots of light. The one to the south, away from Polaris, is M81. It has an obvious oval shape, roughly half again as long as it is wide. The other one, M82, is thin and pencil-shaped, looking like a string of loosely spaced fuzzy beads.

Sky brightness makes a real difference for a small telescope here. These can be a challenge to see under suburban skies, but once you get away from city lights they leap out at you.

In a Dobsonian telescope: M81 and M82 are rather bright, as galaxies go, and they make quite a striking pair if the sky is clear and dark.

M81, the more rounded one, is not perfectly round but rather elongated in the northwest–southeast direction. It's slightly brighter in the center, but without a sharply defined nucleus.

M82, the thin one, is lumpy and irregular. The darker the night, the more you can see of the outer edges of this galaxy and so the more pencil-like it will appear. It may even appear a tiny bit curved, like a shallow bowl open to the northwest. On really good nights, a Dobsonian can show a dark lane of dust across M82. Try observing it with high power.

Move M81 to the western edge of your field of view and look southeast for NGC 3077, a dash of light in a cluster of stars.

A full low-power field to the southwest of M81 is another faint galaxy, NGC 2976.

M81 is an example of a spiral galaxy, at least 50,000 light years across, but in a small telescope we only see the brighter central core. It contains a few hundred billion stars.
M82 is an irregular galaxy. It doesn't have nicely defined spiral arms, but is full of irregular dust clouds and lumpy collections of stars. It's about half the size of M81, but still contains tens of billions of stars.

These galaxies are located about twelve million light years away from us. They are quite close to each other, perhaps some ten times closer together than the Andromeda Galaxy is to us. Astronomers living in one of these galaxies should have a fine view of the other one!

In fact, these two galaxies passed close to each other about 200 million years ago, leading to a stream of hydrogen gas being pulled out from each, from which young star clusters have begun to be formed between the galaxies.

These galaxies, along with several dozen galaxies including NGC 2976 and NGC 3077 make up the *M81 Group*. (Galaxy *groups* have fewer members than galaxy *clusters*; the dividing line is roughly 50 galaxies.) Interestingly, most of the members of the M81 Group are spiral galaxies.

Detailed observations of M82 suggest that an astoundingly enormous explosion occurred in its nucleus, sending shock waves thousands of light years across the galaxy. What could have caused such an explosion? That's still in dispute. In any event, this is undoubtedly one of the strangest galaxies that you can see in a small telescope.

For more about galaxies, see page 109.

find more at: www.cambridge.org/features/turnleft/northern_skies.htm

In Ursa Minor: *Polaris*, a double star, Alpha Ursae Minoris
In Cepheus: NGC 188, an open cluster

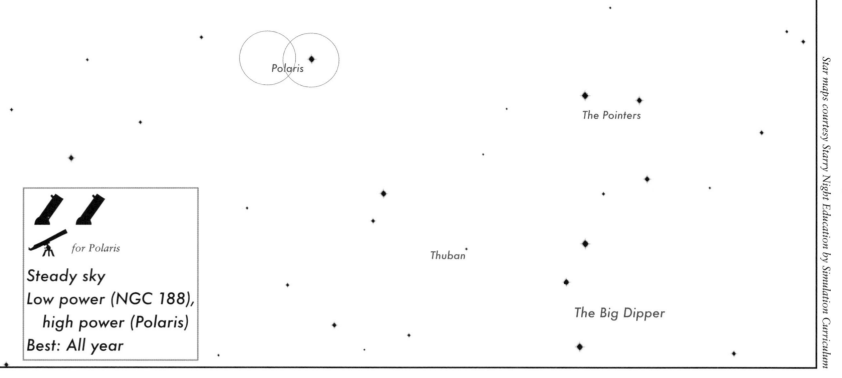

Polaris

The Pointers

Thuban

The Big Dipper

Star maps courtesy Starry Night Education by Simulation Curriculum

for Polaris

Steady sky
Low power (NGC 188),
high power (Polaris)
Best: All year

- Famous star, easy to locate
- Challenging double (faint companion)
- Subtle, ancient open cluster

Where to look: Find the Big Dipper, and look to the two stars in the bowl of the dipper farthest from the handle. These are the Pointer Stars; draw an imaginary line from the bottom to the top, and extend this line until you find a fairly bright star. This is Polaris, the North Star. Face this star and you are always facing due north.

In the finderscope: Polaris, a second-magnitude star, is the brightest of the stars in its area and so difficult to mistake. (Of course, it is not the brightest star in the sky; some 45 other stars are brighter. The North Star derives its fame solely from its position, not its brightness.)

NGC 188: Step in the direction towards Cassiopeia from Polaris to a fourth-magnitude star, HR 285, about four degrees away. Aim here, and then with your low-power eyepiece move slowly to the south.

S

to the Beehive

to Boötes

Celestial North Pole ×

Polaris

S

S

HR 285

NGC 188 ✕

HR 240

to the Pleiades

to Cygnus

S

NGC 188 *in a star diagonal at low power*

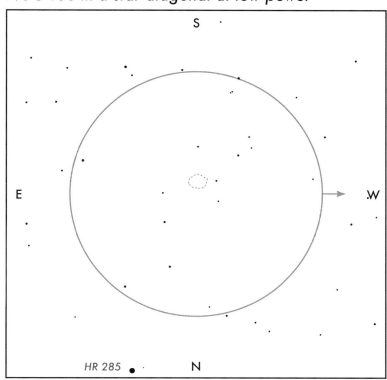

In a small telescope: The cluster is faint; you won't see it in a 3" unless the night is especially dark. The dotted line marks where we did see it in a small telescope, once, observing from the desert in west Texas...

NGC 188 *in a Dobsonian at low power*

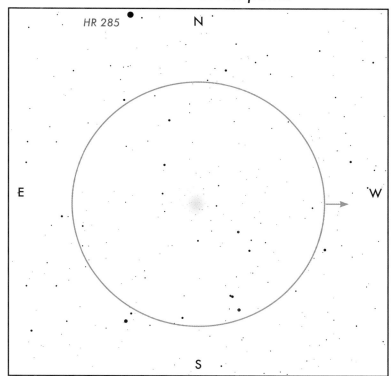

In a Dobsonian telescope: Easy to find, on a dark night the cluster appears as a grainy ball of light with a delightful sprinkling of very dim stars. It's a nice test, available year-round, of how good the night is.

Polaris is considered a good challenge for a 2"–3" telescope. The difficulty with splitting it is not that the two stars are particularly close to each other, but rather that the primary is so much brighter than the other star. High power helps to move the dimmer star out of the glare of the primary.

Polaris is located about 400 light years away from us, and it is at least a triple system. The two stars we can see are separated by more than 2,000 AU, and must take at least 40,000 years to orbit each other. But star A is also a *spectroscopic binary*; judging from oscillations noted in the spectrum of the light coming from star A, it also has a dim companion, hundreds of times closer than B, orbiting A in only thirty years.

An equatorial mount (see page 19) is extremely awkward to use here because normally the equatorial axis is aligned to a point very close to Polaris. To avoid this problem, just reset your telescope so that this axis points somewhere else. And, if you have one, turn off the clock drive – Polaris is the one star that won't move out of your field of view! However, once you've finished looking at Polaris you have to remember to realign your equatorial mount.

Polaris is most noteworthy to residents of planet Earth for being located in that part of the sky where the axis through our north pole happens to be pointing at the moment. As the years go by and the Earth wobbles on its axis, eventually our north polar axis will be pointing elsewhere. During the times

5' circle

of the ancient Egyptians and Babylonians, Polaris was more than 10 degrees away from true north; another star, Thuban, was used as a pole star back then. Nowadays, Polaris is less than a degree away from true north. In the year 2100, it'll be less than half a degree away. That will be the closest Polaris will come to being a true indicator of north.

Polaris A is a giant star, not much hotter than the Sun but much bigger and about 1,500 times as bright. It is a Cepheid-type variable star (see page 196): its brightness varies, slightly, as its stellar atmosphere expands and contracts. More oddly, the amplitude of its brightness variation has decreased with time. A hundred years ago, Polaris changed brightness by 0.12 magnitudes every four days; but its variation today is only a few hundredths magnitude, impossible to detect with the naked eye.

Polaris B is a fairly ordinary star, only a bit bigger and about three times brighter than our Sun. From Polaris, our Sun would appear to be three times dimmer than the dim companion appears to us.

NGC 188: The open cluster is located almost 6,000 light years from us. It is one of the oldest known open clusters, about six billion years old (judged from the ages of a double star seen within it). It may have survived as a cluster because, like M67 (page 94), it orbits well out of the plane of the Galaxy, away from the perturbations of other stars.

Polaris (Alpha Ursae Minoris)			
Star	**Magnitude**	**Color**	**Location**
A	2.1	Yellow	Primary star
B	9.1	Blue	18" from A

find more at: www.cambridge.org/features/turnleft/northern_skies.htm

In Canes Venatici: The *Whirlpool Galaxy*, M51

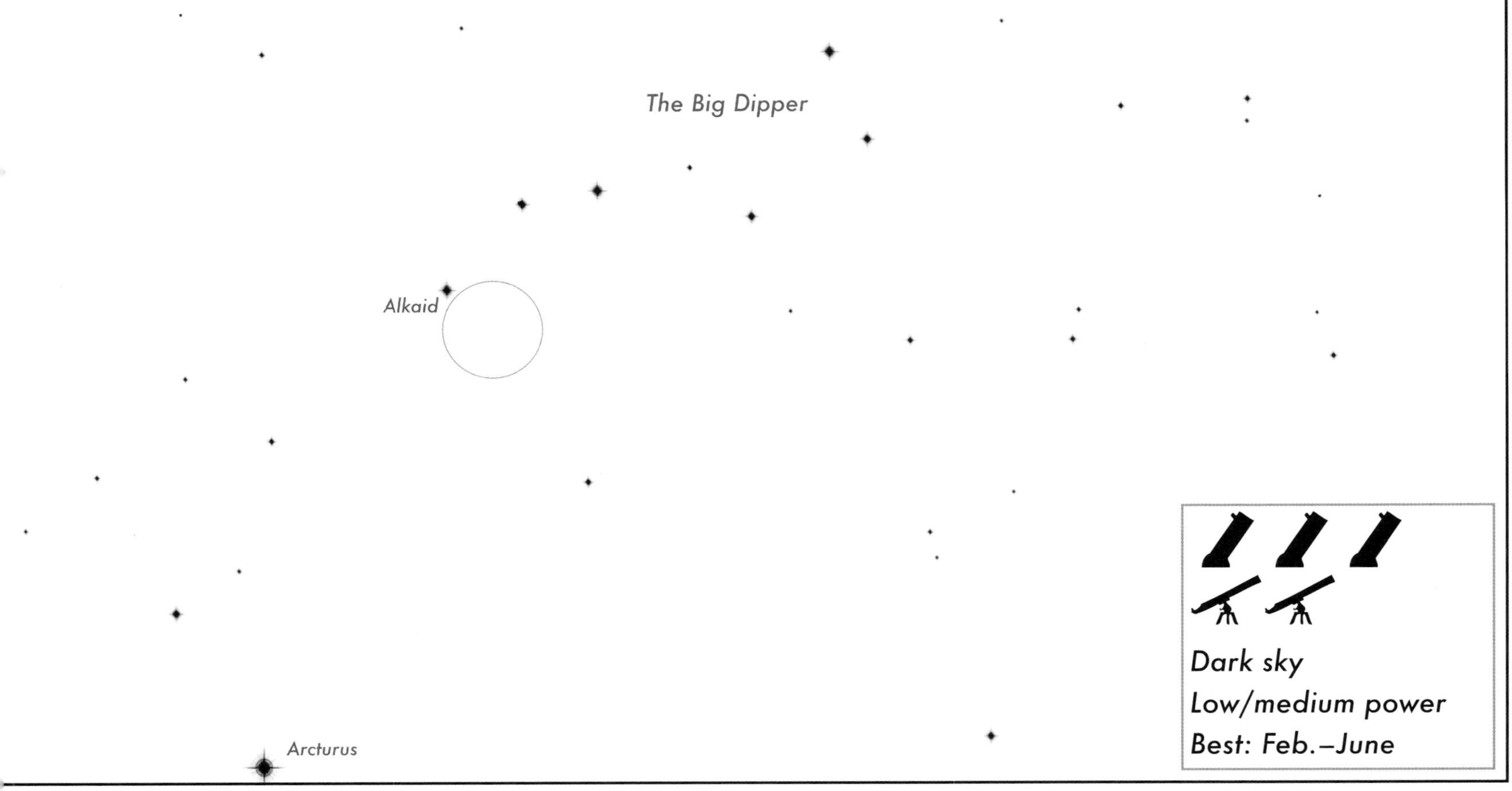

The Big Dipper

Alkaid

Arcturus

Star maps courtesy Starry Night Education by Simulation Curriculum

Dark sky
Low/medium power
Best: Feb.–June

- Spectacular galaxy (if the night is dark)
- Small neighbor galaxy also visible
- Fun history

Where to look: Go to the star at the end of the Big Dipper's handle, called Alkaid.

In the finderscope: You'll see Alkaid and a somewhat fainter star (24 Canum Venaticorum, or 24 CVn for short) off to the west of Alkaid. Center on 24 CVn. Call the finderscope view a clock face, with 24 CVn as the hub. With Alkaid at the 9 o'clock position, look for M51 at the 5 o'clock position.

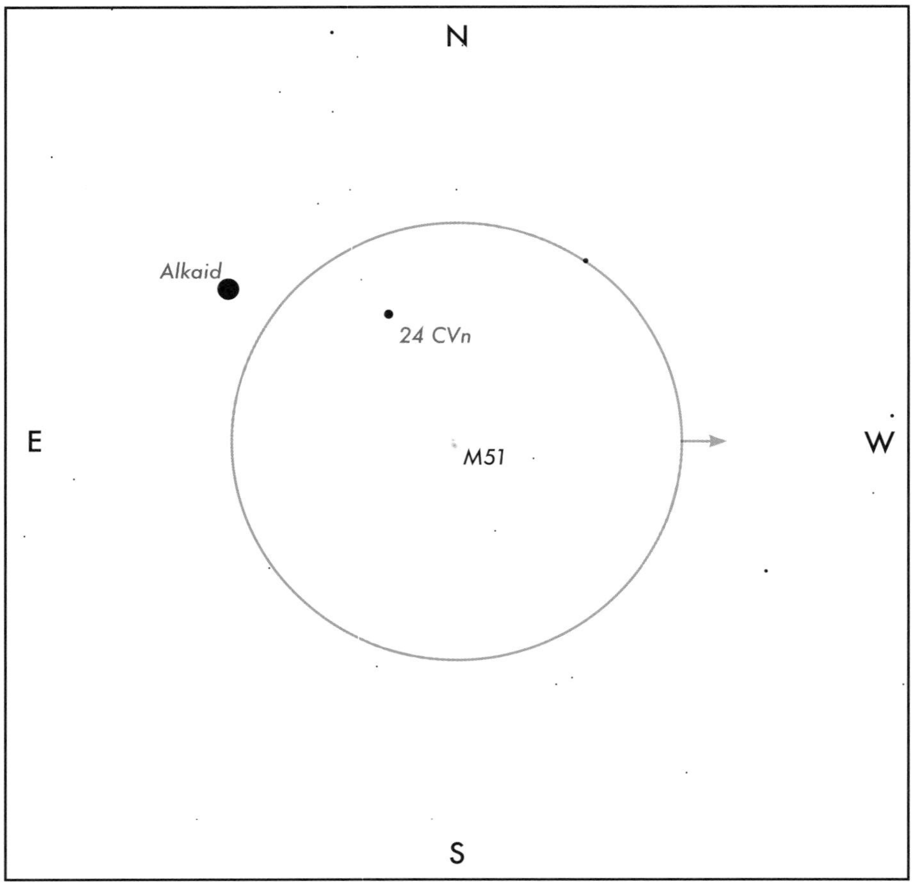

N

Alkaid

24 CVn

E

M51

W

S

M51 in a star diagonal at low power

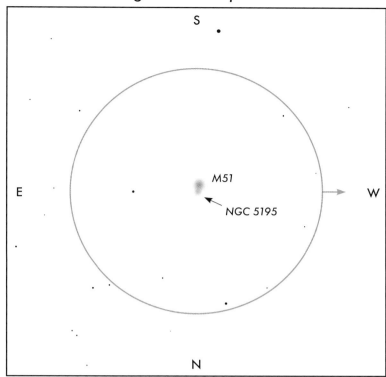

M51 in a Dobsonian at medium power

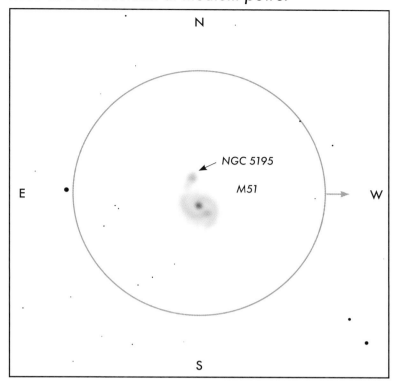

In a small telescope: The galaxy will look like a faint hazy patch of light. A closer look with averted vision shows two separate concentrations of light, something like a badly out-of-focus double star. The larger object is M51, while its smaller companion galaxy, to the north, is designated NGC 5195. It's a bit more intense than the larger, but more diffuse, main galaxy.

In a Dobsonian telescope: Find in low power, observe in medium power. The two individual galaxies can be made out relatively easily on a good night. If the sky is especially dark, the brighter core and spiral arms of M51 can just be made out even in an 8" Dobsonian.

The **Whirlpool Galaxy** is so named because of the beautiful spiral shape seen in large telescopes. In fact, it was the first galaxy seen to have a spiral structure. With perfect conditions, an 8" Dob can just show the spiral arms.

The main galaxy, M51, consists of 160 billion stars in a classic spiral, 40,000 light years in diameter. **NGC 5195** is probably a dense elliptical galaxy and may actually be more massive than its wider companion. Recent measurements of the distance to these galaxies show that they are located about 25 million light years from us, ten times as far from us as the Andromeda Galaxy. Large telescopes reveal a dust cloud in M51 which seems to lie in front of NGC 5195, which indicates that the smaller galaxy is apparently on the other side of M51 from us.

Though M51 was seen and catalogued by Charles Messier in 1774, it was Lord Rosse (born William Parsons) who first noted its spiral structure in 1845. At his estate in Birr Castle, in Parsonstown, Ireland, he had built the largest telescope in the world – a distinction it kept until the 100" telescope was set up on Mt. Wilson, California, in 1917. Rosse's telescope, nicknamed "The Leviathan," had a six-foot (1.8-meter) aperture mirror at the bottom of a Newtonian tube so large that it could not be steered across the sky; instead, it sat between two 40-foot high walls, oriented to the south, with a series of

winches to move it from the horizon to the zenith. Between the two walls, which supported the observer platforms, there was only a 15° (one hour angle) motion available in azimuth.

Ireland is not normally considered an ideal spot for astronomy due both to its famously rainy weather and its location far to the north, where many of the deep-sky objects of Scorpius and Sagittarius will never rise high above the horizon. This far northern location was a blessing for observing the Whirlpool, however. Only from such a northern latitude (53° N) would an object with such a northern declination as M51 (47° N) be visible to a telescope that can only look to the south of the zenith! In fact, M51 is only seven degrees away from straight up when it crosses into this telescope's field of view and, of course, only 6° from the zenith as it transits.

Lord Rosse's telescope still exists, and it has recently been restored to working order. Birr Castle is now a museum and the site of the annual Whirlpool Star Party. Every summer, amateur astronomers can set up their own telescopes in the shadow of the Leviathan and observe the Whirlpool for themselves … weather permitting.

find more at: www.cambridge.org/features/turnleft/northern_skies.htm

In Ursa Major: The *Pinwheel Galaxy*, M101; and *Mizar*, a double star, Zeta Ursae Majoris *(with Alcor)*

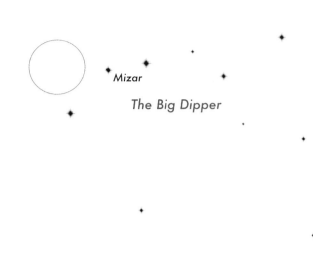

Mizar

The Big Dipper

Arcturus

Star maps courtesy Starry Night Education by Simulation Curriculum

for Mizar

for Alcor

Dark sky (M101),
any sky (Mizar)
Low power
Best: Feb.–June

- M101: Spectacular galaxy (if the night is dark)
- Mizar: Very easy to find and split
- Fun history

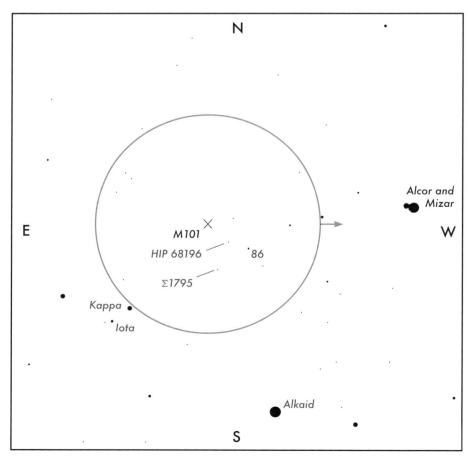

Where to look: Mizar is the middle star in the handle of the Big Dipper. With your naked eye you should be able to see a faint star next to it, called Alcor. This pair of stars was called the Horse and Rider by the Arabs; it was also known as the Puzzle, and they considered it a test of how sharp one's eyesight was. In fact, most nights it's an easy test.

In the finderscope: Mizar and Alcor are easily visible in the finderscope.

M101: Look east from Mizar, past Alcor, to a string of sixth-magnitude stars running to the east-southeast. Past the fourth star of the string (86 Ursae Majoris) are three seventh-magnitude stars, running roughly north–south. Step from 86 Ursae Majoris to the middle of the three fainter stars (HIP 68196); one step farther brings you close to the galaxy. If you can't see the fainter stars in your finder, start at 86 Ursae Majoris and move northeast.

Also in the neighborhood: The southeastern of the three seventh-magnitude stars past 86 Ursae Majoris is **Struve 1795**, *a challenging Dob double: the seventh-magnitude primary has a magnitude 9.8 companion to its north, just 7.9" away.*

Iota and Kappa Boötis, a finder field northeast of Alkaid, are both nice double stars. Kappa, stars of magnitude 4.5 and 6.6 (14" to the west-southwest), shows subtle colors; Iota, magnitude 4.8, has a magnitude 7.4 companion 39" to the northeast.

M101 *in a star diagonal at low power*

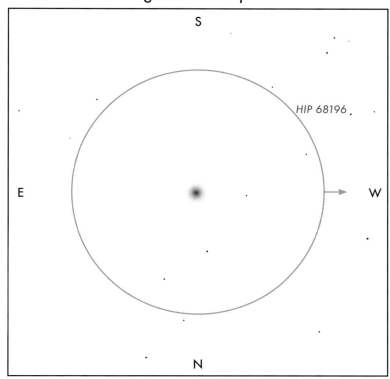

M101 *in a Dobsonian at low power*

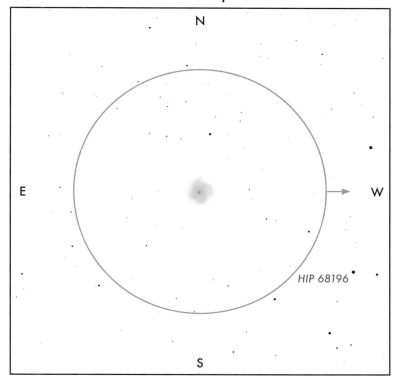

In a small telescope: It's easiest to start at 86 Ursae Majoris and move northeast, past a pair of seventh-magnitude stars, looking for a faint smudge of light. This galaxy is a challenge for a small telescope in suburban skies; it's fainter than the M81/82 pair or the Whirlpool, so if the night isn't good enough to show them, don't expect to see this.

In a Dobsonian telescope: Look for a lumpy disk of light among a scattering of faint stars. If the night is good, you can see a hint of the spiral arms. This galaxy is fainter than the M81/82 pair or the Whirlpool; if the night isn't good enough to show them, don't expect to see this.

The **Pinwheel Galaxy** (M101) is so named for its beautiful spiral shape, seen in photographs from large telescopes. It lies 23 million light years away from us. With a diameter of 180,000 light years (80% bigger than the Milky Way) it's one of the largest disk galaxies known. We normally see only its central core, though (as with its near neighbor M52) under excellent conditions an 8" Dob can just show its spiral arms. It is one of a group of at least eight other galaxies.

Mizar was the first double star to be discovered, and the first to be photographed. Both stars are relatively colorless, however, and so while it is easy to find it lacks some of the charm of other double stars.

Alcor is not an orbiting companion star to Mizar; but the two stars are related. All the stars of the Big Dipper, including Mizar and its companion, as well as Alcor, are members of the *Ursa Major Group*, the leftover remains of an open cluster. All the stars in this group were probably formed at the same time. Now they are slowly drifting past our Solar System, and apart from each other, as they follow their separate orbits about the center of the Galaxy. Most of these stars are less than 100 light years from us, which is why so many of them are bright. In fact,

Sirius, the brightest star in our sky, is a member of this group! Its path just happens to have taken it on the opposite side of our Sun from most of the other Ursa Major Group stars.

Best available estimates put Mizar at 78 light years from us, while Alcor is 81 light years away. Thus, they are quite close to each other. In the skies over a planet orbiting Alcor, Mizar would be a magnificent naked-eye double star. Mizar A would appear as bright as Venus, Mizar B brighter than Jupiter, and their separation would be perhaps seven arc minutes (about a quarter the width of the Moon).

Of the Mizar pair themselves, the smaller star orbits the primary at a distance of roughly 500 AU; a full orbit takes several thousand years. The primary star is two and a half times as massive as our Sun, twice its width, and 25 times as bright. Its companion is twice the Sun's mass, about 60% larger in radius, and 10 times as bright.

Also in the field of view is an eighth-magnitude star with the impressive name of **Sidus Ludovicianum**. An enterprising eighteenth-century German astronomer claimed this star was a new planet he'd discovered, and he named it after Ludwig V, his king. It is certainly one of the dimmest stars to bear a name.

Mizar (Zeta Ursae Majoris)			
Star	**Magnitude**	**Color**	**Location**
A	2.2	White	Primary star
B	3.9	White	14" SSE from A

find more at: www.cambridge.org/features/turnleft/northern_skies.htm

Southern skies

Some of the most astounding deep-sky objects are visible only from the southern hemisphere. Southern hemisphere residents have a rich endowment of sky wonders; it would be a crime to not take advantage of your location. Northerners travelling south of the equator, don't forget to bring your telescope, or at least a good pair of binoculars.

As one travels south, some stars familiar to northerners are lost below the northern horizon, but new stars come into view in the south.

South of latitude 30° N – Florida, southern Texas, the Middle East, southern Japan, and China – you can see objects down to declination 60° S: Omega Centauri is visible from these sites. In the Caribbean, Central America, Hawaii, and India you can see Rigel Kentaurus, the Southern Cross, and other objects down to a declination of 75° S.

At the equator all the sky is visible, at least in theory. However, faint objects near the celestial south pole, like the Magellanic Clouds, are only seen well from deep within the

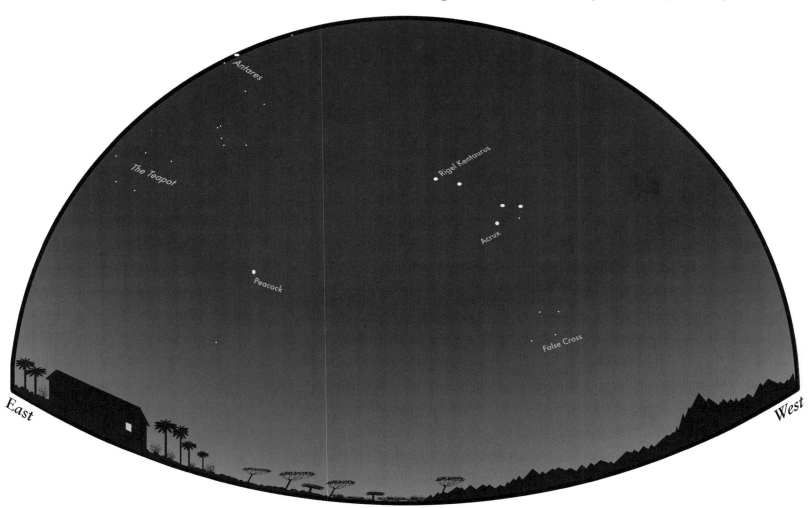

Looking south, June

Finding your way:
Southern sky guideposts
(as seen from 35°S)

The most prominent southern constellations, along with the Milky Way, arch across the southern hemisphere sky in June.

Scorpius, with its red star **Antares**, and the Teapot of *Sagittarius* can be seen rising in the east. Due south, skirting the horizon in Hawaii and the Caribbean but more and more dominating the sky as you travel farther south, are the bright stars **Rigel Kentaurus** (to the left) and **Hadar**, and **Acrux** at the foot of the *Southern Cross*. Earlier in the evening or earlier in the year (or observing from farther south) you may see one or two bright stars setting in the west: **Canopus** and **Sirius**.

In June, the northern winter stars of, say, Australia are the southern summer stars of California. Use the Guidepost pages for July–September (pages 118 ff)…but you'll need to turn the charts upside down!

On the northern horizon (not illustrated here), the *Big Dipper* can still be seen on occasion, even from New Zealand. Follow the arc of its handle to **Arcturus** and **Spica**, now high overhead. **Regulus**, at the foot of the "backwards question mark" (now upside down as well) of *Leo*'s mane, sets in the west; lying on his back along the horizon, Leo may appear unusually large thanks to the horizon Moon illusion. The familiar Summer Triangle – better named a Winter Triangle, here – will just be starting to rise; it won't be easily seen until later in the night (or later in the year).

southern hemisphere itself. Look for them from Chile or Argentina, South Africa, Australia, or New Zealand.

Finally, a caveat: the authors of this book grew up under northern skies. Objects visible from the northern hemisphere are old friends to us, guys that we'd looked at all our lives before we re-observed them again (and again) to pin down our descriptions for this book. The same is, alas, not true for our descriptions of southern hemisphere objects. This means that what follows is not nearly as detailed, nor as complete, as we would like.

And, as always, haze, clouds, light pollution, and a full Moon can change the appearance of even the best deep-sky object. Take our suggestions as a starting point only.

To the north: Visitors to the southern hemisphere will find that many of their familiar "summer" objects take on a whole new appearance when you no longer have to see them through a murky southern horizon. And since June nightfall in the south comes earlier than in the north, they'll stay visible longer. If you travel south, be sure to catch these old favorites:

Object	Constellation	Type	Page
Lagoon Nebula	Sagittarius	Nebula	146
Trifid Nebula	Sagittarius	Nebula	148
M22, M28	Sagittarius	Globular clusters	158
M54, M55	Sagittarius	Globular clusters	160
NGC 247/253	Cetus/Sculptor	Galaxies	174

Looking south, December

In December, look for *Orion* rising in the east (just off the left edge of our chart above). Farther east and south is **Sirius**, the brightest star in the sky.

South of Orion you'll see a bright star, **Canopus**. When Orion is at its highest (February evenings), Canopus can be seen even as far north as the southern USA. It's set in a large oval of second- and third-magnitude stars shaped roughly like a large letter D that make up a group once called the ship Argo, of Jason and the Argonauts fame. It's known nowadays by its constituent members: *Puppis*, the ship's poop deck; *Carina*, the ship's keel; and *Vela*, the sails. Nearby swim *Dorado*, the swordfish, and *Volans*, the flying fish.

Due south in December is the first-magnitude star **Achernar**. To the west is another bright star, **Fomalhaut**, and south of it

is the **Peacock** star. They all seem rather lonely by themselves in an otherwise barren part of the sky. However, the lack of bright southern stars at this time of year is more than made up for by the gems of the southern hemisphere: the *Magellanic Clouds*. If you're far enough south, and it's a dark night, they jump out like two large bright rivals to the Milky Way.

Summer stars in the northern half of these skies (not illustrated here) include all the familiar "winter" objects from *Andromeda* to the *Twins*. In particular, the deep-sky objects in Puppis – M46, M47, and M93 – are not to be missed. You'll find descriptions of them on pages 84 and 88.

On the following pages, we indicate the months when our favorite southern objects are highest in the sky, and how far south in latitude you need to be before you can see them well.

In Tucana: The *Small Magellanic Cloud*

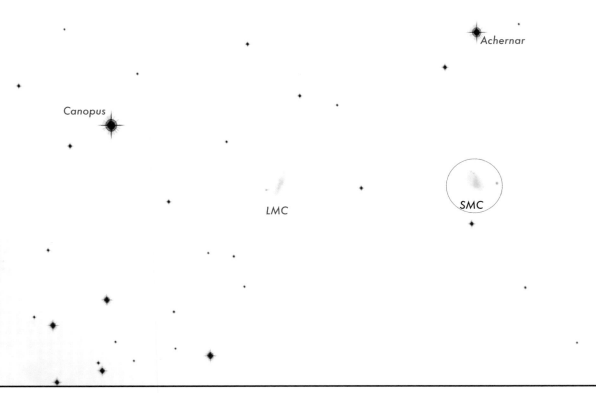

Star maps courtesy Starry Night Education by Simulation Curriculum

Dark sky
Low/medium power
Nebula filter
Best: Nov. > 10° S

Where to look: Draw a line from Canopus to Achernar; the Magellanic Clouds are just south of this line, with the Small Magellanic Cloud (SMC) lying to the west. On a reasonably dark night, if you are located far enough south of the equator the SMC should be very easy to spot, looking like a detached lump of the Milky Way.

In the finderscope: With a diameter of nearly five degrees, the SMC is too big and too diffuse to look particularly interesting in the finderscope. It's easier to see with the naked eye.

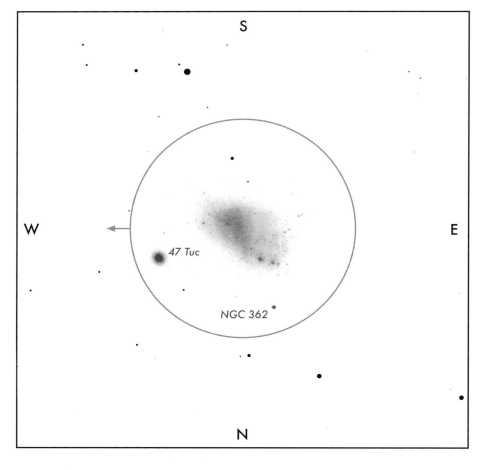

The **Small Magellanic Cloud (SMC)** is a satellite galaxy of our Milky Way. Best estimates give it close to a billion stars, located about 200,000 light years away from us. The bright central core is about 10,000 light years in diameter.

More than a dozen NGC objects can be found in the SMC, as listed below:

Object	Description
NGC 220/222/231	Open clusters
NGC 249	Emission/reflection nebula
NGC 261	Emission/reflection nebula
NGC 265	Open cluster
NGC 330	Open cluster
NGC 346	Cluster and nebula
NGC 371	Open cluster
NGC 376	Open cluster
NGC 395	Open cluster
NGC 419	Open cluster
NGC 458	Open cluster
NGC 456/460/465	Open clusters

The Small Magellanic Cloud in a star diagonal

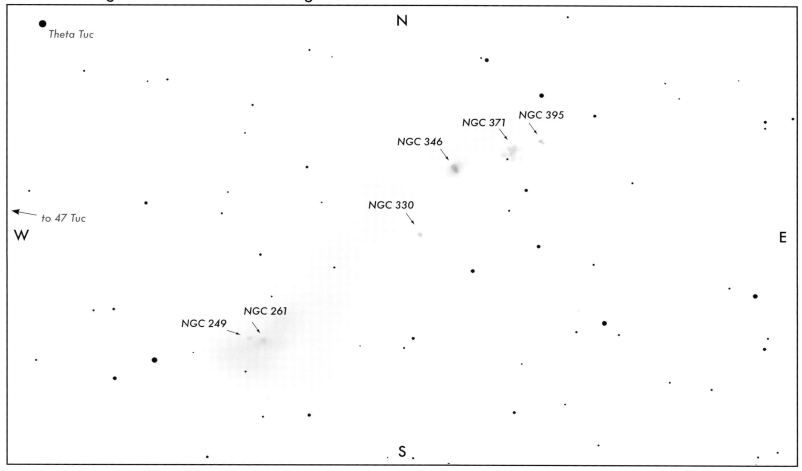

In the telescope: The SMC looks like a very lumpy cloud of light, many times larger than the field of view, with several noticeable lumps of light embedded within. Look especially near the edges, where objects stand out more easily in contrast to the dark sky instead of having to compete with the background of the SMC itself.

The Small Magellanic Cloud in a Dobsonian

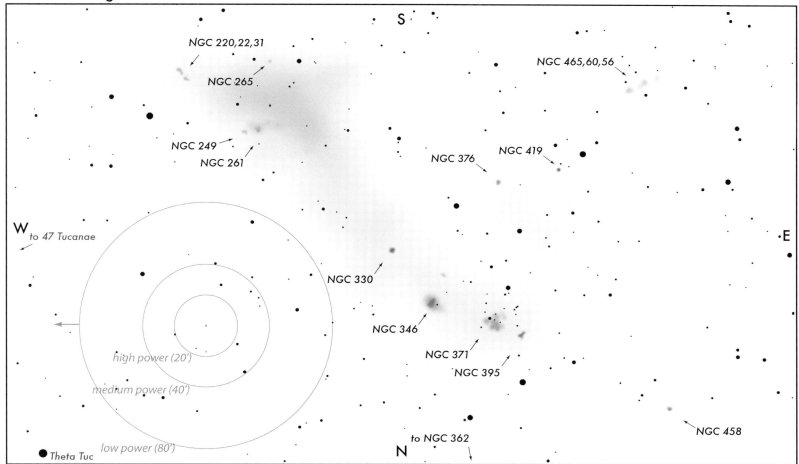

find more at: www.cambridge.org/features/turnleft/southern_skies.htm

In Tucana: *47 Tucanae*, a globular cluster, NGC 104

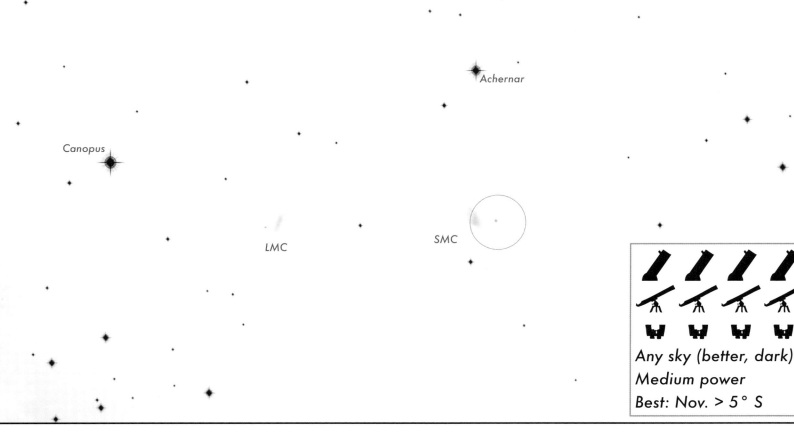

Star maps courtesy Starry Night Education by Simulation Curriculum

Any sky (better, dark)
Medium power
Best: Nov. > 5° S

- Spectacular globular cluster
- Easy to find and see (even with naked eye)
- Unusually young?

Where to look: Find the Small Magellanic Cloud (SMC): draw a line from Canopus to Achernar and you'll find the Magellanic Clouds south of this line. Or, if you're far enough south and the Southern Cross is visible as a circumpolar object, trace a line from the top star, Gacrux, to Acrux, to beyond the south celestial pole, and onwards towards the SMC.

In the finderscope: Aim at a fuzzy spot west and slightly north of the SMC.

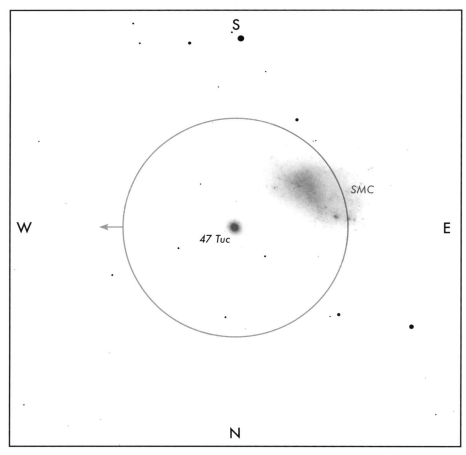

47 Tucanae in a star diagonal at medium power

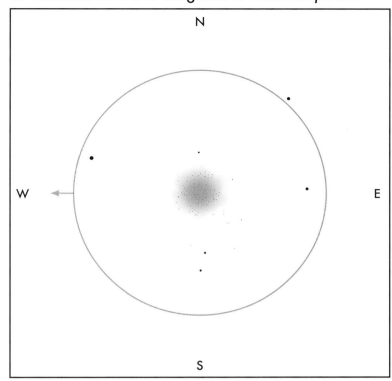

In a small telescope: The globular cluster is a large, bright, and grainy ball of light in a rich field of stars. The outer, somewhat dimmer portion is about three times as wide as the core.

47 Tucanae in a Dobsonian at medium power

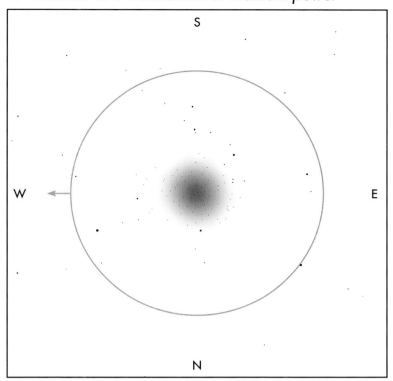

In a Dobsonian telescope: The globular cluster is a large, bright, and very grainy ball of light in a rich field of stars. As one moves away from the bright center, individual stars are more easily resolved.

This globular cluster is the second-brightest globular in the sky, after Omega Centauri (see page 238). It is visible even to the naked eye as a fuzzy dot, if the night is dark, and is easily seen as a cloud of light in the finderscope or in a good pair of binoculars. A four-inch telescope or larger should be enough to start resolving individual stars within this cluster.

It's significantly bigger and brighter than M13, and of similar size but a magnitude brighter than M22. The nucleus is much brighter than the surrounding cloud of light; so bright that, in fact, it is hard to resolve the outer part of the cluster from the glare of the core!

Notice the bright stars also in the field of view. The cluster extends to the south, almost out to a ninth-magnitude field star (not part of the cluster itself), across nearly a third of the low-power field, about two thirds the size of a full Moon.

The cluster is located 15,000 light years from us; its mass has been estimated at over half a million times the mass of our Sun.

Compared to other globular clusters, the stars in 47 Tucanae are unusually rich in heavy elements. As noted in our discussion of globular clusters on page 115, the stars in most globulars are poor in such elements. Since these elements are made deep inside stars, stars with heavy elements at their surfaces must have been formed from the debris of earlier stars.

Thus, while we think most globular clusters were formed very early in the history of the Galaxy, before heavy elements had been produced, by contrast 47 Tucanae may be among the youngest of globulars. It formed after an earlier generation of stars had time to form, make heavy elements, and explode into supernovae.

One other peculiar oddity of this cluster is that careful observations of individual stars shows an unusual lack of double stars. That's telling us something about how double stars are formed; but we're not sure just what, yet.

For more about globular clusters, see page 115.

find more at: www.cambridge.org/features/turnleft/southern_skies.htm

In Tucana: A globular cluster, NGC 362

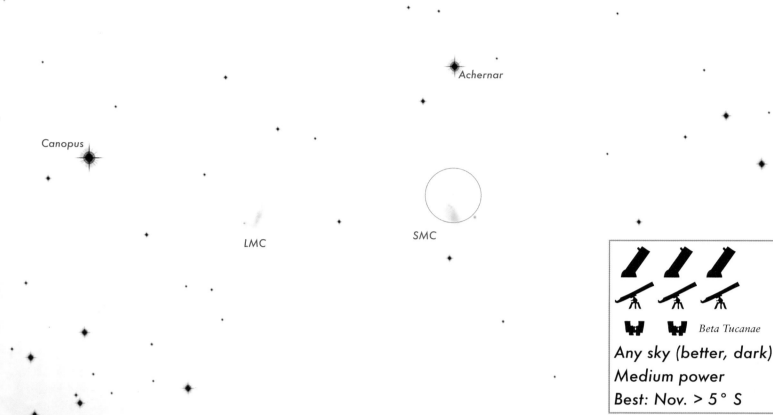

Star maps courtesy Starry Night Education by Simulation Curriculum

Beta Tucanae

Any sky (better, dark)
Medium power
Best: Nov. > 5° S

- Small but bright
- Easy to find
- Near several complex double stars

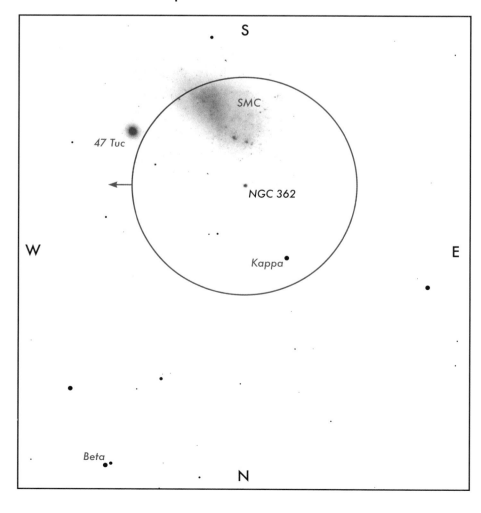

Where to look: Find the Small Magellanic Cloud (SMC): draw a line from Canopus to Achernar and you'll find the Magellanic Clouds just south of this line. Or, if you're far enough south that the Southern Cross is visible as a circumpolar object, trace a line from the top star, Gacrux, to Acrux, to beyond the south celestial pole, and onwards towards the SMC.

In the finderscope: The northeast corner of the SMC should be visible in the finderscope. Look for a magnitude 6.5 "star" just beyond the SMC. That's NGC 362.

NGC 362 in a star diagonal at medium power

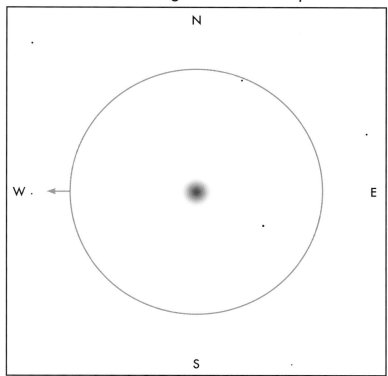

NGC 362 in a Dobsonian at medium power

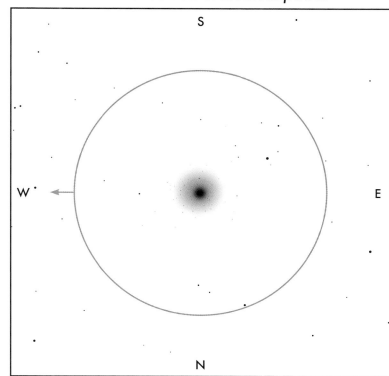

In a small telescope: In a rich field of stars, NGC 362 is a tiny but bright globular cluster. You'll need at least medium power to appreciate NGC 362 well; fortunately, it's bright enough that high power is no problem. The whole cluster is about 13 arc minutes across, but the central core (the part most easily seen) is only about half this total radius.

In a Dobsonian telescope: In a Dob, the field of stars around NGC 362 is especially rich; you'll want at least medium power to see them well. The cluster itself shows a tiny but bright central core, surrounded by a grainy cloud of light about 13 arc minutes across.

Compared to other globular clusters, including its near neighbor 47 Tucanae, **NGC 362** looks small; and, in fact, it probably contains only about 100,000 stars, compared to the half million or more 47 Tucanae or the nearly million stars in a huge cluster like Omega Centauri. Compared to other globular clusters, its stars have a higher content of metals, suggesting that it is a few billion years younger than the typical globular; still, that makes NGC 362 about 10 to 11 billion years old. It is located about 28,000 light years from us, roughly double the distance as Omega Centauri and 47 Tucanae.

For more about globular clusters, see page 115.

Also in the neighborhood: *Two degrees northeast of NGC 362, visible in the finderscope view, is the multiple star **Kappa Tucanae**. The primary is a yellow fifth-magnitude star, while its companion, five arc seconds northwest, is a distinctly orange star of magnitude 7.4. In addition, look for a magnitude 7.9 star about five arc minutes (319") also to the northwest. It is actually a pair of eighth-magnitude stars (magnitudes 7.9 and 8.4) oriented southeast–northwest with a one arc second separation. That's too close for a 3" telescope, but a Dobsonian with high power can just split it.*

*Ten degrees north of NGC 362 and the SMC are a pair of stars very close together, but visible as separate stars even to the naked eye. They are the **Beta Tucanae** stars, a complex system. The brighter of the naked eye pair can itself be split by a small telescope into a pair, Beta-1 and Beta-2, magnitude 4.4 and 4.5, separated by 27 arc seconds; Beta-2 is the one to the south-southeast.*

Beta-3 is the other naked-eye star, magnitude is 5.2, 10 arc minutes away to the southeast. It is just possible to split it from the Beta-1/Beta-2 pair with the naked eye, and fun to see in binoculars; and even at high power (as illustrated here) it should be visible in the same telescope field as Beta-1 and Beta-2.

In fact, all three stars are very close doubles, too close to be split with a small telescope, thus making this a sextuple star. The whole system lies about 150 light years from us.

find more at: www.cambridge.org/features/turnleft/southern_skies.htm

In Dorado: The *Large Magellanic Cloud*

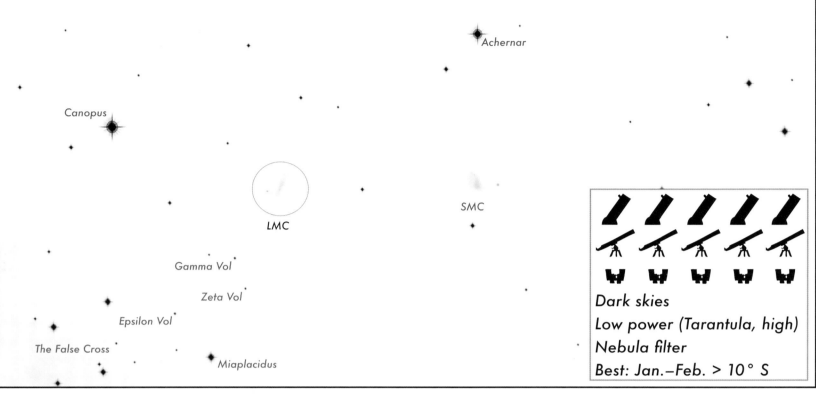

Star maps courtesy Starry Night Education by Simulation Curriculum

Dark skies
Low power (Tarantula, high)
Nebula filter
Best: Jan.–Feb. > 10° S

- Nearby satellite galaxy
- Dozens of superb objects – just roam!
- The Tarantula Nebula: incredible

Where to look: Draw a line from Canopus to Achernar; the Magellanic Clouds lie south of this line. On a reasonably dark night, if you are located far enough south of the equator, the Large Magellanic Cloud (LMC) should be very easy to spot.

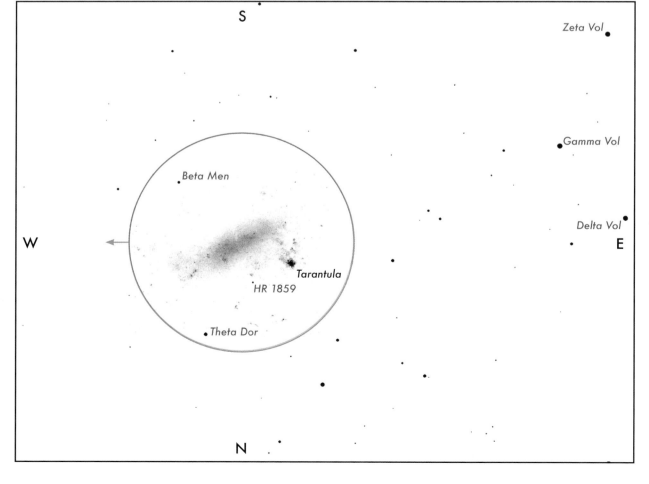

In the finderscope: Seven degrees across at its widest, the LMC will be too large to fit completely into your finderscope. The brightest and most spectacular member of the LMC is the Tarantula Nebula, NGC 2070; it should be easily visible in the finderscope.

The **Large Magellanic Cloud** (LMC) is a satellite galaxy of our Milky Way, about 160,000 light years away – about 15 times closer than the Andromeda Galaxy. It's small, as galaxies go, containing perhaps 10 billion stars, many of which are unusual for being rich in carbon; such carbon stars are far rarer in our Galaxy.

The odd shapes of both the SMC and the LMC are likely result from the gravitational interactions during a past near-collision. In fact, radio telescopes have detected a magnetic field in the space between the Magellanic Clouds, called the Magellanic Bridge. It's described as a filament of gas and dust, 75,000 light years long, between the two clouds.

The Tarantula Nebula (NGC 2070), also known as 30 Doradus, exceeds superlatives. It is the most active star-forming region in any of the galaxies of our Local Group, giving birth to thousands of stars. Some are over 200 times as massive as our Sun and therefore will have very short lifespans. If the nebula were as close to us as the Orion Nebula it would appear over a thousand times brighter, and fill more of the sky than the entire constellation of Orion!

The LMC, with its many levels of intensity, contains dozens and dozens of clusters and emission nebulae against a background haze of light and a foreground peppering of stars. It fills a roughly oval region about five degrees wide and seven degrees long. Even at low power, this represents about fifty telescope fields of view … every one as rich and interesting as any stand-alone nebula! Cloud upon cloud of light, with bright patches and darker lanes, peppered with clusters of stars, it simply has to be seen to be believed.

In the table to the right, we list 47 NGC-catalogued objects that can be seen in the LMC with a Dob or small telescope on a good night. (There are actually more than 47 objects represented here; those marked with plus signs have multiple NGC numbers, but look like single objects in a small telescope.)

On the following two-page spreads we give charts indicating where to look for these sights. The first spread of charts are oriented for a small telescope with a star diagonal; the next spread is oriented as seen in a Dobsonian. Each spread has four charts. Three mark off the north (N), central (C), and south (S) regions of the LMC, while a fourth smaller chart is concentrated around the Tarantula Nebula (T), the richest and brightest part of this galaxy.

Also in the neighborhood: *Halfway between the LMC and the False Cross is a loose collection of four stars, all of roughly third magnitude, in a diamond pattern. Three are multiple stars.*

The one closest to the LMC, **Gamma Volantis**, *is an easy split even in a small telescope, a magnitude 3.8 yellow with a blue magnitude 5.7 companion 13" north-northwest.*

Epsilon Volantis *is more of a Dob challenge, a yellow–blue pair in a busy star field; the primary star is magnitude 4.4, while its companion, magnitude 7.4, is only 5" north-northeast. That's beyond the range of most small telescopes.*

Zeta Volantis *has a faint, orange tenth-magnitude companion, 16" east-southeast of its fourth-magnitude yellow primary. Given such a strong magnitude difference, it is a nice challenge in a small telescope; easier in a Dob.*

Object	Chart	Description
NGC 1711	C	Open cluster
NGC 1727	C	Cluster and nebula
NGC 1743	C	Cluster and nebula
NGC 1755	N	Open cluster
NGC 1786	N	Globular cluster
NGC 1835	C	Globular cluster
NGC 1837	S	Cluster and nebula
NGC 1845	S	Open cluster
NGC 1847	N, C	Open cluster
NGC 1850	N, C	Open cluster
NGC 1854	C	Open cluster
NGC 1856	C	Open cluster
NGC 1858	N, C	Cluster and nebula
NGC 1872	C	Cluster and nebula
NGC 1874+	C	Cluster and nebula
NGC 1901	N	Open cluster
NGC 1910	C	Open cluster; variable S Doratis
NGC 1934+	N	Cluster and nebula
NGC 1962+	N, C	Cluster and nebula
NGC 1967+	C	Open cluster
NGC 1974+	N	Cluster and nebula
NGC 1983	T	Open cluster
NGC 1986	S, C	Open cluster
NGC 1994	T	Open cluster
NGC 2001	N, C, T	Cluster and nebula
NGC 2009	T	Open cluster
NGC 2014	N	Cluster and nebula
NGC 2015	T	Cluster and nebula
NGC 2031	S	Open cluster
NGC 2032+	N	Cluster and nebula
NGC 2033	S, T	Cluster and nebula
NGC 2042	T	Cluster and nebula
NGC 2044	T	Cluster and nebula
NGC 2048	S, T	Emission/reflection nebula
NGC 2055	T	Cluster and nebula
NGC 2060	T	Cluster/supernova remnant
NGC 2070	T	Cluster and nebula (Tarantula)
NGC 2074	T	Cluster and nebula
NGC 2077	T	Cluster and nebula
NGC 2079	S	Cluster and nebula
NGC 2080	S	Cluster and nebula
NGC 2081	T	Cluster and nebula
NGC 2086	S, T	Cluster and nebula
NGC 2098	T	Cluster and nebula
NGC 2100	T	Open cluster
NGC 2103	S	Cluster and nebula
NGC 2122	T	Cluster and nebula

find more at: www.cambridge.org/features/turnleft/southern_skies.htm

Northern LMC in a star diagonal at high power

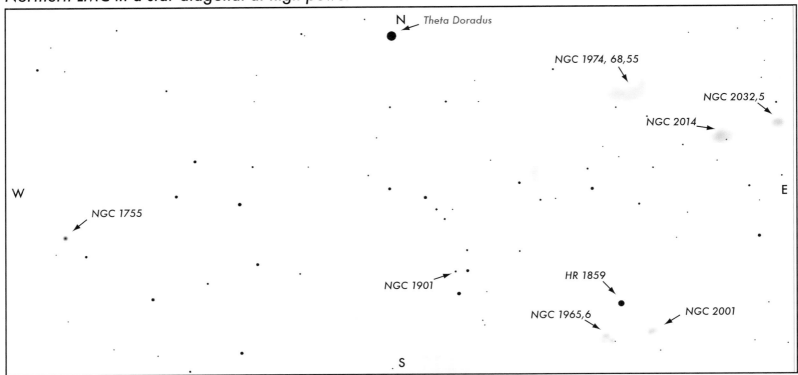

N — Theta Doradus

NGC 1974, 68, 55

NGC 2032,5

NGC 2014

W

E

NGC 1755

NGC 1901

HR 1859

NGC 1965,6

NGC 2001

S

Central LMC in a star diagonal at high power

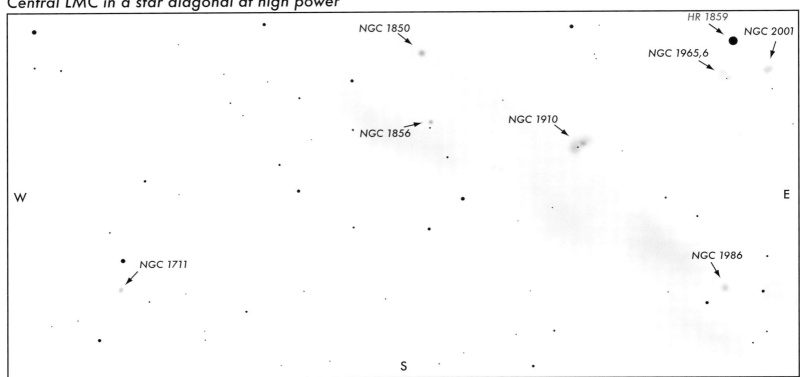

NGC 1850

HR 1859

NGC 2001

NGC 1965,6

NGC 1856

NGC 1910

W

E

NGC 1711

NGC 1986

S

Finding your way around the LMC with a star diagonal:

To the right is a diagram of the Large Magellanic Cloud, mirror-imaged, with north at the top. The brightest part of the nebulosity runs diagonally from the northwest to the southeast. Our charts here follow that line of light: north (N), central (C), and south (S). From the southeast end of this line, an arm of light branches off to the north, ending in a knot of bright nebulosity. That knot is the Tarantula Nebula, in its own square box. Each of the charts here are 2° tall. The wide N, C, and S charts are a bit over 4° wide, and the Tarantula is a 2° square.

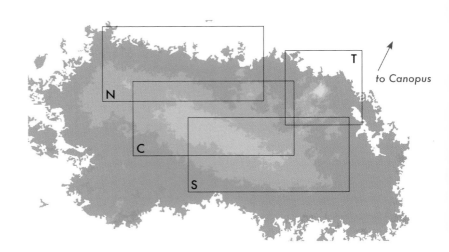

T

to Canopus

N

C

S

In a small telescope: This is a beautiful four square degrees of sky!

Start with the Tarantula; it is the easiest, brightest, and most spectacular field.

If you can pull yourself away from its pleasures, move southwards to the bright central corridor of the LMC; look for knot of light that is NGC 2103.

From there, move west and look for the fifth-magnitude star, Beta Mensae. Put it at the southern edge of your low-powered field, and look for NGC 1837 and NGC 1845. Then move back northeastward until you encounter the bright central corridor again. Follow the corridor past half a dozen NGC objects, and at its end move westward to pick up NGC 1711.

Moving back eastward to the corridor, keep moving to a sixth-magnitude star, HR 1859; just south in your low-power field will be a number of nice NGC nebulae.

Moving northwest from HR 1859, look for the fifth-magnitude star Theta Doradus. Put it in the western edge of your low-power field, and you should be able to find the objects around NGC 1974. Keep moving east to see NGC 2014 and 2032/35.

The mismatch in the scale and number of stars from the three larger charts to the Tarantula chart is intentional – to find the dimmer blobs in the rest of the LMC, you really do want a reasonably dim limiting magnitude. Showing stars to 11.0 (the default standard for the other star-diagonal fields in this book) would have been too busy, so we only show stars here to magnitude 10 for those charts (including some fainter red stars, and dropping some brighter blue stars, since the human eye sees these better than a CCD camera with a V-band filter). For the Tarantula, even that would have been just too busy; thus we show there a somewhat subdued set of stars that don't overload the field.

Around the Tarantula in a star diagonal, low power

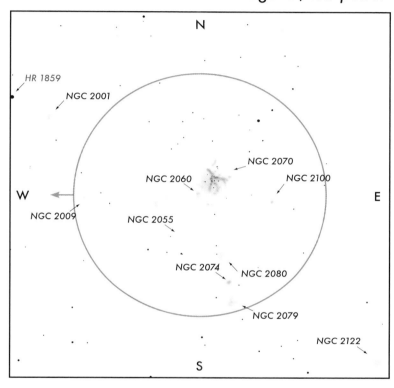

Southern LMC in a star diagonal at high power

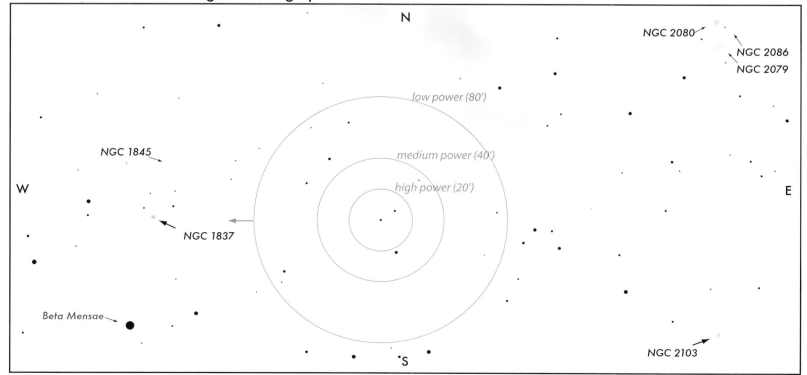

Finding your way around the LMC with a Dobsonian: To the right is a diagram of the Large Magellanic Cloud, with south at the top. The brightest part of the nebulosity runs diagonally from the southeast to the northwest. Our charts here follow that line of light: south (S) to the right, central (C) and north (N) below. From the southeast (upper-right) end of this line, an arm of light branches down to the north, ending in a knot of bright nebulosity. That knot is the Tarantula Nebula, in its own square box (opposite). Each of the charts here are 2° tall. The wide S, C, and N charts are a bit over 4° wide, and the Tarantula is a 2° square.

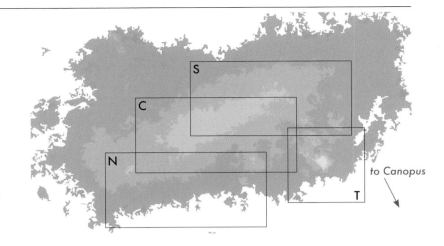

Central LMC in a Dobsonian

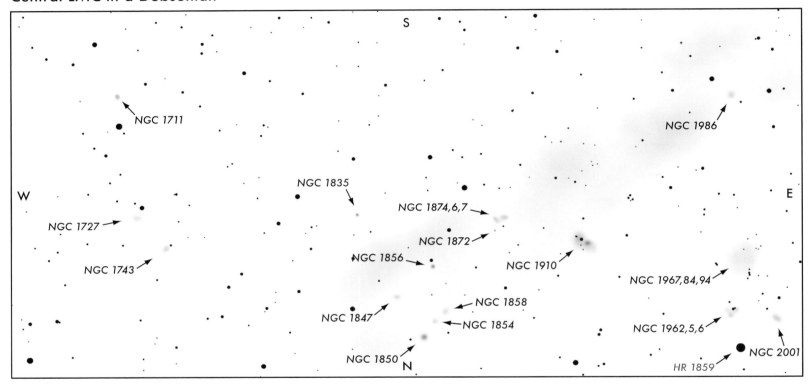

Northern LMC in a Dobsonian

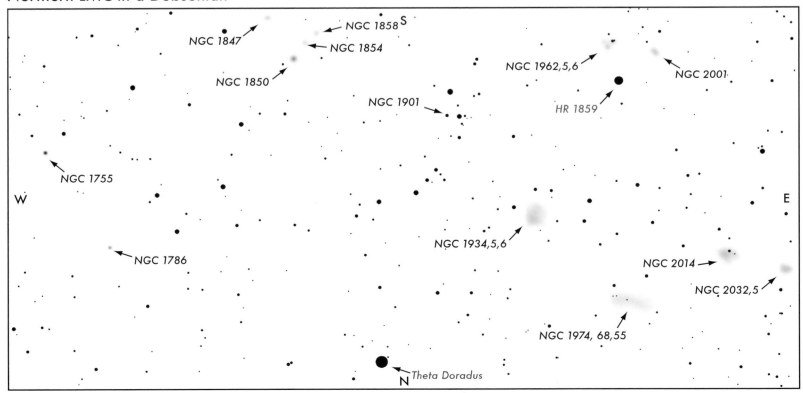

Southern LMC in a Dobsonian

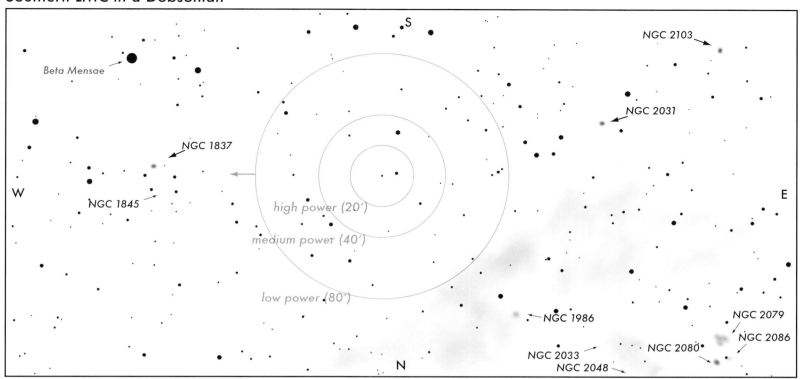

Around the Tarantula in a Dobsonian, low power

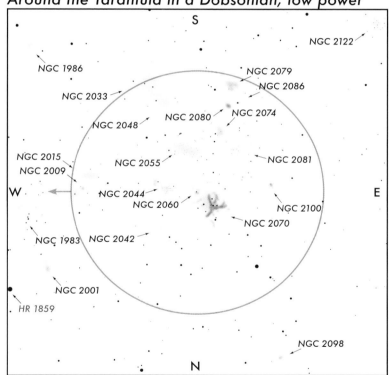

In a Dobsonian telescope: This is one really wild four square degrees of sky!

Start with the Tarantula; it is the easiest, brightest, and most spectacular field.

If you can pull yourself away from its pleasures, move southwards to the bright central corridor of the LMC; look for NGC 2103 and NGC 2031.

From there, move west and look for the fifth-magnitude star, Beta Mensae. Put it at the southern edge of your low-powered field, and look for NGC 1837 and NGC 1845. Then move back northeastward until you encounter the bright central corridor again. Follow the corridor past a dozen NGC objects, and at its end move westward to pick up NGC 1711, 1727, and 1743.

Moving back eastward to the corridor, keep moving to a sixth-magnitude star, HR 1859; just south in your low-power field will be a number of nice NGC nebulae.

Moving northwest from HR 1859, look for the fifth-magnitude star Theta Doradus. Put it in the western edge of your low-power field, and you should be able to find the objects around NGC 1974. Keep moving east to see NGC 2014 and 2032/35. Return back to Theta and now move west until you see NGC 1786; to its southwest is NGC 1755, the end of our tour.

The mismatch in the scale and number of stars from the three larger charts to the Tarantula chart is intentional – to find the dimmer blobs in the rest of the LMC, you really do want a reasonably dim limiting magnitude. Showing stars to magnitude 12.0 (the default standard for the other Dob fields in this book) would have been too busy, so we only show stars here to about magnitude 11.5 for those charts. For the Tarantula, however, even that would be just too busy; thus we show a somewhat subdued set of stars so we don't overload the field.

Remember, these charts are merely designed to help you find the objects in the telescope. Our drawings can't possibly do justice to what you'll actually see for yourself in the eyepiece. And they're geared to what the human eye sees – not to be used for starship navigation!

In Vela: An open cluster, NGC 2547, and Gamma Velorum, a multiple star

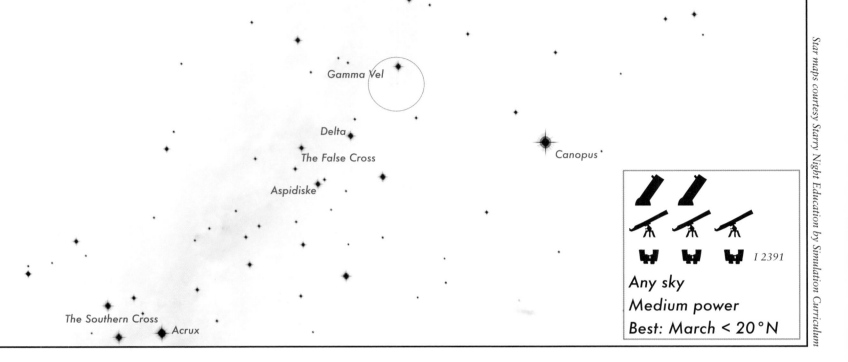

Gamma Vel

Delta

The False Cross

Aspidiske

Canopus

The Southern Cross

Acrux

Star maps courtesy Starry Night Education by Simulation Curriculum

I 2391

Any sky
Medium power
Best: March < 20°N

- Gamma Velorum: easy quadruple
- NGC 2547: nice small telescope cluster
- I 2391: nice nearby cluster for binoculars

Where to look: Find the False Cross, the kite-shaped groups of stars west of the true Cross. Step westward across the stars of the cross-bar, Aspidiske to Delta; one more step brings you to Gamma Velorum.

In the finderscope: From Gamma, look south for a fifth-magnitude star, IS Velorum. One step farther brings you to a pair of slightly fainter stars in a southwest–northeast line; continue a step past them and you should be near the open cluster. On a good night it will be visible in the finderscope.

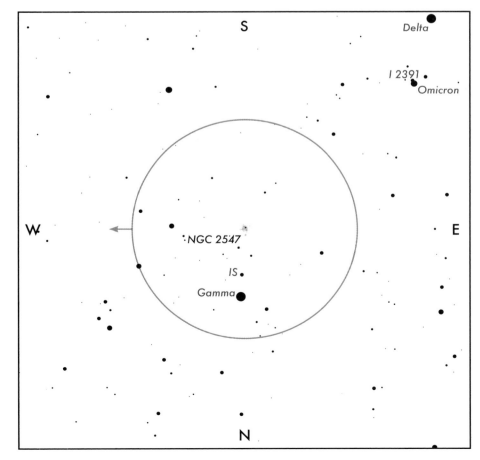

S

Delta

I 2391

Omicron

W

NGC 2547

IS

Gamma

E

N

NGC 2547 in a star diagonal at medium power

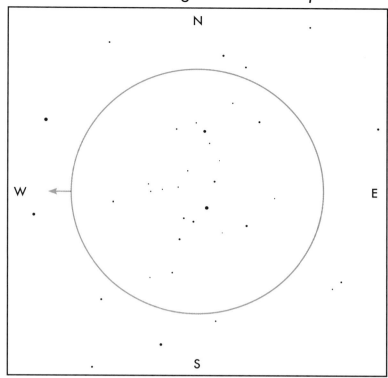

NGC 2547 in a Dobsonian at medium power

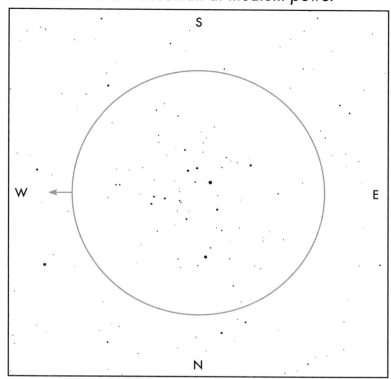

In a small telescope: The open cluster is a pleasant, loose collection of about a dozen stars easily visible in a small telescope, with a hint of a background haze of light.

In a Dobsonian telescope: The cluster is not seen at its best in a Dob; it is too resolved, and it can be hard to make it out against so many background Milky Way stars.

Gamma Velorum is a charming, easy multiple star; the A/B pair are easy to split, ideal for binoculars or a small telescope. The dimmer third star C makes a nice equal-sided triangle with the A/B pair, while the fainter D star lies just past it. The system is located 520 light years from us.

Gamma Velorum A is probably the closest example of a *Wolf–Rayet star*. These are massive stars at a late stage in their evolution with very strong stellar winds that can blow away a Sun's worth of mass in 10,000 years.

The open cluster **NGC 2547**, located 1,500 light years away from us, has been studied in the infrared by the Spitzer space telescope, which found more than 160 members of the cluster. With its lack of red-giant stars and the presence of unburned lithium in many of the smaller stars, the cluster is thought to be quite young, formed only 25–50 million years ago.

Also in the neighborhood: On your way from Delta to Gamma, you will pass a pretty cluster of stars, *I 2391*, centered around the magnitude 3.5 star **Omicron Velorum**. It is a pleasant grouping for binoculars, and nice to see in the finderscope.

Gamma Velorum			
Star	**Magnitude**	**Color**	**Location**
A	1.8	Blue	Primary star
B	4.1	Blue	41" SW from A
C	7.3	Blue	62" SSE from A
D	9.4	Blue	93" SSE from A

In Carina: The *Southern Beehive,* an open cluster, NGC 2516

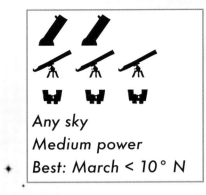

Star maps courtesy Starry Night Education by Simulation Curriculum

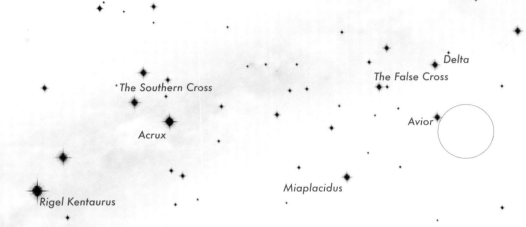

- Rich starfield against a background glow
- Easy to find
- Ideal for small telescopes

Where to look: The Milky Way runs from Orion south, past the first-magnitude star Canopus, and down to the Southern Cross. Roughly halfway between Canopus and the Southern Cross is another cross of four stars, called the *False Cross.* These stars are slightly farther apart and slightly dimmer than those in the true Cross. Point at the brightest (third-magnitude) and westernmost star of the False Cross, Avior (Epsilon Carinae). On a good night it is visible to the naked eye as a clump of light south of the Milky Way.

In the finderscope: From Avior, move the finderscope west and a bit south. Just about the time Avior leaves the finderscope field of view, you'll be centered on the open cluster. It should be easily visible in the finderscope, and quite pretty in binoculars.

NGC 2516 in a star diagonal at medium power

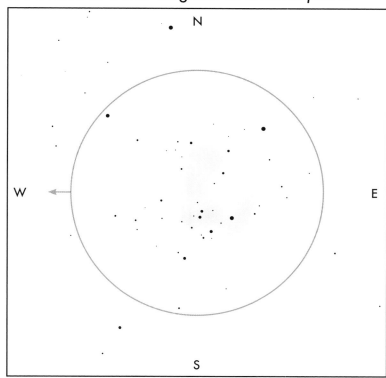

NGC 2516 in a Dobsonian at medium power

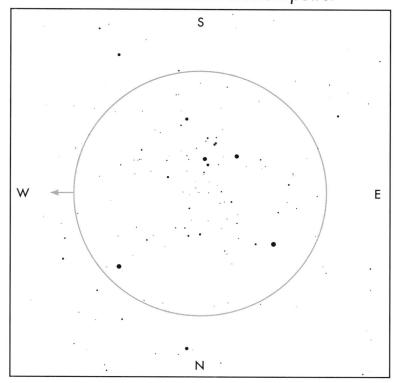

In a small telescope: A small telescope should be able to pick out at least two dozen stars in the center of this cluster. In a 3" telescope, unresolved fainter stars behind them add a fuzzy glow of light. The whole cluster extends more than a degree across; wander about with the telescope to see them all.

Note the bright (fifth-magnitude) reddish star east and slightly north of center, at the end of a tongue of grainy light; and three double stars: pairs of eighth-magnitude, with just under 10 arc seconds separation each.

In a Dobsonian telescope: The nebula is an impressive dense collection of bright stars, standing out against the fainter background of the Milky Way. Rather than an unresolved haze, a Dob resolves countless fainter pinpricks of light between the brighter members of the cluster.

More than 100 stars reside in the **Southern Beehive** in a space a few dozen light years across, located a bit over a thousand light years from us. The fifth-magnitude star in the center is the only prominent red star, and the doubles include B- and A-type stars, indicating that the cluster is still relatively young – perhaps five to ten million years since star formation began.

The large number of 12th- and 13th-magnitude stars produce the fuzzy glow described above; they provide a wonderful, graceful backdrop to this cluster as seen in a small telescope. It's a perspective that bigger telescopes, with their greater resolving power, lose out on.

*Also in the neighborhood: On the southeast edge of your finderscope view, about six degrees from the cluster, is the double star **Rumker 8**; in a Dob you can see the yellow fifth-magnitude primary orbited by an orange eighth-magnitude star, 4" to the east-northeast.*

*East of Avior is a fourth-magnitude star, HIP 42568; a degree south and west from it is the double star **h 4128**. It may appear at first as merely an elongated star; a Dobsonian at very high power reveals this to be pair of stars, not quite a cat's-eye pair: one is just brighter than 7th magnitude, the other just a shade dimmer. They have a 1.2" separation, in a roughly north–south direction.*

find more at: www.cambridge.org/features/turnleft/southern_skies.htm

In Carina: A globular cluster, NGC 2808

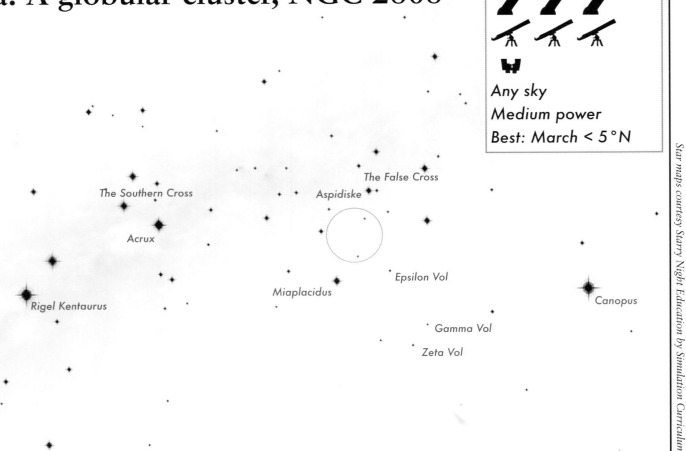

Star maps courtesy Starry Night Education by Simulation Curriculum

Any sky
Medium power
Best: March < 5°N

- Elegant cluster in a rich field of stars
- Easily resolved in a Dob
- Perhaps the core of a captured dwarf galaxy?

Where to look: Find Miaplacidus, the brightest star (second-magnitude) between Canopus and the Southern Cross. Start by aiming your finderscope there.

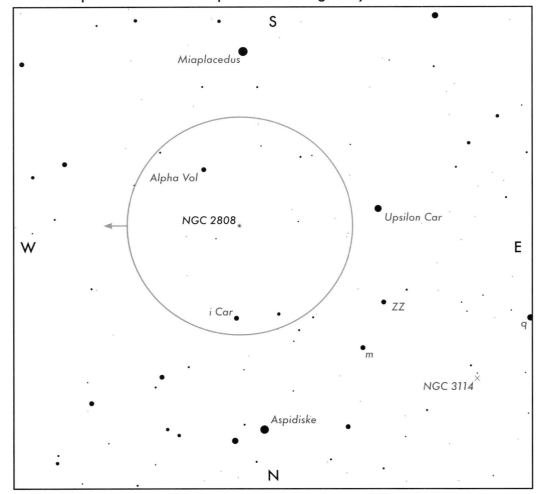

In the finderscope: Move from Miaplacidus north towards Aspidiske, the eastern star of the False Cross. En route you will go past two fourth-magnitude stars, Alpha Volantis and i Carinae. Aim halfway between these stars, and then move the finderscope a degree to the east; the cluster should be visible in the finderscope as a fuzzy star.

Alternately, if you are starting from Eta Carinae and the Keyhole, move south and west from Eta, past q Carinae, to Upsilon Carinae, and then west until you see Alpha Volantis and i Carinae.

NGC 2808 in a star diagonal at medium power

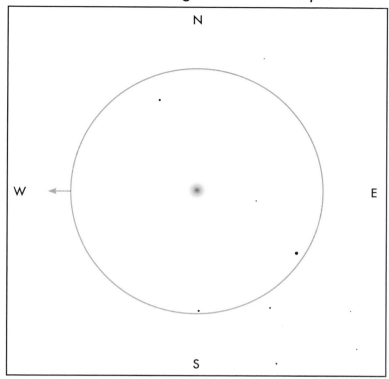

NGC 2808 in a Dobsonian at medium power

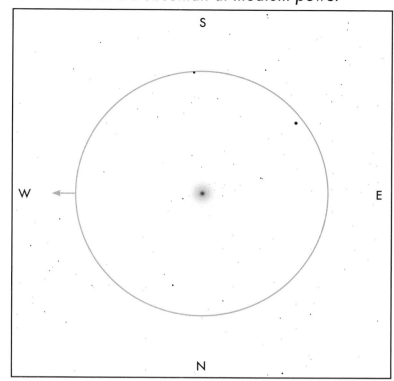

In a small telescope: The globular cluster is small but bright, making an equal-sided triangle with a pair of stars, a seventh-magnitude star to the southeast and an eighth-magnitude star to the south. The background rich field of stars makes it an especially pretty sight.

In a Dobsonian telescope: The globular cluster shows a small bright core surrounded by a partially resolved ball of light. The rich field of background stars adds to its delight.

NGC 2808 is an unusually massive globular cluster, containing more than a million stars. It is located about 30,000 light years from us.

In recent years, this globular cluster has been the focus of intense research because detailed study of the stars within the cluster suggest that it may contain three generations of stars. Globular clusters, as described on page 115, are thought to be among the oldest collections of stars, made up primarily of low-metal first-generation stars; the presence here of stars that contain the metals made inside previous generations of stars was quite surprising. One theory suggests that the cluster's unusual size allowed the rapid formation of several generations of metal-forming stars. Another suggestion, however, is that it isn't a globular cluster at all but rather the core of a dwarf galaxy that has been captured by the Milky Way.

Also in the neighborhood: Four degrees east of NGC 2808 is the second-magnitude star **Upsilon Carinae.** *It's a pair of yellow stars, a third-magnitude primary with a sixth-magnitude secondary 5" to the southeast; a small telescope can just split it. In the same field of view, on a line with this pair, is a pair of ninth-magnitude stars that look like a cat's-eye double; but apparently they are unrelated, and just coincidentally close to each other from our point of view.*

From i Carinae, move your finder east past the pair of fourth-magnitude stars, m and ZZ Carinae, until you see the third-magnitude star q Carinae. Look north of the line connecting m Carinae and q Carinae for a pair of sixth-magnitude stars; aim there and look for an open cluster, **NGC 3114.** *It should be visible as a haze of stars in the finder or a pair of binoculars. In a small telescope it will resolve into dozens of stars half-filling the low-power field of view.*

ZZ Carinae *itself (also known as l Carinae – the letter "l" not the numeral "1") is a Cepheid variable, changing from magnitude 3.4 to 4 with a three-week period. See page 196 for more on this type of variable star.*

Southwest of NGC 2808 are the third-magnitude stars Epsilon, Gamma, and Zeta Volantis. They are multiple stars described on page 217.

find more at: www.cambridge.org/features/turnleft/southern_skies.htm

In Carina: Eta Carinae and the *Keyhole Nebula*, NGC 3372; and four nearby open clusters

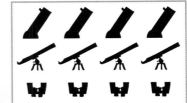

Star maps courtesy Starry Night Education by Simulation Curriculum

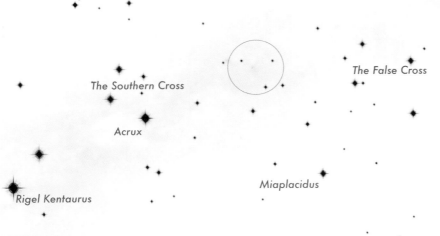

The Southern Cross

Acrux

The False Cross

Rigel Kentaurus

Miaplacidus

Canopus

Any sky; better, dark
Low to high power
Nebula filter
Best: April < 10°N

- Bright, spectacular luminosity
- Pre-supernova star?
- Nearby region rich with bright clusters

Where to look: The Milky Way runs between Orion and the Twins, south past Sirius, past the bright star Canopus, and down to the Southern Cross. Roughly halfway between Canopus and the Southern Cross is another cross of four stars, called the *False Cross*. These stars are slightly farther apart and slightly dimmer than those in the true Cross. Look for a bright knot of light on the southern side of the Milky Way, halfway between the Southern Cross and the False Cross.

In the finderscope: Aim halfway between the Southern Cross and the False Cross for two fuzzy patches of light in a very rich part of the Milky Way: the **Keyhole Nebula**, to the west, and the much larger fuzz of **NGC 3532** to the east.

The cluster **NGC 3293** lies inside a triangle of stars northwest of the Keyhole. The cluster **NGC 3766** is halfway between the Keyhole and the Southern Cross, easily visible in the finder. The **Southern Pleiades** are a finder field due south of the Keyhole.

All these objects are easy to see in the finder; most are naked-eye visible on a good night.

(Also, note q Carinae; it's a stepping stone between here and NGC 2808, described on pages 226–227.)

S

The Southern Pleiades

h 4432

NGC 3766

q

W

The Keyhole

×Feinstein 1

NGC 3532

E

× NGC 3293

j

N

The Keyhole in a star diagonal at low power

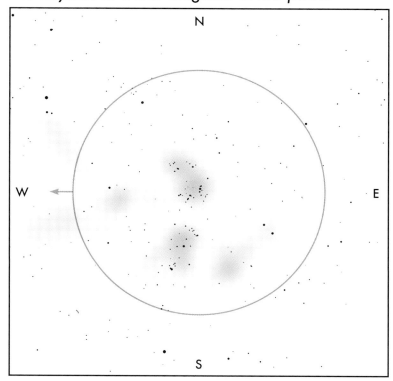

The Keyhole in a Dobsonian at low power

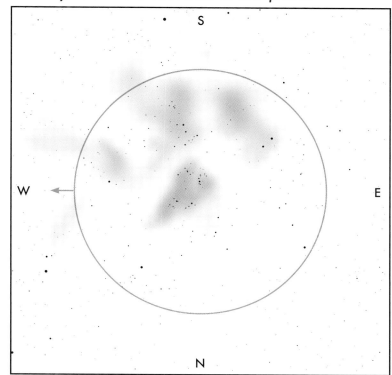

In a small telescope: The nebula is quite large, about a degree and a half across, and it will more than fill even a low-power field of view. If the night is dark, you'll see a highly structured set of bright patches and dark lanes, which gradually fade off into the background Milky Way. Try higher powers for more detail.

Note a bright, distinctly orange star in the center of the field; larger telescopes reveal this to be a small, manikin-shaped nebula, nicknamed the Homunculus.

In a Dobsonian telescope: The nebula will more than fill even a low-power field of view. If the night is dark, you'll see a breathtaking, highly structured set of bright patches and dark lanes that fade off into the background Milky Way. There is a close double in the very bright nebulosity at the center of the field of view. Note a bright, distinctly orange star in the center of the field; larger telescopes reveal this to be a small, manikin-shaped nebula, nicknamed the Homunculus.

Try moderate and high powers, and a nebula filter, for more detail.

The **Keyhole Nebula** is named for the distinct dark hole in the field of light. Though the Orion Nebula's center is brighter, on a dark night the Keyhole shows a size and complexity that's even more fascinating. The contrast between the bright nebulosity and the dark lanes, the differing shades of brightness among the components, and the peppering of bright stars through the nebulosity, make it an unforgettable sight.

This giant cloud of gas and dust, a rich region of star formation, is estimated to be 7,500 light years away from us; but as with the Jewel Box (page 236), correcting for dust between it and us makes calculating a distance difficult. It's been estimated to have more than 1,200 associated stars.

The brightest of these stars, Eta Carinae, has a remarkable history. Today it's a mildly varying seventh- to eighth-magnitude star, but in the 1700s Eta varied between second- and fourth-magnitude. In the early part of the 1800s, it flared up even brighter; by 1827, it had become a first-magnitude star. And in April 1843, it actually reached magnitude –0.8, brighter than any star but Sirius. Given its distance from us, it was probably as bright as a million Suns – amateurs observing from the Andromeda Galaxy could have seen it in a 4" telescope. Astronomers are still arguing about what could have caused this; such increases in brightness can occur with novae or supernovae, but they usually do not last for decades.

The **Gem Cluster**, the open cluster NGC 3293, is perhaps 8,000 light years from us. Determining its age (see page 71) has turned out to be tricky. The colors of the fainter stars in this group suggest it is about 25 million years old, but the bright blue stars appear to be only around six million years old. Perhaps they were formed more recently than the main group. The **Wishing Well**, NGC 3532, resembles a scattering of coins at the bottom of a well. It is notable for being the very first target of the Hubble Space Telescope, in 1990. It lies 1,600 light years from us.

The **Pearl Cluster**, NGC 3766, gets its name from the string of stars across its northern edge. It is 5,500 light years away. The **Southern Pleiades** is less than 500 light years distant, not too different from its northern namesake, M45 (page 64). However, it is older (50 million years) and smaller; you'll find half again as many stars of a given brightness in M45 than here.

Also in the neighborhood: Five degrees east of the Southern Pleiades is h 4432, a pair of fifth magnitude bluish stars 2.3" apart on a northwest–southeast line.

Five degrees northwest of the Keyhole is J Velorum; it's a triple, with an eighth-magnitude star 7" east of the fifth magnitude primary, and a ninth-magnitude companion 37" to the south.

find more at: www.cambridge.org/features/turnleft/southern_skies.htm

The Gem Cluster in a star diagonal at high power

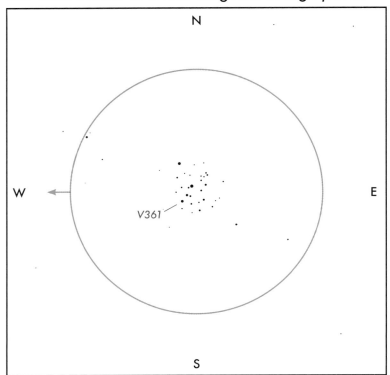

The Gem Cluster in a Dobsonian at high power

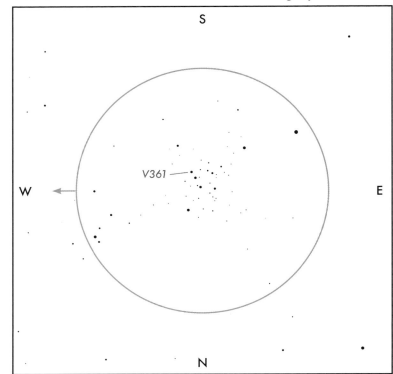

In a small telescope: Note a clump of stars to the northwest of the Keyhole. This is NGC 3293, the **Gem Cluster**, a small sprinkling of faint stars against a background haze. See a red star in the southwest, V361 Carinae, a red giant that varies from magnitude 7.1 to 7.6.

A quite pleasant open cluster, NGC 3532, the **Wishing Well,** lies three degrees due east of the Keyhole, visible to the naked eye as a knot of light in the Milky Way. In a small telescope, it's a loose cluster of sparkles, like the coins at the bottom of a well, filling the field of view. One degree south of it (see the finder view) is Feinstein 1, a loose open cluster seen best at low power.

In a Dobsonian telescope: NGC 3293, the **Gem Cluster**, is the clump of stars northwest of the Keyhole. The aperture of a Dob lets you look in high power, revealing dozens of faint stars. In the southwest, V361 Carinae is a red giant that varies from magnitude 7.1 to 7.6.

NGC 3532 lies three degrees due east of the Keyhole. Like the Keyhole, it's visible to the naked eye as a knot of light in the Milky Way. In a Dob, it's a bright, loose cluster of dozens of stars, with easily a hundred fainter stars filling in the background. Look one degree south (see the finder view) for Feinstein 1, a loose open cluster; use your lowest power.

NGC 3532 in a star diagonal at medium power

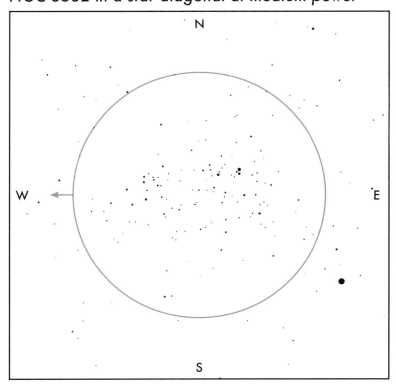

NGC 3532 in a Dobsonian at medium power

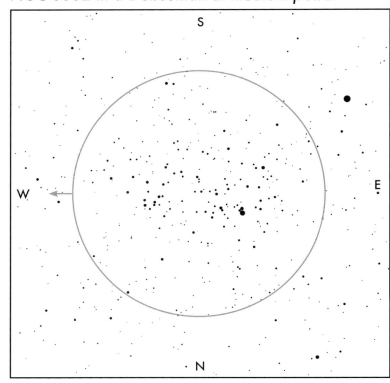

The Southern Pleiades, star diagonal, low power

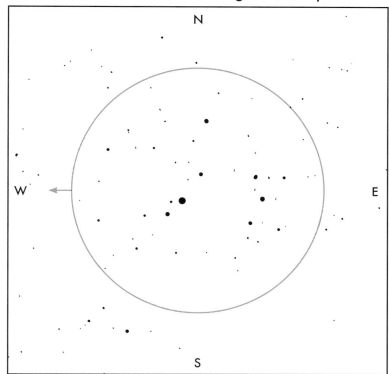

The Southern Pleiades in a Dobsonian at low power

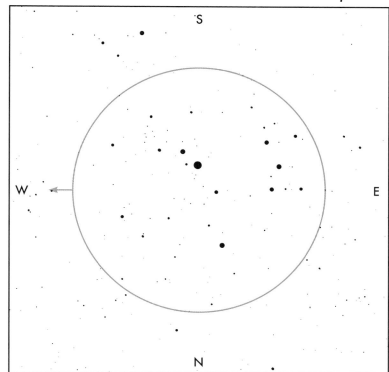

In a small telescope: About five degrees south of the Keyhole is a loose open cluster, IC 2602, popularly known as the **Southern Pleiades**. It's an awkward size for a small telescope – too small for the finderscope, too big for a low-power field. But it is perfect for binoculars.

Almost exactly halfway between the Keyhole Nebula and Acrux is the open cluster NGC 3766, called the **Pearl Cluster**. In a small telescope, it will look like a grainy haze of light flanked to its east by a pair of fifth-magnitude stars. Dominating the cluster are a triangle of stars; the two to the north are red giants.

In a Dobsonian telescope: About five degrees south of the Keyhole is a loose open cluster, IC 2602, popularly known as the **Southern Pleiades** – a big loose gathering of stars in an already busy field of Milky Way stars. Use your lowest power.

Almost exactly halfway between the Keyhole Nebula and Acrux is the open cluster NGC 3766, called the **Pearl Cluster**. In a Dob, it's a tight, compact group faint stars with a pair of fifth-magnitude stars to its east. Notice within the cluster a triangle of brighter stars; the two to the north are red giants.

The Pearl Cluster in a star diagonal, medium power

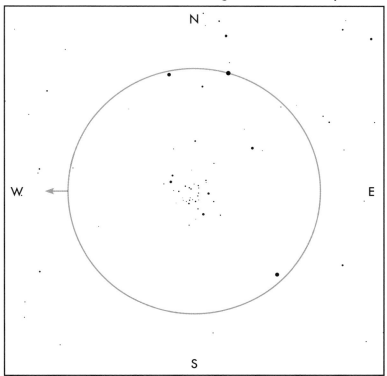

The Pearl Cluster in a Dobsonian at medium power

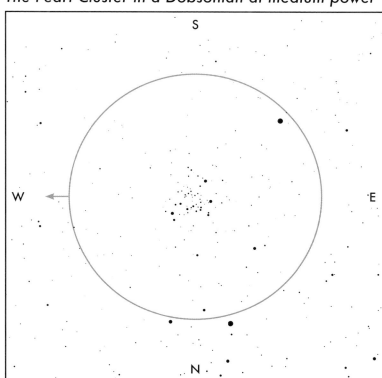

find more at: www.cambridge.org/features/turnleft/southern_skies.htm

By Centaurus and Crux:
Seven double stars

Gacrux

Any sky

Highest power

Best: May–June < 10°N

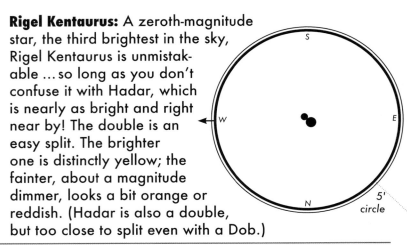

Rigel Kentaurus (Alpha Centauri)

Star	Magnitude	Color	Location
A	0.0	Yellow	Primary star
B	1.3	Orange	6–10" S from A

Dunlop 159

Star	Magnitude	Color	Location
A	5.0	Yellow	Primary star
B	7.6	Blue	8.9" SSE from A

Mu Crucis

Star	Magnitude	Color	Location
A	3.9	Blue	Primary star
B	5.0	White	35" NNE from A

Gacrux (Gamma Crucis)

Star	Magnitude	Color	Location
A	1.6	Orange	Primary star
B	6.5	White	127" NNE from A
C	9.5	White	155" E from A

Alpha Circini

Star	Magnitude	Color	Location
A	3.2	White	Primary star
B	8.5	Red	16" SW from A

Beta Muscae

Star	Magnitude	Color	Location
A	3.5	White	Primary star
B	4.0	White	0.7" ENE from A

Acrux (Alpha Crucis)

Star	Magnitude	Color	Location
A	1.3	Blue	Primary star
B	1.6	Blue	4" W from A
C	4.8	Blue	90" SSW from A

Rigel Kentaurus is a triple star, the brightest (easy) multiple star in the sky, and notable for being the system nearest to our own Solar System – a mere 4.395 light years away. The pair visible to small telescopes has a period of 80 years, and its separation varies from 12 AU (about the distance from Saturn to the Sun) to 36 AU. In our telescopes, this motion translates to a separation ranging from 1.7 to 22 arc seconds. Until about 2035 the companion will be 6" to 10" south of star A; in 2037–2038 it passes within 2" to the west; then it quickly moves away, to the north and east.

The primary star is a near twin to our Sun in color and brightness. Looking at it gives you a good idea of what we would look like to astronomers from Rigel Kentaurus looking at us. (Of course, we don't have a bright, slightly orange companion.) Indeed, there could be such astronomers there looking back at us; any planet within three or four AU of either A or B (or more than 70 AU away from both, a chillier location) would be in a stable orbit, and conceivably habitable.

The third member of this group is 11th magnitude, a tough find for most small telescopes. It orbits about a quarter of a light year, 14,000 AU, from the other two – comparable to the distance from our Sun to the Oort Cloud of comets. From our vantage point that's more than two degrees away from A and B, and well out of the telescope field of view. At its present location in its orbit, it lies slightly closer to us (4.22 light years) than the main pair, and so it has the honor of being the star nearest to our Sun; thus it bears the name **Proxima Centauri**. We've noted its location with a small cross, though actually it should only just barely be visible under the best conditions in a three-inch telescope. In fact, even if we moved ourselves to an Earth-like orbit around Rigel Kentaurus itself, it'd be only barely visible to the naked eye (magnitude 4.5).

The **Acrux** system is in fact at least a quadruple, since star A is itself known to be a spectroscopic binary: its spectral lines shift back and forth with a period of 76 days, indicating the presence of a closely orbiting star tugging it back and forth. This system of stars lies about 500 light years away from us.

Acrux was only the third star to be revealed as a double (after Mizar and Mesarthim); the fourth was Rigel Kentaurus. Both discoveries were made by three friends who were fellow astronomers and Jesuit missionaries from France. Fathers Jean de Fontanay and Guy Tachard split Acrux in 1685, from the Cape of Good Hope, South Africa; and Fr. Jean Richaud split Rigel Kentaurus in 1689 while observing a nearby comet from Pondicherry, India.

Dunlop 159: Aim your finderscope halfway between Rigel Kentaurus and Hadar, and look one and a half degrees north. You'll find a fifth-magnitude star, Dunlop 159, a pleasant yellow–blue double star. (The 10th magnitude star to its north is just a field star.) The pair are located about 300 light years away.

Alpha Circini: Look south from Rigel Kentaurus for a thirdmagnitude star. If Rigel Kentaurus is the hub of a clock face, and Hadar is located at 2:30, then Alpha Circini would be at the 6 o'clock position. The companion is significantly fainter than its primary, making this a nice Dob challenge. This star, located only 53 light years from us, is

Mu Crucis: Step northeast from Acrux to Mimosa; half a step farther arrives at this fourthmagnitude star. It's a wide, easy double, ideal for a small telescope. They lie 360 light years from us.

Gacrux is the northernmost star of the Cross. The two companions are widely separated (try with binoculars) but significantly dimmer than the primary and thus a fun challenge to spot. In fact they are only an optical double pairing; Gacrux A is only 88 light years distant, while B is more than 400 light years away.

also notable for being a rapidly pulsating variable, flickering (a fraction of a magnitude) every 6.8 minutes.

Beta Muscae: Step southeast from Delta Crux to Acrux; one more step brings you to third-magnitude Beta Muscae. The pair, 350 light years away, have a 194 year period, moving notably in our telescopes. For the next 30 years star B will be moving north to within 0.6" of star B; in a Dob, look for an elongated image rather than actual separation.

Acrux: Even in a 3" telescope, the A/C separation is so large that you might think it would be a boring pair to look at; but look closely (or at high power) and you'll notice that the primary star is itself a double, a lovely cat's-eye pair of bright stars. Furthermore, these stars lie in the midst of the Milky Way and so you'll also see a rich field of background stars about a magnitude dimmer than Acrux C. Seeing a bright multiple star standing out in front of these other stars gives a unique and lovely illusion of depth.

Star maps courtesy Starry Night Education by Simulation Curriculum

In Musca: Two globular clusters, NGC 4833 and NGC 4372

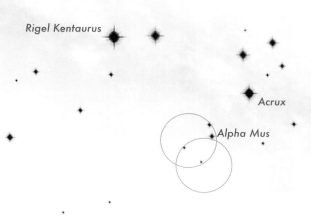

Rigel Kentaurus

Acrux

Alpha Mus

The False Cross

Star maps courtesy Starry Night Education by Simulation Curriculum

Any sky
Medium power
Best: May > 5° S

Where to look: Look south of the Southern Cross for a pair of second-magnitude stars, Alpha and Beta Muscae. Alpha is farther to the south. Aim the finderscope there.

In the finderscope, NGC 4833: From Alpha, move south and east towards the third-magnitude star Delta Muscae. The cluster is found just north of that line, about a degree north of Delta. It may be visible in the finderscope as a faint, fuzzy star.

NGC 4372: From Alpha, move south (or, from Delta, move southwest) to the third-magnitude star Gamma Muscae. The cluster is about one degree southwest of Gamma, near a faint star designated HIP 60561. On a good night the cluster itself may be visible in the finderscope.

Also in the neighborhood: Step from Alpha to Beta Muscae (towards Hadar) and look for a trio of fifth-magnitude stars in a north–south line; the northernmost of these, m Centauri, has a little sixth-magnitude companion visible in the finderscope. From m Centauri, follow a trail of sixth- and seventh-magnitude stars that curl to the south and east, to a small smudge of light at the end of the curl. That's **NGC 5189,** *the Spiral Planetary Nebula. In larger telescopes one can see the internal detail that gives it its name, but that's a challenge in a typical Dob.*

Beta Muscae itself is a nice double star, as described on page 233.

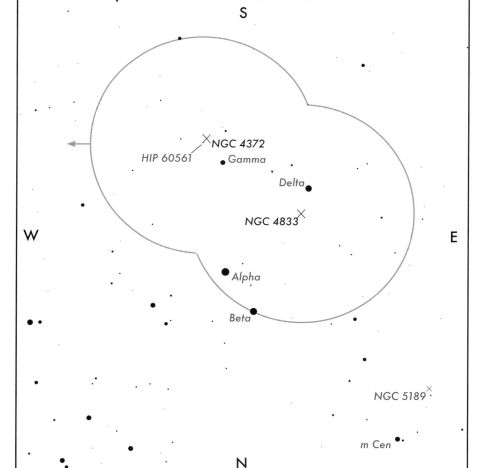

NGC 4833 in a star diagonal at medium power

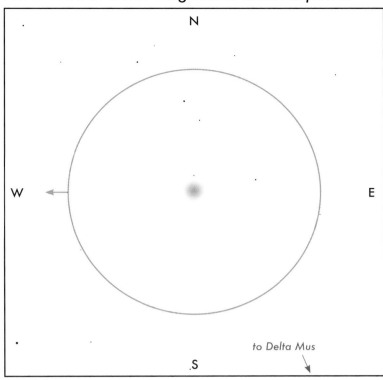

NGC 4833 in a Dobsonian at medium power

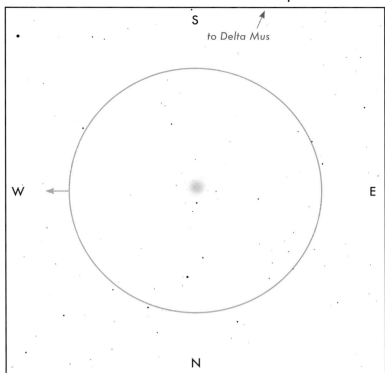

In a small telescope: Both clusters look remarkably alike: small featureless disks of light with a star at one edge. However, NGC 4372 is larger and more diffuse, and its companion star is much brighter than that of NGC 4833. To keep track of which is which, note that NGC 4833 has three stars to its north in the field of view.

In a Dobsonian telescope: Both clusters look oddly pale compared to other globulars. Both are disks of light, just starting to be resolved, with a slight central concentration and a prominent star at one edge. NGC 4372 is the larger and more diffuse globular, and its companion star is much brighter than that of NGC 4833.

NGC 4372 in a star diagonal at medium power

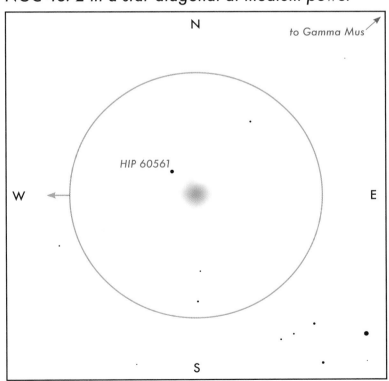

NGC 4372 in a Dobsonian at medium power

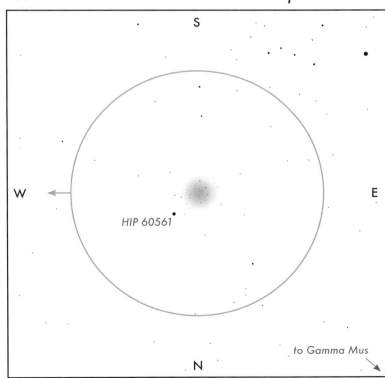

These two globular clusters, **NGC 4833** and **NGC 4372**, are in fact separated by only about 2,500 light years; NGC 4833 is 21,000 light years from us, while the distance to NGC 4372 is 18,900 light years. The light from both clusters must pass through a cloud of cold gas and dust lying between them and us, which scatters much of their blue light (just as our atmosphere turns sunsets red). The result is the subdued color of these clusters, making them look rather yellowish compared to the blue–white stars seen in unobscured globular clusters.

find more at: www.cambridge.org/features/turnleft/southern_skies.htm

In Crux: *The Jewel Box*, an open cluster, NGC 4755.

Star maps courtesy Starry Night Education by Simulation Curriculum

Rigel Kentaurus   *Hadar*

Mimosa

Acrux

Any sky
Medium, high power
Best: May < 10°N

- Beautiful, colorful, bright cluster
- Easy to find
- Nearby dark cloud, the Coal Sack

Where to look: Find the Southern Cross. The southern-most star, at the bottom of the cross, is Acrux (Alpha Crucis). The left star of the cross-piece, the star nearest Rigel Kentaurus and Hadar, is Mimosa (Beta Crucis, also sometimes called Becrux). Aim there.

In the finderscope: Look for an equal-sided triangle of Mimosa and two fourth-magnitude stars, Lambda and Kappa Crucis. Lambda is to the north, Kappa to the south (towards Acrux). Aim at Kappa.

Note that Mimosa and this cluster can both fit in the same low-power field of view; that's handy for finding it. However, since the Jewel Box is small but relatively bright, it is better seen at medium or high power.

S

The Coal Sack

Acrux   BZ Cru

W   E

Kappa

Mimosa

Lambda

Gamma   Mu

N

The Jewel Box in a star diagonal at medium power

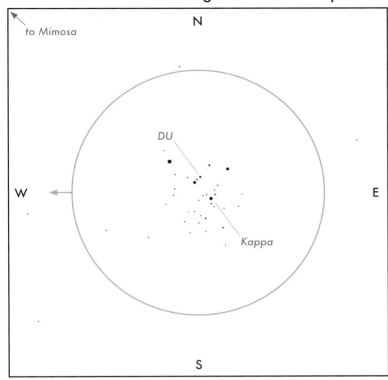

The Jewel Box in a Dobsonian at high power

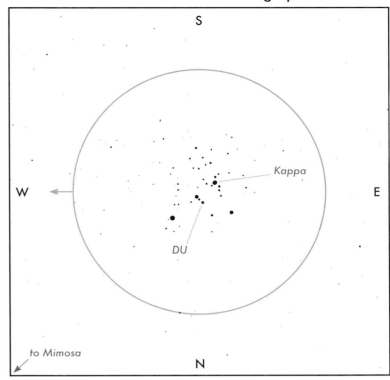

In a small telescope: What looked like a single fourth-magnitude star to the naked eye is revealed to be a cluster of about half a dozen sixth- to eighth-magnitude stars, with perhaps another two dozen fainter stars among them and a very faint haze of light behind them all. You'll first notice a wedge-shaped group centered on three bright stars in a triangle pointing west-northwest; further observing reveals the rest of the cluster behind and around them. The southeast end of the line of three bright stars near the center of the cluster is the sixth-magnitude star officially designated Kappa Crucis.

In a Dobsonian telescope: Notice a wedge-shaped group centered on three bright stars in a triangle pointing west-northwest, with the rest of the cluster behind and around them. The Dob's larger aperture means that you can stand to observe this cluster at higher power, bringing out more of the fainter stars. The southeast end of the line of three bright stars near the center of the cluster is the sixth-magnitude star officially designated Kappa Crucis.

The **Jewel Box** gets its name from the way the brighter members jump out against a background of fainter stars, like large gems nestled in a box of jewels: "a casket of variously coloured precious stones," to quote Sir John Herschel's nineteenth-century description. Quite nice even in a three-inch telescope, it gets even more impressive in a larger scope as the background nebulosity increases in brightness and starts to become resolved.

The Jewel Box is a collection of at least 50 stars located perhaps about 7,500 light years from us. There's quite a large uncertainty with this distance measurement, however. The standard technique is to take stars of a known spectral type, of known absolute brightness, and then compute how far away they must be to have the brightness we actually see. However, the region around Crux is heavily obscured with dust clouds (the Coal Sack is right near by) and guessing how to correct for this dust is tricky.

Of the ten brightest stars in this cluster, the one in the center of our bright triangle is a red supergiant, DU Crucis; Herschel described it as a "ruby in a set of diamonds and sapphires." As indicated by its name, it's a variable star; its brightness goes from magnitude 7.1 to 7.5. The other nine brightest stars are blue B types, each more than ten thousand times as bright as our Sun. (Given a 7,500 light year distance, the brightest may shine like 80,000 of our Suns.) Such bright stars tend to consume their fuel quite rapidly, turning themselves into red giants; the fact that they aren't all red giants yet indicates that this cluster is quite young, only a few million years old. (Learn more about open clusters on page 71.)

*Also in the neighborhood: The Southern Cross, the smallest of the constellations, lies entirely within the band of the Milky Way. However, on a good dark night when the Milky Way is easy to see, you'll notice with the naked eye that a large region to the east is significantly darker than the rest of the Galaxy. This is the **Coal Sack**, a large dark cloud of gas and dust lying only about 600 light years from us, blocking out the stars behind it. There's quite literally nothing to see in a telescope; but it's a beautiful naked-eye object. A fun naked-eye challenge: look for the fifth-magnitude (slightly variable) star, **BZ Crucis**, which sits in the middle of the dark region. It is actually located behind the dark cloud, 1,000 light years from us. Apparently it is shining through a rift in the cloud.*

Gacrux and Mu Crucis are lovely double stars; we talk about them on page 233.

find more at: www.cambridge.org/features/turnleft/southern_skies.htm

Antares

Scorpius

Alpha Lup

Zeta Cen

Epsilon Cen

Gamma Cen

Rigel Kentaurus

Hadar

Mimosa

Acrux

Any sky
Medium power
Best: May < 30°N

Star maps courtesy Starry Night Education by Simulation Curriculum

Where to look: South and west of Scorpius is the clump of second-magnitude stars of Lupus; southernmost of them is Alpha Lupi. To its west is the northernmost bright star of Centaurus, Zeta Centauri. Step from Alpha Lupi to Zeta Centauri; half a step farther brings you to a fourth-magnitude starlike object, Omega Centauri.

Alternately, step north from Acrux (foot of the Southern Cross) to Mimosa (the left arm) and continue three more steps.

In the finderscope: Even in the finderscope it should be obvious that you're looking at an extended ball of light, not a point star. Likewise, it shows up wonderfully in binoculars.

Also in the neighborhood: Put Omega Centauri in the northeast corner of your finder and look south-west for the stars Xi-1 and Xi-2; between them is an edge-on galaxy, **NGC 4945,** *a nice Dob challenge. Stretching 150,000 light years across, it lies about 12 million light years from us.*

South-southeast from Omega Centauri is Epsilon Centauri. Two degrees to its east is the double star **N Centauri:** *two blue stars, magnitudes 5.4 and 7.6, oriented east–west, separated by 18 arc seconds. A degree south and slightly east of Epsilon is* **Q Centauri,** *a double whose components are a bit*

S

Q

Epsilon Cen

N

M

NGC 5286

Xi-2

Xi-1

NGC 4945

W

Zeta Cen

E

Omega

NGC 5128
(Cen-A)

N

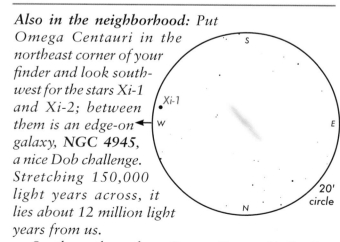

S

Xi-1

W

E

N

20'
circle

Omega Centauri, star diagonal at medium power

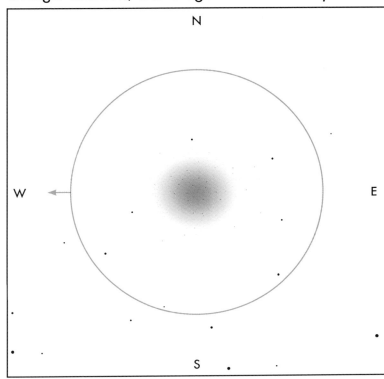

Omega Centauri in a Dobsonian at medium power

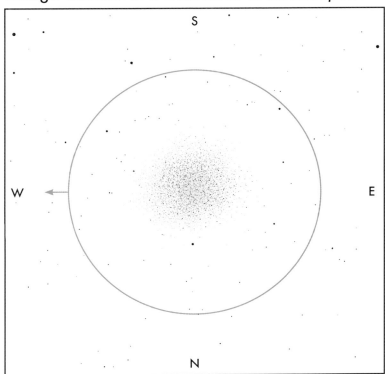

In a small telescope: The globular cluster sits in a rich field of stars only slightly less bright than the background stars shown here. It fills a significant fraction of your field of view, even at low power. Its bright central condensation extends more than halfway across the whole nebula. In a 3" telescope it will appear grainy; on a good night you may actually resolve individual stars.

In a Dobsonian telescope: The globular cluster fills the field of view, even at low power, in a rich field of Milky Way stars. Its bright central condensation extends more than halfway across the whole nebula. In a Dob you will have no problem resolving individual stars.

brighter than N but harder to split; the magnitude 6.6 companion lies only 5" south-southeast from the 5.3 primary. It's a fun challenge for a small telescope, easier in a Dob.

About five degrees southeast of Omega Centauri is the globular cluster **NGC 5286.** *It's just northwest of a relatively bright (fifth-magnitude) yellow star, M Centauri. It's easy in a Dob, but a challenge in a small telescope.*

The galaxy **NGC 5128,** *containing the strong radio source Centaurus-A, is five degrees north of Omega Centauri. (It can just peep above the horizon even as far north as New York; we've seen it from the south shore of Long Island on a perfect night!) Move your finder due north; stop when you see a group of sixth magnitude stars shaped like a numeral "7" at the western edge of your field. Sometimes it is called the "Hamburger Galaxy" – in a Dob, look for a dark dust lane across its disk, splitting it like a hamburger patty between two buns.*

This galaxy is the namesake of the NGC 5128 *Galaxy Cluster, located about 22 million light years from us.* NGC 4945 *is also a member.*

Omega Centauri is the biggest, brightest, most spectacular globular cluster in the sky. More than half a degree across (bigger than a full Moon!) it's half again as large as its closest rival, M22, and more than three times as bright. And compared to M13, the so-called "Great Cluster" in Hercules, the best globular of the northern sky, it's twice the angular diameter and more than two full magnitudes brighter. Nothing else in the northern hemisphere even comes close (but compare it with 47 Tucanae on page 212).

Higher power in a small telescope doesn't seem to gain you much; the central core may look bigger, but then the fainter outer edges get more spread out and harder to see.

Omega Centauri was known (as a star) in antiquity; due to the wobbling of the Earth's spin axis, which causes the locations of the constellations to appear to shift over time, it was visible as far north as Egypt some 2,000 years ago. It was labelled and given its Greek name as a star in Renaissance star charts. Then in the 1670s, Edmund Halley (of comet fame) travelled to South Africa and became one of the first telescope observers of the southern sky; in 1677 he discovered that this "star" was in fact a globular cluster.

Or indeed is it a cluster? Some have speculated that it's the core of a small galaxy that was captured into the Milky Way.

Located about 15,000 light years from us, it contains more than a million stars. The core of the cluster, as seen here, is about 50 light years across.

For more on globular clusters, see page 115.

find more at: www.cambridge.org/features/turnleft/southern_skies.htm

In Pavo: A globular cluster, NGC 6752

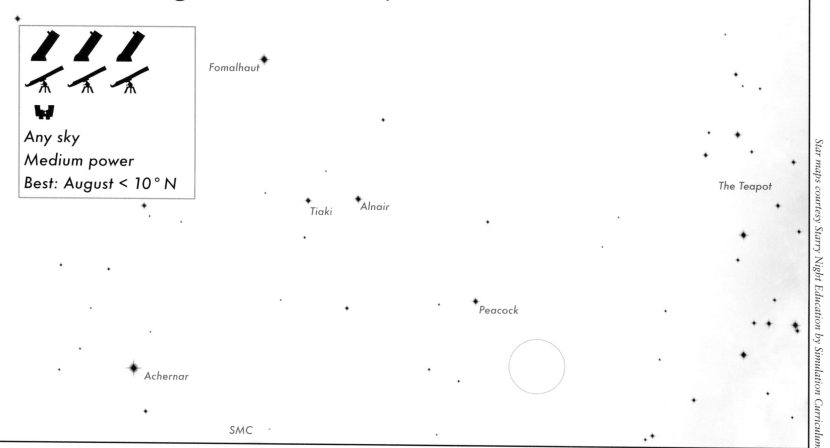

Any sky
Medium power
Best: August < 10° N

Star maps courtesy Starry Night Education by Simulation Curriculum

- Easy to find in any sky
- Darker skies reveal more detail at edges
- Resolved even in small telescopes

Where to look: Find Fomalhaut. South of it are a pair of second-magnitude stars, Alnair and Tiaki. Step from Fomalhaut to Alnair; one step farther brings you to the Peacock Star, Alpha Pavonis. At magnitude 1.9, it's the brightest star in its corner of the sky.

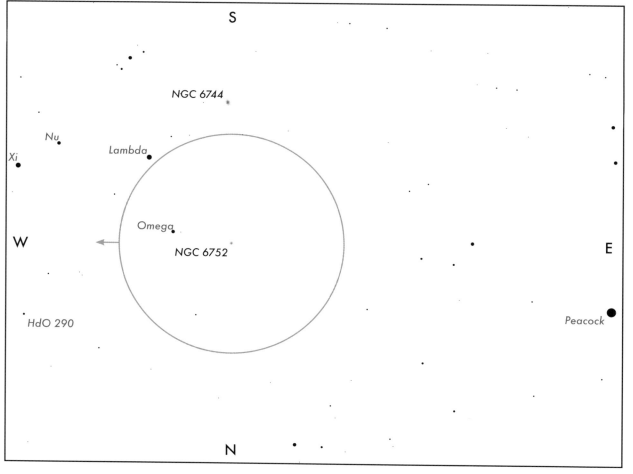

In the finderscope:
From Peacock, move west past a sixth-magnitude star to a trio of fifth-magnitude stars, looking like a clock reading a quarter to one. Follow the minute hand one finder field farther and look for the cluster as a faint fuzzy star.

If you overshoot, you'll reach a collection of fourth-magnitude stars: Lambda, Nu, and Xi Pavonis. From Lambda move northward to fifth-magnitude Omega; aim there and move eastward, back towards Peacock, looking for the cluster in your lowest-power telescope field.

NGC 6752 in a star diagonal at medium power

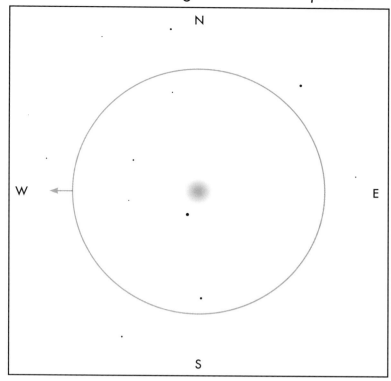

NGC 6752 in a Dobsonian at medium power

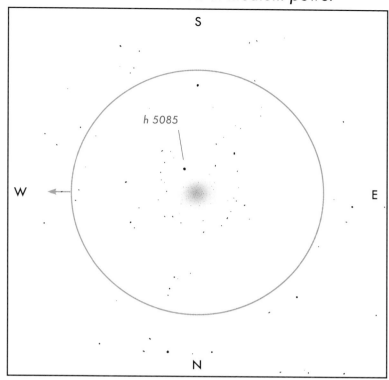

In a small telescope: The cluster is a bright, sweet sight in a small telescope. The central core is easy to see and resolve even when the Moon is up; the darker the night, the more detail you will see and the larger the cluster will appear to be.

In a Dobsonian telescope: The cluster is easy to see and resolve into stars. If the night is dark enough, you will see the bright and the strongly resolved central core surrounded by a fainter haze of light, three times as wide.

Note a double star at the southern edge of the cluster; that's **h 5085**, a 7.6 primary with a 9.1 companion 2.7" to the west.

NGC 6752 is reckoned as one of the best globular clusters in the sky, on a par with M22 and M13. Because it has a strong central concentration, small telescopes underestimate its size; in fact, long-exposure images show that the cluster extends to nearly three quarters of a degree in diameter, larger than a full Moon and almost as large as Omega Centauri or 47 Tucanae.

As with many objects in this part of the sky, the presence of dust clouds between us and this cluster adds some complication to estimating its intrinsic brightness and distance, but with modern corrections the current estimate is that the cluster lies about 13,000 light years from us, comparable to the distances to Omega Centauri, 47 Tucanae, and NGC 362.

The star Peacock only got its name in the mid-twentieth century. In the 1930s, when the British Nautical Almanac Office was preparing an almanac for pilots and navigators flying south of the equator, the Royal Air Force decided that referring to such a prominent star only as Alpha Pavonis was potentially confusing, so the star was given the name of its constellation. (*Pavo* is Latin for peacock.) Even today it is often referred to simply as "the Peacock Star." Avior (Epsilon Carinae – see page 224) got its name at the same time, in honor of the aviators who used it for navigation.

Also in the neighborhood: About five degrees due south of the globular cluster NGC 6752 is **NGC 6744**, *a spiral galaxy seen almost face on. In a small telescope you should be able to make out the bright central nucleus of the galaxy; a Dob begins to hint at its spiral structure.*

There are a number of nice Dob doubles in the area. **Peacock** *itself is an interesting case, with two faint companions (ninth and tenth magnitude) at a large separation – four arc minutes – to the east.* **Xi Pavonis** *has a magnitude 4.4 primary with a magnitude 8.1 companion 3.4" to the south-southeast – a good Dob challenge. To its north, the sixth-magnitude star* **HdO 290** *has a magnitude 10.5 companion 30" to the east-southeast; though the separation is wide, the brightness contrast makes this a challenge.*

find more at: www.cambridge.org/features/turnleft/southern_skies.htm

About this Fifth Edition...

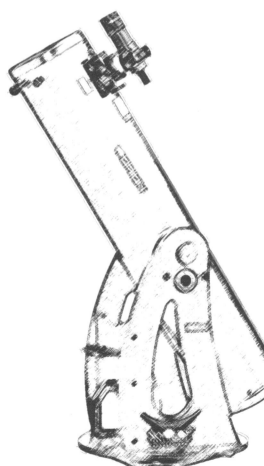

A lot has happened in the 30-odd years since Dan first showed me Albireo. Back then, we would have to tip-toe when coming in late, so as not to wake up his kids; now Léonie gets mad if we wake up their grandkids. In 1983, I had given up a research job at MIT to join the Peace Corps; in 1989, I gave up a professorship at Lafayette College to enter the Jesuits, who assigned me to the Vatican Observatory in Rome. For my sins, in 2015 Pope Francis named me its director.

I now have a number of Vatican instruments (and other professional telescopes) to use. Meanwhile, Dan's collection of telescopes, large and small, now outnumbers even the number of cats in their household.

It's not only our personal lives that have changed. In the last thiry years, major developments have occurred in the world of amateur astronomy.

Amateur telescopes have changed: the Dobsonian design has put 8" mirrors in the price range of nearly everyone. Meanwhile, computer-controlled production techniques have made small catadioptrics better than ever, while holding the line on prices. So why settle for a little three-incher any more?

Personal computers, and the astronomy software that runs on them, have changed the way most of us find objects in the nighttime sky. For the first edition, we drew our star charts by hand, in ink; now we are also guided by *Voyager* and *Starry Night* for finding star positions. In fact, you can buy a computer-controlled telescope; just punch in some numbers, and the scope slews itself to the pre-programmed object. With all this convenience, who needs a book?

And finally, if you really want to see spectacular astronomical sights you can just log on to the internet. So, for that matter, who needs a telescope at all?

And yet... for the sheer fun of the hunt, not to mention portability to dark sites, it's hard to beat a three-incher!

The second edition of this book fixed a lot of typos; the third edition added objects visible in the southern hemisphere; for the fourth edition we re-observed every object with both a small Cat and a Dob, and added a number of additional objects including many new "neighborhood" objects, objects beyond the typical 3" but within the range of a Dob. To accommodate all this new material, we went to a bigger page size (and a spiral binding). Now, for this fifth edition, we are working to completely integrate these printed pages with the power of the internet.

For instance, there's a wide range of telescope models and designs available nowadays. Many telescope finderscopes now come with a star diagonal (mirror flipping what we show you here). Likewise, an 8 inch Cat, quite a popular design, will show the mirror image of an 8 inch Dob. Even with bigger pages we can't possibly show all the different telescope configurations. So we've made flipped versions of all our telescope drawings and put them online at **www.cambridge.org/turnleft** for you to download and print out before you go observing.

Our website also gives links to up-to-date eclipse tables and the location of planets – and not just bright planets! Uranus, Neptune, asteroids, and comets... they're faint, and they move; no book chart alone could show you enough detail to let you find them. But the internet charts can! With that in mind, we've completely rewritten our Planets chapter.

And let's face it, a lot of stuff that was true a third of a century ago has changed with time. The science has changed, for certain: what we know of what's out there has grown, and even some of the names we now use have changed. Plus, several double stars have visibly moved in their orbits since the first edition.

Equipment's changed, too. Modern lenses show a much wider field of view, and now they can stand up to cleaning, especially with the latest cleaning kits. And skies have changed. Given the problems of light pollution, we now use brighter stars in our finderscope images. (The fact that our eyes are also 30 years older may have played a role there, as well.)

But for all that, our basic philosophy has remained unchanged. We still assume you have a small telescope, a few hours' spare time, and a love of the nighttime sky. *Turn Left at Orion* is still the book I need beside me at the telescope. We're delighted that other folks have found it to be a faithful companion, as well.

— *Brother Guy Consolmagno SJ (Vatican City State; 2018)*

So, you've made it through the seasons and you've seen most of the objects in this book that are visible from where you live. Don't stop here! You've probably found that some of these objects have turned into old friends, and you'll look forward to seeing them again and again. Furthermore, the references listed below can lead you to plenty more fun objects to observe that we haven't included in this book.

The fact is that our choice of objects was rather idiosyncratic, emphasizing our personal favorites. Guy is particularly fond of colorful double stars, while I'm partial to partially resolved open clusters and globulars. These preferences are reflected in the large numbers of such objects in the book. We hope that you have come to disagree strongly with at least a few of our object ratings. That would mean that you've developed your own discerning palate when it comes to stargazing. Not everybody gets excited by the same types of objects.

Where do you go from here? You'll want to keep informed of what is going on in the sky. Below, we suggest some excellent magazines and web links to help you do so. We also list some of our favorite planetarium software and star atlases (essential if you're going to do really serious stargazing), as well as reference books on everything from deep-sky objects to the planets, Moon, and Sun. In observing the Sun, make sure you have safe filters and use them properly – take no risks observing our very potent daytime star.

There are now professional-level databases available on the internet for all to use – we'll guide you to some of them as well. Advanced stargazers' references rely on stellar coordinates in order to locate objects. The coordinate system for the sky is really very simple, but it can appear intimidating to the beginner. However, once you've gotten used to the sky and how it moves through the night and the seasons, the coordinate systems (and the more advanced charts and atlases) will start to make sense. If you'll never want to do more than stargaze on occasion, then our book should be all you'll ever need. But if (as we hope) we've whetted your appetite for amateur astronomy, all the better. Clear skies!

— Dan Davis (Stony Brook, New York, 2018)

Where do you go from here?

In figuring out where to go from here, you can start with our Turn Left at Orion *web resources (**www.cambridge.org/ turnleft**). There you can find customized charts that match your equipment and hemisphere, which you can print out for yourself; tables for when to observe the planets; and more!*

Reference sources

Here are some starting places for where to go next! If you have suggestions for items we could add to the links below, feel free to contact us at turnleft@ cambridge.org

Periodicals
Astronomy:
  www.astronomy.com
Astronomy Now:
  www.astronomynow.com
Sky & Telescope:
  www.skyandtelescope.com

Planetarium software
Aladin Sky Atlas: aladin.u-strasbg.fr
Sky Safari: skysafariastronomy.com
Starry Night: www.starrynight.com
Stellarium: www.stellarium.org

Other websites
Astronomy Picture of the Day:
  apod.nasa.gov/apod/
Heavens Above (artificial satellite tracking): www.heavens-above.com

For further reading
M. Bratton: *The Complete Guide to the Herschel Objects*
I. Cooper, J. Kay, and G. R. Kepple: *The Night Sky Observer's Guide* (three volumes: *vol. 1 Autumn & Winter*, *vol. 2 Spring & Summer*, and *vol. 3 The Southern Skies*)
J. Kanipe and D. Weaver: *Annals of the Deep Sky series* (a growing set of volumes, alphabetical by constellation)
C. Luginbuhl and B. Skiff: *Observing Handbook and Catalogue of Deep-Sky Objects*
S. J. O'Meara: the *Deep-Sky Companions* series, including *The Messier Objects*, *The Caldwell Objects*, *The Secret Deep*, and *Hidden Treasures*
J. Pasachoff and D. Menzel: *A Field Guide to the Star and Planets*
H. A. Rey: *The Stars*
I. Ridpath: *Norton's 2000.0 Star Atlas*
I. Ridpath and W. Tirion: *Stars and Planets*
A. Rükl: *Atlas of the Moon*
P. Taylor: *Observing the Sun*

Atlases
B. MacEvoy and W. Tirion: *The Cambridge Double Star Atlas*
J. Mullaney and W. Tirion: *The Cambridge Atlas of Herschel Objects*
R. Stoyan and S. Schurig: *Interstellarum Deep Sky Atlas*
W. Tirion and R.W. Sinnott: *Sky Atlas 2000.0*
W. Tirion, B. Rappaport, and W. Remaklus: *Uranometria 2000.0*
C. E. Scovil: *The AAVSO Variable Star Atlas*

Organizations
AAVSO, the American Association of Variable Star Observers (www.aavso.org)
Astronomical Society of Australia (astronomy.org.au)
Astronomical Society of the Pacific (www.astrosociety.org)
British Astronomical Association (www.britastro.org)
International Dark-Sky Association (www.darksky.org)
Royal Astronomical Society of Canada (www.rasc.ca)

Accessories

Over the years, we've accumulated all sorts of wonderful gizmos to go along with our telescopes. Some of these items we now feel we could never live without; others are just mighty convenient to have around. You can find most of these wherever you bought your telescope. Here's a list of some of our favorites.

A set of slow-motion gears between the tripod and your Cat makes it easy to center the telescope and track objects.

Tool kit: Keep a separate set of the Allen wrenches and screwdrivers you need to adjust your telescope. Don't let them stray from your telescope.

Mirror alignment laser: A specially designed laser tool lets you align your Dobsonian optics quickly and painlessly. We use one all the time. But they can be expensive.

Good tripod: Most small catadioptrics need their own tripod. A good one can be as expensive as the telescope itself. You'll want something with heft to keep the telescope from shaking when you track an object as it moves across the sky, yet small and light enough to carry with you when you travel.

Slow-motion gears: This small gizmo with worm gears fits between the tripod and your Cat. It makes it easy to center the telescope and track objects.

Telrad: For a lot of folks, not just beginners, it's embarrassingly difficult to get the telescope pointed close enough to even be able to see the appropriate stars in the finderscope. A simple device called a Telrad lets you look at the sky through a clear screen while it projects a red bullseye to show where the telescope is pointed.

Finder: Some otherwise wonderful telescopes come with really inferior finderscopes. Consider upgrading your finder; it makes all the difference in the world to be able to see the nearby stars that guide you to your object. But you may have to adjust your telescope to balance out the extra weight of a bigger finderscope.

Star diagonal: As we mention in the text, you'll want a real 90° star diagonal, not a 45° angle "erecting prism." More intriguingly, you can bring new life to an old telescope by replacing an old star diagonal designed to use the smaller eyepieces common in years past with one that can handle the newer 1¼" eyepieces.

Solar filter: You absolutely, positively can't observe the Sun without one. Use only the kind that covers the whole objective mirror or lens, not just the eyepiece.

Dew control: There are many reasonably priced battery-powered little heaters on the market that can protect your optics from dew. They can be invaluable: losing an otherwise wonderful stargazing session to dew can be very frustrating!

Extra eyepieces: After you know your skies and your favorite objects, save up for a really good eyepiece – wide-angle or high-powered, depending on your taste. Looking through a good eyepiece is like sticking your head up into the stars themselves.

Barlow lens: Small telescopes from department stores often come with cheap Barlow lenses to help boost their "power" to unrealistic levels. Don't let them discourage you from getting a *good* Barlow to fit behind your high-power eyepiece. If the night is steady enough, the extra power can make challenging double stars pop out!

Coma corrector: If you have a low *f*-ratio Dob and enjoy looking at big extended objects like open clusters at low power, then this attachment will sharpen up your viewing by reducing the distortion at the edge of your field of view.

Table: Find a lightweight table, like a card table, to hold your books, charts, and spare eyepieces. Use one that can stand to get dewy wet.

Filters

Filters increase contrast, bringing out features in interesting colors by removing light with less interesting colors. But they only work by subtraction; and thus, they work better with bigger telescopes, like Dobs, where you have more aperture and so more light to spare.

Neutral density: If you like looking at the Moon in a Dob (and who doesn't?) this is essential; without it, the brightness of the full Moon can hurt your eye. (In a pinch, try sunglasses!)

Nebula (or OIII): Diffuse and planetary nebulae emit most of their visible light in a certain green color emitted by ionized oxygen atoms. These filters let only that narrow color pass, cutting away the rest of the light from the sky (and surrounding stars). It turns faint nebulae like the Rosette into whole new objects. But it won't help for most galaxies or double stars.

Light pollution: If your city uses mostly yellow sodium vapor lighting, then a filter that specifically cuts away that color of light is very helpful.

Red/blue colors: Enjoy looking at planets? A red filter makes the features on Mars stand out; a blue filter enhances your views of Venus and the clouds on Jupiter.

Clothing: Cargo pants have lots of extra pockets for storing lens caps (or lenses); buy a size that fits over your regular trousers, for extra warmth. Fleece vests work remarkably well at keeping you warm when standing outside for long times in the cold and dark. Fisherman's vests have lots of pockets for eyepieces, filters, lens caps – all easy to misplace in the dark. And thin gloves let you focus on focusing, not freezing.

Observer's chair: Special ones for amateur astronomers have seats that move up and down to fit wherever you are pointing your telescope.

Flashlight: A regular flashlight with its lens painted red with nail polish works fine, but nowadays you can get a purer red with LED flashlights. Some come on straps to hold them to your forehead (or, better, to hang around your neck), keeping your hands free. Beware that some are too bright for dark-sky use; keep a few dying batteries handy to reduce their brightness.

Thinking of upgrading your telescope to a bigger or more portable model?

There's no doubt that the Dobsonian telescope, a Newtonian set in a simple, effective, and very inexpensive mounting, has revolutionized amateur astronomy. On the other hand, Cats can beat Dobs when it comes to portability and – inch for inch of aperture, if not dollar for dollar in cost – in optical quality. Like everything in life, nothing is perfect. There are a number of important issues with Cats and Dobs that every observer should be aware of.

The truth about Cats and Dobs

A low f-ratio telescope like a Dob, which is usually around f/4.5, does a great job of gathering lots of light in a relatively small length. But at very low power you may begin to see the shadow of the secondary mirror in the image itself. It shows up as a characteristic darker spot in the center of your view. It's not completely black, just dimmer than the edges of your field of view.

A parabolic mirror will focus light just fine along its axis – at the center of your field of view. But the farther away you get from the center, the poorer the image becomes. Cat telescopes correct for this with carefully shaped secondary mirrors. However, an inexpensive Dob just uses a flat secondary mirror. This is no problem when you're observing at moderate to high power, but at very low power – a very wide field of view – you'll notice that the stars at the outer edge of the field of view are no longer crisp points but more like little V-shaped splotches. This distortion is called *coma*.

Every time a beam of light goes through a lens or bounces off a mirror, a little bit of light is lost. Typically mirrors reflect about 90% of the light they receive. Lenses and prisms lose light both by reflecting some of it (anti-reflection coatings help) or by absorbing it within the glass; they are at best around 95% efficient, per lens. A Dob bounces light off a primary and secondary mirror, so that's 90% of 90%, or 81% efficiency total by the time the light gets to the eyepiece. But a Cat not only has two mirrors, it also passes the light through a corrector plate and a star diagonal, so its efficiency is worse – a couple tenths of a magnitude dimmer. (Refractors also face this issue. The great Lick refractor was measured to be only 60% efficient by the time all its lenses were accounted for.)

In a Cat, the secondary mirror is glued to the corrector lens; for a Dob, it's held in place by thin wires, called the *spider*. Those wires, however, cause the waves of the starlight to diffract into faint streaks spreading out perpendicular to those wires. (Surprisingly, the thinner the wires, the more pronounced the effect.) That means that, unlike in a Cat, the image of a bright star in a Dob will have characteristic spikes of light. It looks kind of cool and dramatic... until you're trying to look for a faint double-star companion right where one of those spikes is getting in your way.

Dob mirrors are big – wonderful! But that means they need more time to cool down than a small Cat.

Dobsonians are remarkably inexpensive for their size, and extremely easy to use. But they're not easily portable. It can be hard to take them camping, or even just carry them out into the back yard.

Don't try observing the Sun in a Dob...unless you're willing to pay. That big mirror gathers way too much sunlight, enough to do damage to whatever eyepiece you use for projecting the Sun's image and certainly enough to break many in-the-eyepiece solar filters (which you shouldn't trust in any event). The only safe way to observe the Sun with a Dob – or *any* telescope – is to use a *full aperture solar filter*. And a filter big enough to fit the 8" aperture of a Dob will run you well over $100.

The low cost of a Dob comes from its clever mounting, which holds the telescope using a simple pair of friction pads. But that means the telescope has to be in good balance, or the pads can't hold it in place. Heavy eyepieces or Barlows, or a heavy finderscope attached at the end of the long Dob tube, can cause a Dob to become top-heavy. To solve this problem, some Dobs allow you to adjust the pivot point; others let you add counterweights. Be sure you can adjust yours, before investing in a heavy new lens or finder.

M. Skirvin

Tables

Page	Name	Constellation	RA		Dec		Magnitudes	Separation	Chapter
188	Struve 3053	*Cassiopeia*	0	2.6	66°	6'	6.0, 7.3	15"	North
215	Beta Tucanae	*Tucana*	0	31.5	−62°	58'	4.3, 4.5	27"	South
188	Eta Cassiopeiae	*Cassiopeia*	0	49.0	57°	49'	3.4, 7.4	13"	North
188	Burnham 1	*Cassiopeia*	0	52.8	56°	38'	8.6, 9.3, 8.9, 9.7	1.1", 3.7", 8.5"	North
215	Kappa Tucanae	*Tucana*	1	15.8	−68°	53'	5.0, 7.7, 7.8, 8.4	5.0", 319", 320"	South
194	Struve 131	*Cassiopeia*	1	33.2	60°	41'	7.3, 9.9	14"	North
195	Struve 153	*Cassiopeia*	1	46.6	61°	16'	9.4, 10.4	7.7"	North
188	Struve 163	*Cassiopeia*	1	51.2	64°	51'	6.8, 9.1, 10.7	34", 114"	North
181	1 Arietis	*Aries*	1	53.4	22°	17'	6.3, 7.2	2.9"	October–December
180	Gamma Arietis, *Mesarthim*	*Aries*	1	53.4	19°	18'	4.5, 4.6	7.5"	October–December
180	56 Andromedae	*Andromeda*	1	56.2	37°	15'	5.6, 6.1	201"	October–December
180	Lambda Arietis	*Aries*	1	58.0	23°	36'	4.8, 6.7	37"	October–December
180	Gamma Andromedae, *Almach*	*Andromeda*	2	3.9	42°	20'	2.3, 5.0	9.8"	October–December
180	59 Andromedae	*Andromeda*	2	10.9	39°	3'	6.1, 6.7	17"	October–December
180	6 (Iota) Trianguli	*Triangulum*	2	12.4	30°	18'	5.3, 6.7	3.7"	October–December
188	Iota Cassiopeiae	*Cassiopeia*	2	29.0	67°	24'	4.6, 6.9, 9.1	2.9", 7.0"	North
202	Alpha Ursae Minoris, *Polaris*	*Ursa Minor*	2	31.8	89°	16'	2.1, 9.1	18"	North
183	HJ 1123	*Perseus*	2	41.9	42°	47'	8.4, 8.5	20"	October–December
183	Otto Struve 44	*Perseus*	2	42.2	42°	41'	8.3, 9.0	1.4"	October–December
183	HJ 2155	*Perseus*	2	48.8	42°	49'	8.3, 10.3	17"	October–December
186	Eta Persei, *Miram*	*Perseus*	2	50.7	55°	54'	3.8, 8.5	29"	North
61	39 Eridani	*Eridanus*	4	14.4	−10°	15'	5.0, 8.5	6.3"	January–March
60	Omicron-2 Eridani, *Keid*	*Eridanus*	4	15.3	−7°	39'	4.5, 10.2, 11.5	83", 7"	January–March
70	Omega Aurigae	*Auriga*	4	59.3	37°	53'	5.0, 8.2	4.8"	January–March
58	Beta Orionis, *Rigel*	*Orion*	5	14.5	−8°	12'	0.3, 6.8	9.3"	January–March
70	14 Aurigae	*Auriga*	5	15.4	32°	41'	5, 9, 8, 8	10", 14", 2"	January–March
63	h 3752	*Lepus*	5	22.5	−24°	46'	5.4, 6.6	3.4"	January–March
58	Eta Orionis	*Orion*	5	24.5	−2°	24'	3.6, 4.9	1.8"	January–March
66	Struve 118	*Taurus*	5	29.3	25°	9'	5.8, 6.7	4.4"	January–March
58	32 Orionis	*Orion*	5	30.8	5°	57'	4.4, 5.7	1.2"	January–March
58	Delta Orionis, *Mintaka*	*Orion*	5	32.0	0°	20'	2.2, 6.8	53"	January–March
54	Struve 745	*Orion*	5	34.7	−6°	0'	8.5, 8.5	29"	January–March
54	Struve 747	*Orion*	5	35.0	−6°	0'	4.7, 5.5	36"	January–March
58	Lambda Orionis, *Meissa*	*Orion*	5	35.1	9°	56'	3.5, 5.4	4.1"	January–March
55	Theta-1 Orionis, *Trapezium*	*Orion*	5	35.3	−5°	25'	5.1, 6.6, 6.6, 6.3, 11.1, 11.5	13", 17", 12", 5", 5"	January–March
54	Iota Orionis	*Orion*	5	35.4	−5°	54'	2.9, 7.0, 9.7	11", 49"	January–March
55	Theta-2 Orionis	*Orion*	5	35.4	−5°	26'	5.2, 6.5	52"	January–March
58	42 Orionis	*Orion*	5	35.4	−4°	50'	4.6, 7.5	1.1"	January–March
65	Struve 742	*Taurus*	5	36.4	22°	0'	7.1, 7.5	4.1"	January–March
56	Struve 761	*Orion*	5	38.6	−2°	34'	7.9, 8.4, 8.6	67", 8.7"	January–March
70	26 Aurigae	*Auriga*	5	38.6	30°	30'	5.5, 8.4	12"	January–March
56	Sigma Orionis	*Orion*	5	38.7	−2°	36'	3.7, 8.8, 6.6, 6.3	11", 13", 41"	January–March
58	Zeta Orionis, *Alnitak*	*Orion*	5	40.8	−1°	59'	1.9, 3.7	2.5"	January–March
58	52 Orionis	*Orion*	5	48.0	6°	27'	6.0, 6.0	1.1"	January–March
73	Eta Geminorum, *Propus*	*Gemini*	6	14.9	22°	30'	3.5, 6.2	1.7"	January–March
77	Epsilon Monocerotis	*Monoceros*	6	23.8	4°	36'	4.4, 6.4	12"	January–March
73	15 Geminorum	*Gemini*	6	27.8	20°	47'	6.6, 8.2	25"	January–March
78	Beta Monocerotis	*Monoceros*	6	28.8	−7°	2'	4.6, 5.0, 5.3	7.2", 2.9"	January–March
81	Alpha Canis Majoris, *Sirius*	*Canis Major*	6	45.1	−16°	43'	−1.4, 8.5	8.4"	January–March
81	Mu Canis Majoris	*Canis Major*	6	56.1	−14°	3'	5.3, 7.1	3"	January–March
87	Epsilon Canis Majoris, *Adhara*	*Canis Major*	6	58.6	−28°	58'	1.5, 7.5	7"	January–March
217	Gamma Volantis	*Volans*	7	8.7	−70°	30'	3.8, 5.7	13"	South
86	Herschel 3945, *Winter Albireo*	*Canis Major*	7	16.6	−23°	19'	5.0, 5.8	26"	January–March
75	Lambda Geminorum	*Gemini*	7	18.1	16°	32'	3.6, 10.7	9.7"	January–March
75	Delta Geminorum, *Wasat*	*Gemini*	7	20.1	21°	59'	3.6, 8.2	5.6"	January–March
75	Struve 1083	*Gemini*	7	25.6	20°	30'	7.3, 8.1	6.8"	January–March
75	Struve 1108	*Gemini*	7	32.8	22°	53'	6.6, 8.2	11"	January–March
74	Alpha Geminorum, *Castor*	*Gemini*	7	34.6	31°	53'	1.9, 3.0, 9.8	4.8", 70"	January–March
89	k Puppis	*Puppis*	7	38.8	−26°	48'	4.4, 4.6	9.8"	January–March
81	Alpha Canis Minoris, *Procyon*	*Canis Minor*	7	39.3	5°	13'	0.4, 10.8	4.7"	January–March
217	Zeta Volantis	*Volans*	7	41.8	−72°	36'	4.0, 9.8	17"	South
85	2 Puppis	*Puppis*	7	45.5	−14°	41'	6.1, 6.9	17"	January-March
75	Kappa Geminorum	*Gemini*	7	77.4	24°	24'	3.7, 8.2	7"	January–March
217	Epsilon Volantis	*Volans*	8	7.9	−68°	37'	4.4, 8.0	6.1"	South
222	Gamma Velorum	*Vela*	8	9.5	−47°	21'	1.9, 4.2, 8.2, 9.1	41", 62", 94"	South
92	Zeta Cancri, *Tegmine*	*Cancer*	8	12.2	17°	39'	5.3, 6.3	1.1"	April–June
225	Rumker 8	*Carina*	8	15.3	−62°	55'	3.6, 5.3	3.7"	South
93	24 Cancri	*Cancer*	8	26.7	24°	32'	6.9, 7.5	5.6"	April–June
97	Phi-2 Cancri	*Cancer*	8	26.8	26°	56'	6.2, 6.2	5.2"	April–June
225	h 4128	*Carina*	8	39.2	−60°	19'	6.8, 7.5	1.1"	South
97	Struve 1266	*Cancer*	8	44.4	28°	27'	8.8, 10.0	24"	April–June
96	Iota Cancri	*Cancer*	8	46.7	28°	46'	4.1, 6.0	31"	April–June
95	Epsilon Hydrae	*Hydrae*	8	46.8	6°	25'	3.5, 6.7	2.8"	April–June
97	53 Cancri	*Cancer*	8	52.5	28°	16'	6.2, 9.7	43"	April–June

Page	Name	Constellation	RA		Dec		Magnitudes	Separation	Chapter
96	57 Cancri	*Cancer*	8	54.2	30°	35'	6.1, 6.4	1.5"	April–June
96	Kappa Leonis	*Leo*	9	24.7	26°	11'	4.6, 9.7	2.4"	April–June
227	Upsilon Carinae	*Carina*	9	47.1	–65°	4'	3.0, 6.0	5.0"	South
96	Gamma Leonis, *Algieba*	*Leo*	10	20.0	19°	51'	2.4, 3.6	4.6"	April–June
229	J Velorum	*Vela*	10	20.9	–56°	3'	4.5, 7.2, 9.2	7.1", 36"	South
101	54 Leonis	*Leo*	10	55.6	24°	45'	4.5, 6.3	6.6"	April–June
229	h 4432	*Musca*	11	23.0	–64°	57'	5.4, 6.6	2.4"	South
113	2 Canum Venaticorum	*Canes Venatici*	12	16.1	40°	40'	5.9, 8.7	11"	April–June
233	Alpha Crucis, *Acrux*	*Crux*	12	26.6	–63°	6'	1.3, 1.6, 4.8	4.0", 91"	South
232	Gamma Crux, *Gacrux*	*Crux*	12	31.2	-57°	7'	1.8,2.5,9.7	125", 167"	South
106	24 Comae Berenices	*Coma Berenices*	12	35.1	23°	40'	5.1, 6.3	20"	April-June
104	Gamma Virginis, *Porrima*	*Virgo*	12	41.7	–1°	27'	3.5, 3.5	0.4" – 6"	April–June
232	Beta Muscae	*Musca*	12	46.3	–68°	6'	3.5, 4.0	0.7"	South
110	Struve 1685	*Coma Berenices*	12	52.0	19°	10'	7.3, 7.8	16"	April–June
232	Mu Crucis	*Crux*	12	54.6	–57°	11'	3.9, 5.0	35"	South
105	Struve 1689	*Virgo*	12	55.5	11°	30'	7.1, 9.1	30"	April–June
112	Alpha CVn, *Cor Caroli*	*Canes Venatici*	12	56.0	38°	19'	2.9, 5.5	19"	April–June
113	Struve 1702	*Canes Venatici*	12	58.5	38°	17'	8.7, 9.4	35"	April–June
206	Zeta Ursae Majoris, *Mizar*	*Ursa Major*	13	23.9	54°	56'	2.2, 3.9	14"	North
238	Q Centauri	*Centaurus*	13	31.7	–54°	17'	5.2, 6.5	5.5"	South
238	N Centauri	*Centaurus*	13	52.0	–52°	35'	5.2, 7.5	18"	South
206	Struve 1795	*Ursa Major*	13	59.0	53°	6'	6.9, 9.8	8.0"	North
206	Kappa Boötis	*Boötes*	14	13.5	52°	0'	4.4, 6.6	14"	North
206	Iota Boötis	*Boötes*	14	16.2	51°	22'	4.8, 7.4	39"	North
232	Dunlop 159	*Centaurus*	14	22.6	–58°	28'	5.0, 7.6	8.9"	South
232	Alpha Cen., *Rigel Kentaurus*	*Centaurus*	14	39.6	–60°	50'	0.1, 1.2	1.7" – 22"	South
117	Pi Boötis	*Boötes*	14	40.7	16°	23'	4.9, 5.8	5.6"	April–June
232	Alpha Circini	*Circinus*	14	42.5	–64°	59'	3.2, 8.5	15"	South
116	Epsilon Boötis, *Izar*	*Boötes*	14	45.0	27°	4'	2.6, 4.8	3.0"	April–June
117	Xi Boötis	*Boötes*	14	51.4	19°	6'	4.8, 7.0	6.4"	April–June
116	Mu Boötis, *Alkalurops*	*Boötes*	15	24.5	37°	23'	4.3, 7.1, 7.6	109", 2.2"	April–June
117	Zeta Coronae Borealis	*Corona Borealis*	15	39.4	36°	38'	5.0, 5.9	6.3"	April–June
151	Xi Lupi	*Lupus*	15	56.9	–33°	58'	5.1, 5.6	10"	July–September
150	Eta Lupi	*Lupus*	16	0.1	–38°	24'	3.4, 7.5, 9.4	14", 116"	July–September
151	Xi Scorpii	*Scorpius*	16	4.4	–11°	22'	5.2, 4.9, 7.3	0.9", 7.9"	July–September
151	Struve 1999	*Scorpius*	16	4.5	–11°	26'	7.5, 8.1	12"	July–September
151	Beta Scorpii, *Acrab (Graffias)*	*Scorpius*	16	5.4	–19°	51'	2.6, 4.5	12"	July–September
150	Nu Scorpii	*Scorpius*	16	12.0	–19°	28'	4.4, 5.3, 6.6, 7.2	1.3", 41", 42" (2.3")	July–September
121	Sigma Coronae Borealis	*Corona Borealis*	16	14.7	33°	51'	5.6, 6.5	7.1"	July–September
150	Alpha Scorpii, *Antares*	*Scorpius*	16	29.4	–26°	26'	1.0, 5.4	2.6"	July–September
150	24 Ophiuchi	*Ophiuchus*	16	56.8	–23°	9'	6.3, 6.3	1.0"	July–September
124	Alpha Herculis, *Rasalgethi*	*Hercules*	17	14.6	14°	23'	3.5, 5.4	4.9"	July–September
121	Delta Herculis	*Hercules*	17	15.0	24°	50'	3.1, 8.3	11"	July–September
150	36 Ophiuchi	*Ophiuchus*	17	15.4	–26°	35'	5.1, 5.1	4.9"	July–September
150	Omicron Ophiuchi	*Ophiuchus*	17	18.0	–24°	17'	5.2, 6.6	10"	July–September
123	Rho Herculis	*Hercules*	17	23.7	37°	9'	4.5, 5.4	4.0"	July–September
199	Psi Draconis	*Draco*	17	41.9	72°	8'	4.6, 5.6	30"	North
149	HN 40	*Sagittarius*	18	2.3	–23°	2'	7.6, 10.4, 8.7	6.1", 11"	July–September
241	Xi Pavonis	*Pavo*	18	23.2	-61°	30'	4.4, 8.1	3.4"	South
241	HdO 290	*Pavo*	18	29.9	-57°	31'	5.8, 10.5	30"	South
129	Epsilon Lyrae, *Double-Double*	*Lyra*	18	44.4	39°	38'	5.2, 6.1,5.3, 5.4	208", 2.3", 2.3"	July–September
129	Zeta Lyrae	*Lyra*	18	44.8	37°	36'	4.3, 5.6	44"	July–September
141	Struve 2391	*Scutum*	18	48.7	–6°	2'	6.5, 9.6	38"	July–September
131	Beta Lyrae	*Lyra*	18	50.1	33°	22'	3.3-4.3, 6.7	46"	July–September
131	Otto Struve 525	*Lyra*	18	54.9	41°	36'	6.1, 9.1, 7.6	1.8", 45"	July–September
160	Zeta Sagittarii, *Ascella*	*Sagittarius*	19	2.6	–29°	53'	3.3, 3.5	0.2" – 0.6"	July–September
141	15 Aquilae	*Aquila*	19	5.0	–4°	2'	5.5, 7.0	39"	July–September
129	Struve 2470, *D-D's double*	*Lyra*	19	8.8	34°	46'	7.0, 8.4	14"	July–September
129	Struve 2474, *D-D's double*	*Lyra*	19	8.8	34°	46'	6.8, 7.9	16"	July–September
132	Beta Cygni, *Albireo*	*Cygnus*	19	30.7	27°	58'	3.2, 4.7	37"	July–September
139	Epsilon Sagittae	*Sagitta*	19	37.3	16°	28'	5.8, 8.4	87"	July–September
139	HN 84	*Sagitta*	19	39.4	16°	34'	6.4, 9.5	28"	July–September
128	16 Cygni	*Cygnus*	19	41.8	50°	31'	6.0, 6.2	40"	July–September
139	Zeta Sagittae	*Sagitta*	19	49.0	19°	8'	5.0, 9.0	8.1"	July–September
139	Theta Sagittae	*Sagitta*	20	9.9	20°	55'	6.6, 8.9	12"	July–September
128	Struve 2725	*Delphinus*	20	46.2	15°	54'	7.5, 8.2	6.0"	July–September
128	Gamma Delphini	*Delphinus*	20	46.7	16°	8'	4.4, 5.0	9.1"	July–September
134	61 Cygni	*Cygnus*	21	6.6	38°	42'	5.2, 6.1	31"	July–Septeber
197	Struve 2815	*Cepheus*	21	37.7	57°	34'	8.6, 10.0	0.9"	North
196	Struve 2816	*Cepheus*	21	39.0	57°	29'	5.7, 7.5, 7.5	12", 20"	North
196	Struve 2819	*Cepheus*	21	40.0	57°	35'	7.4, 8.6	13"	North
173	KV Aquarii	*Aquarius*	22	25.7	–20°	14'	7.1, 8.0	6.9"	October–December
196	Delta Cephei	*Cepheus*	22	29.2	58°	25'	4.2, 6.1	41"	North
188	Sigma Cassiopeiae	*Cassiopeia*	23	59.0	55°	45'	5.0, 7.2	3.2"	North

Tables

Open clusters

Page	Name	Constellation	RA		Dec.		Chapter
193	NGC 129	Cassiopeia	0	29.8	60°	14'	North
210	NGC 220	Tucana	0	40.5	−73°	24'	South
210	NGC 222	Tucana	0	40.7	−73°	23'	South
210	NGC 231	Tucana	0	41.1	−73°	21'	South
192	NGC 225	Cassiopeia	0	43.4	61°	47'	North
210	NGC 265	Tucana	0	47.2	−73°	29'	South
202	NGC 188	Cepheus	0	47.5	85°	15'	North
210	NGC 330	Tucana	0	56.3	−72°	28'	South
210	NGC 371	Tucana	1	3.4	−72°	4'	South
210	NGC 376	Tucana	1	3.9	−72°	49'	South
210	NGC 395	Tucana	1	5.1	−72°	0'	South
210	NGC 460	Tucana	1	14.6	−73°	17'	South
210	NGC 458	Tucana	1	14.9	−71°	33'	South
193	NGC 436	Cassiopeia	1	15.5	58°	49'	North
210	NGC 465	Tucana	1	15.7	−73°	19'	South
193	NGC 457	Cassiopeia	1	19.0	58°	20'	North
194	M103	Cassiopeia	1	33.2	60°	42'	North
179	NGC 604, in M33	Triangulum	1	34.5	30°	47'	October–December
195	NGC 637	Cassiopeia	1	41.8	64°	2'	North
195	NGC 654	Cassiopeia	1	43.9	61°	54'	North
195	NGC 659	Cassiopeia	1	44.2	60°	43'	North
195	NGC 663	Cassiopeia	1	46.0	61°	16'	North
178	NGC 752	Andromeda	1	57.8	37°	41'	October–December
186	NGC 869, Double	Perseus	2	19.3	57°	9'	North
186	NGC 884, Double	Perseus	2	22.4	57°	7'	North
182	M34	Perseus	2	42.0	42°	47'	October–December
64	M45, Pleiades	Taurus	3	46.9	24°	7'	January–March
216	NGC 1711	Mensa	4	50.6	−69°	59'	South
216	NGC 1727	Dorado	4	52.3	−69°	21'	South
216	NGC 1755	Dorado	4	55.2	−68°	12'	South
216	NGC 1845	Mensa	5	5.7	−70°	35'	South
216	NGC 1847	Dorado	5	7.1	−68°	58'	South
216	NGC 1850	Dorado	5	8.7	−68°	46'	South
216	NGC 1854	Dorado	5	9.3	−68°	51'	South
216	NGC 1856	Dorado	5	9.5	−69°	8'	South
216	NGC 1872	Dorado	5	13.2	−69°	19'	South
216	NGC 1876	Dorado	5	13.3	−69°	22'	South
216	NGC 1877	Dorado	5	13.6	−69°	23'	South
216	NGC 1901	Dorado	5	18.2	−68°	27'	South
216	NGC 1910	Dorado	5	18.6	−69°	14'	South
216	NGC 1962	Dorado	5	26.3	−68°	50'	South
216	NGC 1967	Dorado	5	26.7	−69°	6'	South
216	NGC 1968	Dorado	5	27.4	−67°	28'	South
216	NGC 1986	Mensa	5	27.6	−69°	58'	South
216	NGC 1984	Dorado	5	27.7	−69°	8'	South
216	NGC 1983	Dorado	5	27.8	−67°	59'	South
216	NGC 1974	Dorado	5	28.0	−67°	25'	South
71	NGC 1907	Auriga	5	28.1	35°	19'	January–March
216	NGC 1994	Dorado	5	28.4	−69°	8'	South
68	M38	Auriga	5	28.7	35°	51'	January–March
216	NGC 2001	Dorado	5	29.0	−68°	46'	South
216	NGC 2009	Dorado	5	31.0	−69°	11'	South
216	NGC 2015	Dorado	5	32.1	−69°	15'	South
216	NGC 2031	Mensa	5	33.7	−70°	59'	South
216	NGC 2033	Dorado	5	34.5	−69°	47'	South
53	NGC 1981	Orion	5	35.1	−5°	25'	January–March
68	M36	Auriga	5	35.3	34°	9'	January–March
216	NGC 2044	Dorado	5	36.1	−69°	12'	South
216	NGC 2042	Dorado	5	36.2	−68°	55'	South
216	NGC 2055	Dorado	5	37.0	−69°	26'	South
216	NGC 2100	Dorado	5	42.1	−69°	13'	South
216	NGC 2098	Dorado	5	42.5	−68°	17'	South
216	NGC 2122	Mensa	5	48.9	−70°	4'	South
68	M37	Auriga	5	52.3	32°	34'	January–March
73	NGC 2129	Gemini	6	1.1	23°	19'	January–March
72	M35	Gemini	6	8.8	24°	20'	January–March
78	NGC 2232	Monoceros	6	27.2	−4°	45'	January–March
76	NGC 2244	Monoceros	6	31.9	4°	56'	January–March
82	M41	Canis Major	6	47.0	−20°	45'	January–March
80	M50	Monoceros	7	2.9	−8°	20'	January–March
86	NGC 2362	Canis Major	7	18.7	−24°	57'	January–March
85	NGC 2414	Puppis	7	33.2	−15°	27'	January–March
84	M47 (NGC 2422)	Puppis	7	36.6	−14°	29'	January–March
85	NGC 2423	Puppis	7	37.1	−13°	52'	January–March

Tables

Planetary nebulae

Page	Name	Constellation	RA	Dec.	Chapter
60	NGC 1535	Eridanus	4 14.3	−12° 44'	January–March
74	NGC 2392, *Clown Face*	Gemini	7 29.2	20° 55'	January–March
85	NGC 2438	Puppis	7 41.8	−14° 44'	January–March
98	NGC 3242, *Ghost of Jupiter*	Hydra	10 24.8	−18° 39'	April–June
234	NGC 5189, *Spiral*	Musca	13 33.7	−65° 58'	South
198	NGC 6543, *Cat Eye*	Draco	17 58.6	66° 38'	North
130	M57, *Ring*	Lyra	18 53.6	33° 2'	July–September
134	NGC 6826, *Blinking*	Cygnus	19 44.8	50° 31'	July–September
136	M27, *Dumbbell*	Vulpecula	19 59.6	22° 43'	July–September
168	NGC 7009, *Saturn*	Aquarius	21 4.2	−11° 22'	October–December
172	NGC 7293, *Helix*	Aquarius	22 29.6	−20° 50'	October–December

Globular clusters

Page	Name	Constellation	RA	Dec.	Chapter
212	NGC 104, *47 Tucanae*	Tucana	0 24.1	−62° 58'	South
174	NGC 288	Sculptor	0 52.8	−26° 35'	October–December
214	NGC 362	Tucana	1 3.2	−70° 51'	South
210	NGC 419	Tucana	1 8.3	−72° 53'	South
216	NGC 1786	Dorado	4 59.1	−67° 45'	South
216	NGC 1835	Dorado	5 5.1	−69° 24'	South
62	M79	Lepus	5 24.2	−24° 31'	January–March
226	NGC 2808	Carina	9 12.0	−64° 52'	South
234	NGC 4372	Musca	12 25.8	−72° 40'	South
234	NGC 4833	Musca	12 59.6	−70° 52'	South
110	M53	Coma Berenices	13 12.9	18° 10'	April–June
111	NGC 5053	Coma Berenices	13 16.5	17° 42'	April–June
238	NGC 5139, *Omega Centauri*	Centaurus	13 26.8	−47° 29'	South
114	M3	Canes Venatici	13 42.2	28° 23'	April–June
239	NGC 5286	Centaurus	13 46.4	−51° 22'	South
124	M5	Serpens	15 18.5	2° 5'	July–September
152	M80	Scorpius	16 17.1	−22° 59'	July–September
152	M4	Scorpius	16 23.7	−26° 31'	July–September
120	M13, *Great Cluster*	Hercules	16 41.7	36° 27'	July–September
123	NGC 6229	Hercules	16 47.0	47° 32'	July-September
126	M12	Ophiuchus	16 47.2	−1° 57'	July–September
154	M62	Oph.-Sco.	16 54.3	−30° 8'	July–September
126	M10	Ophiuchus	16 57.1	−4° 7'	July–September
154	M19	Ophiuchus	17 2.6	−26° 15'	July–September
122	M92	Hercules	17 17.1	43° 9'	July–September
147	NGC 6544	Sagittarius	18 07.3	−22° 00'	July–September
147	NGC 6553	Sagittarius	18 09.2	−25° 54'	July–September
158	M28	Sagittarius	18 24.6	−24° 52'	July–September
158	M22	Sagittarius	18 36.4	−23° 56'	July–September
160	M54	Sagittarius	18 55.2	−30° 28'	July–September
240	NGC 6752	Pavo	19 10.8	−59° 59'	South
132	M56	Lyra	19 16.6	30° 10'	July–September
160	M55	Sagittarius	19 40.1	−30° 56'	July–September
138	M71	Sagitta	19 53.7	18° 47'	July–September
168	M72	Aquarius	20 53.5	−12° 32'	October–December
164	M15	Pegasus	21 30.0	12° 10'	October–December
166	M2	Aquarius	21 33.5	0° 50'	October–December
170	M30	Capricornus	21 40.4	−23° 11'	October–December

Galaxies

Page	Name	Constellation	RA	Dec.	Chapter
176	M110 (NGC 205), *with M31*	Andromeda	0 40.3	41° 41'	October–December
176	M31, *Andromeda Galaxy*	Andromeda	0 42.7	41° 16'	October–December
176	M32, *companion of M31*	Andromeda	0 42.7	40° 52'	October–December
174	NGC 247	Sculptor	0 47.1	−20° 46'	October–December
174	NGC 253	Sculptor	0 47.5	−25° 17'	October–December
210	Small Magellanic Cloud	Tucana	0 53.0	−72° 50'	South
178	M33, *Triangulum Galaxy*	Triangulum	1 33.9	30° 39'	October–December
216	Large Magellanic Cloud	Dorado	5 20.0	−69° 0'	South
201	NGC 2976	Ursa Major	9 47.3	67° 55'	North
200	M81	Ursa Major	9 55.6	69° 4'	North
200	M82	Ursa Major	9 56.1	69° 42'	North
201	NGC 3077	Ursa Major	10 3.3	68° 44'	North
102	M95	Leo	10 44.0	11° 42'	April–June
102	M96	Leo	10 46.8	11° 49'	April–June
102	M105	Leo	10 47.8	12° 35'	April–June
102	NGC 3384	Leo	10 48.3	12° 38'	April–June
102	NGC 3389	Leo	10 48.5	12° 32'	April–June
102	NGC 3412	Leo	10 50.9	13° 25'	April–June
101	NGC 3607	Leo	11 16.9	18° 3'	April–June
101	NGC 3608	Leo	11 17.0	18° 9'	April–June

Page	Name	Constellation		RA		Dec.		Chapter
100	M65	*Leo*		11	18.9	13°	7'	April–June
101	NGC 3632	*Leo*		11	20.1	18°	21'	April–June
100	M66	*Leo*		11	20.2	13°	1'	April–June
100	NGC 3628	*Leo*		11	20.3	13°	37'	April–June
108	M98	*Coma Berenices*		12	13.8	14°	54'	April–June
108	NGC 4216	*Virgo*		12	15.9	13°	9'	April–June
108	M99	*Coma Berenices*		12	18.8	14°	25'	April–June
107	M84	*Virgo*		12	25.1	12°	53'	April–June
107	M86	*Virgo*		12	26.2	12°	57'	April–June
107	NGC 4435	*Virgo*		12	27.7	13°	5'	April–June
107	NGC 4438	*Virgo*		12	27.8	13°	0'	April–June
107	NGC 4461	*Virgo*		12	29.0	13°	11'	April–June
106	NGC 4469	*Virgo*		12	29.5	8°	45'	April–June
106	M49	*Virgo*		12	29.8	8°	0'	April–June
107	NGC 4473	*Virgo*		12	29.8	13°	26'	April–June
107	NGC 4477	*Virgo*		12	30.0	13°	38'	April–June
107	M87	*Virgo*		12	30.8	12°	23'	April–June
107	M88	*Coma Berenices*		12	32.0	14°	25'	April–June
106	NGC 4526	*Virgo*		12	34.0	7°	42'	April–June
106	NGC 4535	*Virgo*		12	34.3	8°	12'	April–June
107	M91	*Coma Berenices*		12	35.4	14°	30'	April–June
107	M89	*Virgo*		12	35.7	12°	33'	April–June
107	M90	*Virgo*		12	36.8	13°	10'	April–June
107	M58	*Virgo*		12	37.7	11°	49'	April–June
107	M59	*Virgo*		12	42.0	11°	39'	April–June
107	NGC 4638	*Virgo*		12	42.8	11°	27'	April–June
107	M60	*Virgo*		12	43.7	11°	33'	April–June
112	M94	*Canes Venatici*		12	50.9	41°	7'	April–June
105	NGC 4754	*Virgo*		12	52.3	11°	19'	April–June
105	NGC 4762	*Virgo*		12	52.9	11°	14'	April–June
110	M64, *Black Eye Galaxy*	*Coma Berenices*		12	56.8	21°	31'	April–June
238	NGC 4945	*Centaurus*		13	5.4	–49°	28'	South
113	M63	*Canes Venatici*		13	15.8	42°	2'	April–June
239	NGC 5128	*Centaurus*		13	25.5	–43°	1'	South
204	M51, *Whirlpool Galaxy*	*Canes Venatici*		13	29.9	47°	12'	North
204	NGC 5195, *companion of M51*	*Canes Venatici*		13	30.0	47°	16'	North
206	M101, *Pinwheel*	*Ursa Major*		14	3.2	54°	21'	North
121	NGC 6207	*Hercules*		16	43.1	36°	50'	July–September
240	NGC 6744	*Pavo*		19	9.7	–63°	51'	South

Page	Name	Constellation		RA		Dec.		Chapter
181	Upsilon Andromedae	*Andromeda*	Exoplanets	1	36.8	41°	24'	October–December
187	FZ Persei	*Perseus*	Variable	2	21.0	57°	9'	North
187	AD Persei	*Perseus*	Variable	2	20.5	57°	0'	North
188	RZ Cassiopeiae	*Cassiopeia*	Variable	2	48.8	69°	39'	North
188	SU Cassiopeiae	*Cassiopeia*	Variable	2	51.9	68°	53'	North
183	Rho Persei	*Perseus*	Variable	3	5.2	38°	51'	October–December
183	Beta Persei, *Algol*	*Perseus*	Variable	3	8.2	40°	57'	October–December
71	LY Aurigae	*Auriga*	Variable	5	29.7	35°	23'	January–March
55	BM Orionis	*Orion*	Variable	5	35.3	–5°	23'	January–March
55	V1016 Orionis	*Orion*	Variable	5	35.3	–5°	23'	January–March
94	VZ Cancri	*Cancer*	Variable	8	40.9	9°	49'	April–June
96	Gamma Leonis, *Algieba*	*Leo*	Exoplanets	10	20.0	19°	51'	April–June
227	ZZ Carinae	*Carina*	Variable	10	54.4	–58°	47'	South
237	BZ Crucis	*Crux*	Variable	12	42.8	–63°	3'	South
237	DU Crucis	*Crux*	Variable	12	53.6	–60°	20'	South
113	Y CVn, *La Superba*	*Canes Venatici*	Carbon	14	41.1	13°	44'	April–June
116	Xi Boötes	*Boötis*	Exoplanets	14	51.4	19°	6'	April–June
155	RR Scorpii	*Scorpius*	Variable	16	56.6	–30°	35'	July–September
124	Alpha Herculis, *Rasalgethi*	*Hercules*	Variable	17	14.6	14°	23'	July–September
157	BM Scorpii	*Scorpius*	Variable	17	40.9	–31°	13'	July–September
145	U Sagittarii	*Sagittarius*	Variable	18	31.8	–19°	22'	July–September
141	R Scuti	*Scutum*	Variable	18	47.5	–5°	43'	July–September
131	Beta Lyrae, *Sheliak*	*Lyra*	Variable	18	50.1	33°	22'	July–September
141	V Aquilae	*Aquila*	Carbon	19	4.4	–5°	41'	July–September
135	R Cygni	*Cygnus*	Variable	19	36.8	50°	12'	July–September
128	16 Cygni	*Cygnus*	Exoplanets	19	41.8	50°	31'	July–September
196	Mu Cephei, *Garnet Star*	*Cepheus*	Carbon	21	43.5	58°	47'	North
196	Delta Cephei	*Cepheus*	Variable	22	29.2	58°	25'	North
165	51 Pegasi	*Pegasus*	Exoplanets	22	57.5	20°	46'	October–December
190	Rho Cassiopeiae	*Cassiopeia*	Supergiant	23	54.4	57°	30'	North

Index

What, where, and when to observe *for observers 30° – 55° north*

7 p.m.		Jan.	Feb.	Mar.						Sept.	Oct.	Nov.	Dec.
9 p.m.		Dec.	Jan.	Feb.	Mar.	Apr.	May	June	July	Aug.	Sept.	Oct.	Nov.
11 p.m.		Nov.	Dec.	Jan.	Feb.	Mar.	Apr.	May	June	July	Aug.	Sept.	Oct.
1 a.m.		Oct.	Nov.	Dec.	Jan.	Feb.	Mar.	Apr.	May	June	July	Aug.	Sept.
3 a.m.		Sept.	Oct.	Nov.	Dec.	Jan.	Feb.	Mar.	Apr.	May	June	July	Aug.
5 a.m.			Sept.	Oct.	Nov.	Dec.	Jan.	Feb.	Mar.				

Find the time of the evening (standard time) when you'll be observing, and look across that row for the current month. Down that column you'll find where to look for each of the constellations listed to the left. A "+" indicates the object is high up in the sky; a "−" says it is to be found near the horizon.

Constellation	pages	Winter			Spring			Summer			Autumn		
Orion	52–63	SE	S+	SW+	W								E−
Taurus–Auriga	64–71	++	++	W+	W	W−						E−	E
Gemini	72–75	E	++	++	W+	W	W−						E−
Monoceros	76–81	E−	SE	S+	SW	W							
Canis Majoris	82–87	SE−	SE	S	SW								
Puppis	88–89		SE−	S−	S−	SW−							
Cancer	92–97	E−	E	E+	++	W+	W−						
Leo	98–103		E−	E	++	++	W+	W					
Virgo/Coma B.	104–111			E−	E	++	++	W+	W	W−			
Canes Venatici	112–115		NE−	NE	++	++	++	NW+	NW	NW−			
Boötes	116–117				E−	E	++	++	W	W			
Hercules	120–123					E−	E	E+	++	W+	W	W	
Serpens/Oph.	124–127						E	SE	S	SW	SW		
Lyra/Cygnus	128–135						E	E	E+	++	W+	W+	W
Vul/Sagitta/Del	136–139							E	E	++	++	W	W
Scutum	140–141							E−	SE	S	SW	W−	
Scorpius	150–157							SE−	S−	SW−			
Sagittarius	142–161								SE−	S−	SW−		
Pegasus/Aqua	164–169	W	W−						E−	E	E	++	W+
Aqua/Cap/Sculp	170–175									SE−	S−	S−	SW−
And/Tri/Aries	176–181	W+	W	W−						E	E	E+	++
Perseus	182–187	++	++	W+	NW						NE	E	++

Finding geostationary satellites in the telescope field

Stars drift west in a telescope (unless it has a clock drive). But when you're looking in just the right direction you might spot a dot of light that doesn't move: a geosynchronous satellite, appearing to stand still in the sky as it orbits the Earth at the same rate the Earth spins. It's surprising and fun to spot one!

Where these satellites appear in the sky depends on your location on Earth. Use this chart to determine where you're likely to see them. Find your latitude on the horizontal axis of this chart, and then read off from the curve the declination angle where geosynchronous satellites are likely to be seen. (The line is deliberately fuzzy, to account for satellite drift and other small uncertainties in the calculations.) For example if you live at 40°N (say, New York, or Rome, or Tokyo) then whenever you point your telescope at a declination of just over 6° S there's a chance you might encounter a satellite in your field of view. As it happens, that's the declination of the open cluster M11.

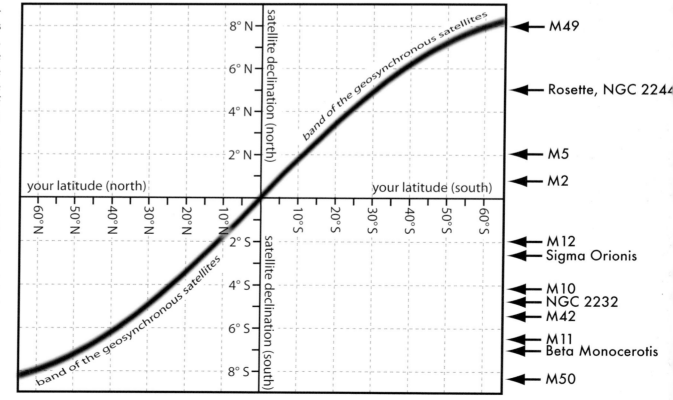